MOUNTAIN SHEEP OF NORTH AMERICA

MOUNTAIN SHEEP OF NORTH AMERICA

Edited by
Raul Valdez and Paul R. Krausman

THE UNIVERSITY OF ARIZONA PRESS   TUCSON

The University of Arizona Press
© 1999
The Arizona Board of Regents
First Printing
All rights reserved

♾ This book is printed on acid-free, archival-quality paper.
Manufactured in the United States of America

04   03   02   01   00   99   6   5   4   3   2   1

Library of Congress Cataloging-in-Publication Data
Mountain sheep of North America / edited by Raul Valdez and
Paul R. Krausman.
p. cm.
Includes bibliographical references (p.   ) and index.
ISBN 0-8165-1839-4 (cloth : acid-free paper)
1. Mountain sheep—North America.   I. Valdez, Raul.
II. Krausman, Paul R., 1946–
QL737.U53 M75 1999   98-25333
599.649—ddc21   CIP

British Cataloguing-in-Publication Data
A catalogue record for this book is available from the British Library.

*Dedicated to
Adrian Sada Treviño,
wild sheep aficionado and benefactor;
Terri L. Steel and Rick F. Seegmiller,
students who devoted their lives
to the study of mountain sheep;
and
Rachel L. Krausman and Shane C. Wegner,
future wild sheep aficionados*

CONTENTS

List of Illustrations viii
List of Tables x
Preface xi
Acknowledgments xiii
List of Abbreviations xiv

1 Description, Distribution, and Abundance
of Mountain Sheep in North America
RAUL VALDEZ AND PAUL R. KRAUSMAN 3

2 Natural History of Thinhorn Sheep
LYMAN NICHOLS AND FRED L. BUNNELL 23

3 Natural History of Rocky Mountain and California Bighorn Sheep
DAVID M. SHACKLETON, CHRISTOPHER C. SHANK, AND BRIAN M. WIKEEM 78

4 Natural History of Desert Bighorn Sheep
PAUL R. KRAUSMAN, ANDREW V. SANDOVAL, AND RICHARD C. ETCHBERGER 139

5 Adaptive Strategies in American Mountain Sheep:
Effects of Climate, Latitude and Altitude, Ice Age Evolution,
and Neonatal Security
VALERIUS GEIST 192

6 Diseases of North American Wild Sheep
THOMAS D. BUNCH, WALTER M. BOYCE, CHARLES P. HIBLER, WILLIAM R. LANCE,
TERRY R. SPRAKER, AND ELIZABETH S. WILLIAMS 209

7 Management of Bighorn Sheep
CHARLES L. DOUGLAS AND DAVID M. LESLIE JR. 238

Appendix: Cytogenetics and Genetics
THOMAS D. BUNCH, ROBERT S. HOFFMANN, AND CHARLES F. NADLER 263

References 277
About the Contributors 335
Index 341

ILLUSTRATIONS

1.1. Past and present distribution of wild sheep in North America 4
1.2. Age classes of bighorn 7
1.3. First published illustration of North American wild sheep 10
1.4. Pachyceriform, moufloniform, and argaliform 12
1.5. Bering land bridge 16
2.1. Full-curl Dall's sheep ram in summer pelage 24
2.2. Nearly ⅞-curl Stone's sheep ram in summer pelage 25
2.3. Estimated mean spring and fall weights of Dall's sheep ewes 37
2.4. Estimated mean spring and fall weights of Dall's sheep rams 38
2.5. Estimated mean shoulder heights of Dall's sheep rams 38
2.6. Estimated mean shoulder heights of Dall's sheep ewes 39
2.7. Mean annual horn growth of Dall's and Stone's sheep rams and Dall's sheep ewes 41
2.8. Dall's sheep ram horn sectioned along the midline 42
2.9. Cumulative horn growth of Dall's and Stone's sheep rams and Dall's sheep ewes 43
2.10. Dall's sheep ewe nursing her lamb 56
2.11. Sheep's ability and inability to dig through snow 63
2.12. Examples of mortality curves for both sexes of Dall's sheep 72
2.13. Sigmoid growth pattern of four Dall's sheep populations 74
2.14. Relationship between lamb production, lamb survival, and population status 75
3.1. Male group of Rocky Mountain bighorn sheep foraging 82
3.2. Weights of Rocky Mountain and California bighorn lambs 99
3.3. Weights of male bighorn sheep 100
3.4. Weights of female bighorn sheep 101

3.5. Typical pattern of annual horn sheath growth in bighorn males 102

3.6. Variation in annual horn growth between populations 104

3.7. Adult Rocky Mountain female and yearling male showing similar horn and body sizes 106

3.8. Group of mature Rocky Mountain bighorn males 111

3.9. Percent frequency of group sizes of Rocky Mountain and California bighorn 113

3.10. Typical sizes of male, maternal, and mixed groups of Rocky Mountain bighorn 114

3.11. Horn clash between bighorn males 117

3.12. Fecundity of female Rocky Mountain bighorn and domestic sheep 127

3.13. Percentage age-specific mortality rates for Rocky Mountain bighorn 135

4.1. Desert bighorn mountainous habitat 140

4.2. Ram horns 157

4.3. Distinguishing characteristics of yearling males and adult ewes 158

4.4. Variation in mating and lambing seasons of wild sheep 167

4.5. Relationship between lamb crop and survival of lambs 174

4.6. Survivorship curves of desert bighorn rams 179

A.1. Typical G-banded karyotype 264

A.2. Specific fusions of acrocentric chromosomes to form biarmed chromosomes 265

TABLES

1.1. Anatomical and life history characteristics of mountain sheep 6
1.2. Sequence of tooth eruption in lower jaw in wild sheep 7
1.3. Record horn measurements in mountain sheep 8
1.4. Classification of North American mountain sheep 14
1.5. Estimated number of mountain sheep in North America 19
1.6. Estimated number of Dall's and Stone's sheep in North America 20
1.7. Estimated number of Rocky Mountain and California bighorn in North America 20
1.8. Estimated number of desert bighorn in North America 21
2.1. Activity budgets during lambing season 45
2.2. Correlation of lambing period and phenology 54
2.3. Mean dates for onset, end, duration, and peak of lambing period 55
3.1. Number of grasses, forbs, and shrubs in bighorn diets 90
3.2. Geographic variation in percent composition of forage groups in bighorn diets 92
3.3. Seasonal variation in percent composition of forage groups in bighorn diets 93
3.4. Age-specific mortality in Rocky Mountain bighorn males 134
5.1. Limb-bone measurements of late Pleistocene and Recent bighorn, giant sheep, and ibex 196
5.2. Dimensions of male Dall's sheep, Altai argalis, and megasheep 203
6.1. Survey of chronic sinusitis in endemic bighorn 222
A.1. Transferrin allelic frequencies in North American wild sheep 270
A.2. Transferrin allelic frequencies in Old World wild sheep 272
A.3. Red-cell blood-group factors of Mexican and Nelson's desert bighorn and European mouflon 274

PREFACE

Prior to the publication of this book, there was no single reference available where one could find a synthesis of information on the various aspects of the life histories and management of the wild sheep of North America. This volume is intended to fill that void. It represents the efforts of wildlife biologists who have dedicated their energies toward investigating life histories of wild sheep and their management applications.

Perhaps no single individual merits more special recognition or is more synonymous with North American wild sheep than the late Charles Sheldon. His pioneering work with wild sheep and other big game animals published in *The Wilderness of the Upper Yukon* and the subsequent posthumous publication relating to his field studies of wild sheep in the Denali region of Alaska, *The Wilderness of Denali,* are truly classics in their own right. Biologists are indebted to him for his pioneering contributions toward furthering the knowledge of North American wild sheep and their conservation.

Today, however, due to many baffling problems, the very survival of North American wild sheep is in jeopardy. When comparing the data available for wild sheep to data for other ungulates such as white-tailed deer, mule deer, elk, and pronghorn, one realizes that wild sheep research is in its infancy. Of all the hoofed mammals in North America, wild sheep present the greatest management problems to ungulate biologists. Many questions, even basic ones, remain unresolved. The reader will notice that there are occasional contradictory views expressed. This is to be expected, considering the present state of knowledge regarding fundamental questions.

This is an exciting period for biologists studying wild sheep; a great research frontier confronts us. It is also a critical period in wild sheep research and management. Unless urgent questions are answered soon, wild sheep face extinction in many areas. Long-term, multidisciplinary research is needed to solve these complex problems. The short-term, piecemeal efforts of the past have been ineffectual. This has been recognized by biologists, but adequate funding to support long-term, multidisciplinary research has not been available. Wild sheep biologists must pool their knowledge and resources and make a concentrated effort to solve critical problems.

This effort was begun in 1957, when wildlife biologists in the southwestern United States initiated an innovative, cooperative effort, involving state and federal wildlife agencies, to seek answers to pressing problems related to wild sheep management. An outcome was the development of the

Desert Bighorn Council. Biologists studying desert bighorns meet annually to share research findings and discuss solutions to problems, which are published in the *Desert Bighorn Council Transactions*. The biologists who began the council established a pattern that continues to the present and laid the foundation for concentrated, long-term efforts to ensure the perpetuity of wild sheep populations. A similar, cooperative approach resulted in the Northern Wild Sheep Conference in 1976, which became the Northern Wild Sheep and Goat Conference in 1978. These conferences also published results in proceedings.

Another influential group was established in 1977: the Foundation for North American Wild Sheep, a nonprofit sportsmen's group that initiated a private-sector effort to promote and enhance the welfare of wild sheep in North America. The foundation has raised funds for research, management, and education. Because of the past and continuing efforts of sportsmen and U.S., Mexican, and Canadian biologists under the aegis of federal and state agencies, wild sheep in North America face an optimistic future. We hope this book will contribute toward the well-being of mountain sheep throughout their range.

ACKNOWLEDGMENTS

We are indebted to numerous people who made this work possible. The largest group includes the wild sheep biologists around North America who provided the information through their work and publications. Their efforts will ensure the conservation of wild sheep.

C.P.P. Reid, Director, School of Renewable Natural Resources, the University of Arizona; C. C. Kaltenbach, Director, Arizona Agricultural Experiment Station, the University of Arizona; J. C. Owens, Dean, College of Agriculture and Home Economics, New Mexico State University; and G. Cunningham, Director, New Mexico Agricultural Experiment Station, New Mexico State University, provided time and funding for the project. Val Catt, Sue Klein, and Lisa Fox assisted with typing various parts of the document. Rick Spaulding and Lisa Fox assisted with various aspects of editing, and Rick Spaulding assisted in compiling the literature cited.

Funding for editing and final printing was provided by the School of Renewable Natural Resources and the Provost's Author Support Fund, the University of Arizona. Finally, we acknowledge Lisa DiDonato, copyeditor; Christine Szuter and the editors at the University of Arizona Press; anonymous reviewers; and others who helped with the production of this work.

ABBREVIATIONS

ADF acid detergent fiber
BLM Bureau of Land Management
B.P. before the present
CI confidence interval
CP crude protein
CPK creatine pyrokinase
DNA deoxyribonucleic acid
ELISA enzyme-linked immunosorbent assay
EP erythrocyte protein
FDA Food and Drug Administration
GIS geographic information system
GOT glutamic oxalacetic transaminase
HPLC high-performance liquid chromatography
HSI habitat suitability index
HSM harvestable surplus model
LA long-acting antibiotic
LDH lactate dehydrogenase
MVP minimum viable population
PATREC pattern recognition system
PCM partial compensation model
PI parainfluenza
RFLP restriction fragment length polymorphism
SD standard deviation
SE standard error
SI selection index
USGS United States Geological Survey

MOUNTAIN SHEEP OF NORTH AMERICA

one

DESCRIPTION, DISTRIBUTION, AND ABUNDANCE OF MOUNTAIN SHEEP IN NORTH AMERICA

Raul Valdez and Paul R. Krausman

Introduction

Mountain sheep *(Ovis canadensis)* are one of the most striking large mammals in North America. Simply observing them is an exciting and gratifying aesthetic experience. Mountain sheep epitomize wilderness. They occupy some of the most inaccessible, rugged, and spectacular habitats in North America. Their ability to negotiate precipitous terrain is legendary. They inhabit some of the hottest, most inhospitable regions of North American deserts (i.e., desert bighorn) and the highest and coldest alpine regions of the American and Canadian Rockies (i.e., Rocky Mountain bighorn) and Alaskan and Canadian massifs (i.e., Dall's and Stone's sheep). Only the most adventurous and hardy outdoor enthusiasts dare to tread in such hostile habitats of temperature extremes and rugged terrain.

North American mountain sheep belong to one of the most widely distributed genera *(Ovis)* of hoofed mammals (i.e., ungulates) in the world. Mountain sheep range from Europe (originally only on the Mediterranean Islands of Corsica and Sardinia, but they are now widely introduced in the continent) through Anatolia, Southwest Asia, the arid and semiarid regions of the Indian subcontinent, Central Asia, Tibet, Mongolia, and Siberia (Valdez 1982). In North America, mountain sheep are found from Alaska through western Canada and the western United States to northwestern Mexico and Baja California (fig. 1.1).

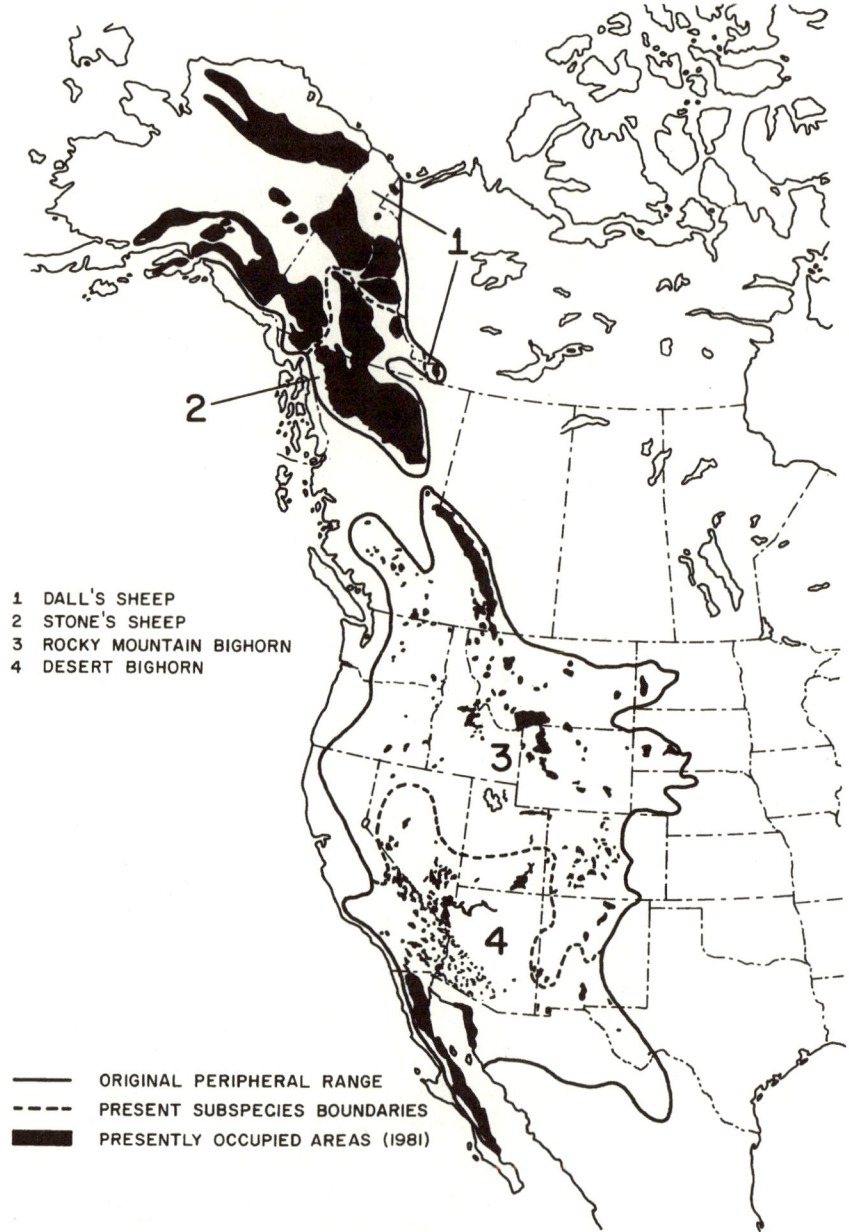

Figure 1.1. Past and present distribution of wild sheep in North America (modified from Trefethen 1975, Hall 1981). Solid lines delineate original peripheral boundaries; only areas of preferred habitat were occupied within the boundaries.

Mountain sheep in North America can be divided into two types: bighorn and thinhorn. Dall's *(Ovis dalli)* and Stone's *(O. dalli stonei)* compose thinhorn. Rocky Mountain *(O. c. canadensis)*, California *(O. c. californiana)*, and desert races of mountain sheep *(O. c. nelsoni, O. c. mexicana, O. c. weemsi, O. c. cremnobates)* compose bighorn. Thinhorn are characterized by horns that are widely curled and by smaller cranial and body measurements than bighorn (Cowan 1940). Unlike horns of thinhorn rams, horns of bighorn rams are often broomed; that is, the tips are broken off, which results principally during clashes when horn tips splinter on impact (Shackleton and Hutton 1971).

North American mountain sheep are adapted to negotiate rocky, precipitous terrain; they generally avoid areas of tall vegetation that obstruct their vision (Risenhoover and Bailey 1985). The diet of mountain sheep is primarily grasses and forbs, although they can shift their diet to shrubs depending on availability. Being ruminants, or cud chewers, they possess a four-chambered stomach in which most of the cellulose of vegetation is digested with the aid of bacteria and protozoans. Sheep are able to regurgitate a cud, or bolus, of food from the rumen to the mouth and further grind the cud. If surface water is available during dry periods, they may water regularly; mountain sheep require a minimum of 4–5% of their body weight of water per day. Some populations are able to obtain moisture from vegetation (Krausman et al. 1985). During summer, some desert bighorn may go without drinking for 5 to 15 days, resulting in the loss of more than 30% of total body water. Like the camel, they are able to consume sufficient water at one drinking session to restore losses. Turner (1970, 1979*a,b*) and Turner and Weaver (1980) discussed water balance and fatty acid physiology in mountain sheep.

Mountain sheep in America are gregarious and extremely loyal to their home ranges; knowledge of home range is passed from generation to generation. Ewes usually bear their first young at 2 or 3 years of age; twinning is rare. Although rams may be sexually mature at 1.5 to 2.5 years, they normally do not participate in the rut until several years later, when they are socially mature and have attained adult body size. In the wild, rams rarely live more than 10 years. However, P. R. Krausman (unpubl. data) documented a 13-year-old wild female sheep that continued to bear young, and Geist (1971) recorded a 19-year-old wild ram.

Description

Anatomical and life history characteristics of mountain sheep reflect adaptations to a mountain environment (table 1.1, fig. 1.2). Age of younger sheep can be determined by the sequence of mandibular tooth eruptions (table 1.2).

Table 1.1. Anatomical and life history characteristics of North American mountain sheep. Measurements are in inches and weights are in pounds (standard units of measurement).

COLORATION	Back of legs, (distinct) rump patch, and usually part of muzzle white. Thinhorn are white (Dall's) or grayish to blackish (Stone's), whereas bighorn usually are some shade of brown. Desert bighorn possess a dark stripe middorsally of dark body hair across rump patch to tip of tail.
HORNS	Bighorn possess more massive, less rugose, and usually broomed and less widely expanded horns than thinhorn. Desert bighorn exhibit more diverging horns than northern sheep (table 1.3). Desert bighorn ewe horns distinctly more expanded and curved than those of other ewes. Horns amber-brown (yellowish) in color in thinhorn and usually darker in bighorn.
WEIGHTS	Thinhorn, birth weights 7–9, adult males ($>$ 4 years) in fall average 180 (up to 300), adult females ($>$ 2 years) average 125; northern bighorn, birth weights 6–10, adult males average 175 (up to 320), females average 130; desert bighorn, birth weights 6–9, adult males average 150 in late summer (up to 230), adult females average 115.
ADULT EXTERNAL MEASUREMENTS	For bighorn males, total length: 52–77; tail: 4–6; hind foot: 14–19; shoulder height: 32–44; females, 46–74, 4–5, 11–17, 30–36. For thinhorn males, 51–70, 3–5, 15–18, 32–42; females, 52–54, 3–4, 11–16, 31–35.
TOOTH FORMULA	For half of upper and lower jaw: (incisors $0/3$, canines $0/1$, premolars $3/3$, molars $3/3$) $\times$ 2 = 32.
TOOTH ERUPTION	Refer to table 1.2.
MATING SEASON	Thinhorn and northern bighorn mate November–December; desert bighorn mate July–December.
GESTATION PERIOD	170–180 days.
LAMBING SEASON	Thinhorn and northern bighorn lamb May–June; desert bighorn lamb January–June.

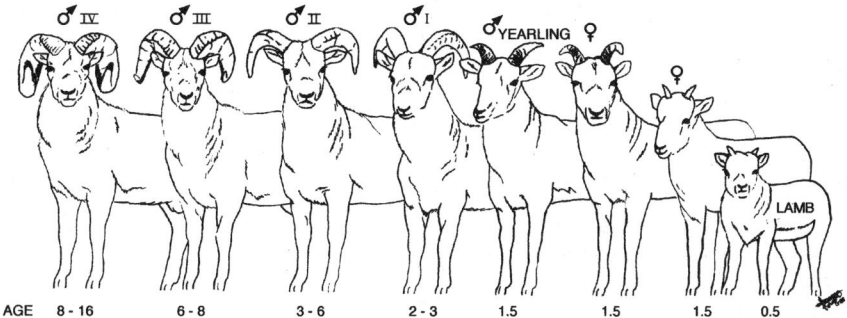

Figure 1.2. Age classes of bighorn (redrawn from Geist 1968:199–215). Note that the classes form a cline in body and horn size.

Table 1.2. Sequence of tooth eruption in lower jaw in North American wild sheep.

	INCISORS			CANINE	PREMOLARS			MOLARS		
AGE	1	2	3	1	2	3	4	1	2	3
6 months	D[a]	D	D	D	D	D	D	(P)		
12 months	(P)	D	D	D	D	D	D	P	(P)	
18 months	P	D	D	D	D	D	D	P	P	
24 months	P	D	D	D	D	D	P	P	P	
36 months	P	P	(P)	D	P	P	P	P	P	(P)
42 months	P	P	(P)	D	P	P	P	P	P	P
44 months	P	P	P	P	P	P	P	P	P	P

Note: Refer to Taber (1969) and Hemming (1969) for variations.
[a]D = deciduous; P = permanent; (P) = erupting.

Table 1.3. Means, standard deviations (SD), and ranges (in inches; standard unit of measurement) of horn measurements of the 25 longest-horned specimens of mountain sheep.

	HORN LENGTH		BASAL CIRCUMFERENCE		TIP-TO-TIP SPREAD	GREATEST SPREAD
	RIGHT	LEFT	RIGHT	LEFT		
Dall's Sheep						
X̄	44.9	44.5	14.3	14.3	26.9	27.2
SD	2.9	2.6	0.6	0.6	4.0	3.6
Range	38.750–49.500	39.000–47.500	13.000–15.250	13.000–15.250	20.875–34.375	20.875–34.375
Stone's Sheep						
X̄	45.0	44.9	14.8	14.8	24.5	26.6
SD	2.2	1.9	0.6	0.6	3.1	2.9
Range	42.125–50.125	41.875–51.625	13.500–16.250	13.500–16.250	19.000–31.375	22.000–31.500
Rocky Mountain and California Bighorn						
X̄	43.7	43.7	15.9	15.9	22.0	23.4
SD	2.3	2.2	0.7	0.7	2.8	1.9
Range	39.125–49.500	40.500–49.250	14.750–17.500	14.750–17.375	18.125–28.875	21.500–28.875
Desert Bighorn						
X̄	40.6	40.8	15.7	15.7	22.1	22.9
SD	2.1	2.3	0.6	0.6	2.9	2.2
Range	37.000–45.625	36.000–46.250	14.500–16.750	14.625–17.000	16.750–27.375	17.750–27.375

Source: Reneau and Reneau (1993).

Horn growth of rams is minimal after ages 7 or 8. Horns are fairly distinct between thinhorn and bighorn races of sheep, as evidenced by the mean measurements of the 25 highest-ranked trophy males of each race (table 1.3). Dall's and Stone's sheep exhibit the longest mean horn lengths, followed by Rocky Mountain bighorn and desert bighorn. Rocky Mountain bighorn exhibit the greatest mean basal circumferences, followed by desert bighorn then Stone's and Dall's sheep. However Dall's and Stone's sheep exhibit the longest means of tip to tip and the greatest spread, followed by Rocky Mountain and desert bighorn (table 1.3). The longest horns ever recorded in a North American wild sheep (right = 50.125 in, left = 51.625 in) are those of the Chadwick ram, a Stone's sheep shot by L. S. Chadwick along the Muskwa River, British Columbia, in 1936.

The largest number of record-sized Rocky Mountain bighorn rams recorded since 1975 originated from introduced populations in Montana (Boone 1988, Gilchrist 1990). Twenty-six of the 100 highest-scoring Rocky Mountain bighorn rams listed in the tenth edition of the Boone and Crockett Club record book are from Montana (Reneau and Reneau 1993).

Interaction with Aboriginal Humans and Discovery by Europeans

Humans and mountain sheep have coexisted in North America for more than 30,000 years (Hopkins 1967, Hopkins et al. 1982). By 12,500 B.P., both occupied the American Southwest (A. H. Harris 1977). Mountain sheep provided a source of meat protein for early humans in North America and for aboriginals in Asia before their immigration to the New World. The preferred methods of hunting bighorn were entrapment using winged corrals to channel sheep, ambushing them near watering holes, or driving them near waiting hunters along sheep trails or escape routes. Initially, the primary weapon was the spear. This was followed by the atlatl, or spear thrower, a narrow, 0.55-m-long stick used to propel the spear (Grant 1980). Some 2,000 years ago, the atlatl was replaced by the more efficient bow and arrow. Unlike their ancestors in the Old World, humans in the New World never domesticated wild sheep.

Mountain sheep were commonly depicted in rock art by Native Americans in the western United States (Grant 1980). These paintings were involved in ceremonies intended to aid hunters in successfully hunting their elusive quarry and to ensure the survival of bighorn, an important animal in Native American cultures. The horns, bones, and hides of bighorn were utilized in several ways. Horns were used to make sickles, scrapers, blades, pendants, utensils, ornaments, and other items. Bones were used primarily for basket-making awls by southwestern tribes, and tanned hides provided material for robes and other clothing items.

Spanish explorers and missionaries in the American Southwest were the first Europeans to record observations of mountain sheep in North America. In 1540, Coronado observed wild sheep when he made his famous journey through New Mexico and Arizona. The first published account of mountain sheep in North America is that of Father Francisco Maria Piccolo, who in

Figure 1.3. The first published illustration of North American wild sheep (Venegas 1757).

1702 published the observations of Jesuit missionaries in Baja California. The first illustration of a wild sheep, labeled the Taye or California deer (fig. 1.3), was published in Madrid, Spain, in 1757 in Miguel Venegas's book *Noticia de la California*. This was translated into English in 1759 under the title *A Natural and Civil History of California* (J. A. Allen 1912).

The first zoological specimen of a Rocky Mountain bighorn was not secured until 1800. It was captured in Alberta Province, virtually at the east gate of Banff National Park (V. Geist, pers. comm.), Canada, by Duncan McGuillivray, an agent for the Northwest Fur Company who accompanied a surveying and exploring expedition to the upper Bow River of Canada in the autumn of 1800. In 1803, McGuillivray sent a specimen to the British Museum. This specimen was formally described in 1804 by George Shaw, a British zoologist, in the London monthly publication *Naturalists' Miscellany* under the name *Ovis canadensis* (J. A. Allen 1912). Mountain sheep were subsequently collected by the Lewis and Clark expedition in 1805 and, based on their specimens, illustrated in Godman (1826).

Dall's sheep, named after the American naturalist and explorer W. H. Dall, were not formally named until 1884. These specimens were collected south of Fort Yukon on the west bank of the Yukon River, probably on the Tanana

Hills, Alaska (Nelson 1884). Stone's sheep were described and named in 1897 from specimens collected by A. J. Stone on the headwaters of the Stikine River, British Columbia, Canada (J. A. Allen 1897). North American wild sheep were encountered by several European and American explorers, trappers, naturalists, and early native and foreign big game hunters (Valdez 1988).

Classification

Sheep are classified in the order Artiodactyla (even-toed ungulates), suborder Ruminantia (ruminating or cud-chewing mammals), and the family Bovidae, which includes cattle, antelopes, and goats. Sheep and their closest relatives (i.e., true goats, *Capra;* aoudad, *Ammotragus;* bharal, *Pseudois;* and tahrs, *Hemitragus*) are further groups in the tribe Caprini. All true sheep are assigned to the genus *Ovis*. True sheep are characterized by the presence of interdigital, inguinal (groin), and preorbital glands and the absence of subcaudal glands and a chin beard (Valdez 1982).

Based on body confirmation and habitat preferences, mountain sheep can be divided into three basic types: moufloniforms, argaliforms, and pachyceriforms (fig. 1.4). Moufloniforms include the European mouflon *(O. musimom)* and Asiatic mouflons and urials *(O. gmelinii* and *O. vignei);* argaliforms comprise the argalis of Central Asia *(O. ammon);* and pachyceriforms include Siberian snow sheep *(O. nivicola)* and North American Dall's, Stone's, and bighorn sheep. Moufloniforms and argaliforms are characterized by long legs and supple, antelopean bodies adapted to running from predators in the montane, undulating habitats they prefer. Pachyceriforms are stockier, muscular types adapted to evading predators by seeking refuge in precipitous terrain. North American mountain sheep occupy habitat similar to that occupied by wild goats in Eurasia (Valdez 1982).

Early zoologists recognized many species of mountain sheep. They used a broad definition of the species concept based on geographic variation. The modern definition of a species requires that forms must be reproductively isolated. Many species that will not breed in the wild readily breed in captivity. However, species status must be based on the ability of animals in the wild to produce fertile offspring that can interbreed with parental types. Because many wild animal populations do not form series of overlapping populations, it is sometimes necessary to make subjective judgements regarding species status in disjunct populations because the detection of natural reproductive barriers is not possible under such circumstances.

A case could be made for classifying all North American wild sheep as a single species. The fact that thinhorn and bighorn are distinct phenotypically

Figure 1.4. (A) Pachyceriform (Ovis canadensis), *(B) moufloniform* (O. gmelinii), *and (C) argaliform* (O. ammon). *(Photographs by Raul Valdez and Thomas Bunch.)*

(i.e., external anatomy) does not preclude the possibility that they would interbreed if their ranges overlapped. Behavioral and chromosomal data (Geist 1971, Bottrell et al. 1978) indicate that there would be no barriers to successful interbreeding if their ranges overlapped. Indeed, they have bred successfully in captivity (T. Bunch, Utah State Univ., pers. comm.). However, we follow the traditional classification of Cowan (1940) and recognize thinhorn and bighorn as distinct species (table 1.4). J. A. Allen (1912), Osgood (1913, 1914), and Cowan (1940) presented a detailed historical treatment of the nomenclature of North American mountain sheep.

Subspecies, or geographic races, are more controversial categories than species because their classification is somewhat subjective. Some zoologists recognize many subspecies, differentiating various races on anatomical features such as size, coloration, and horn shape. Other zoologists are conservative and recognize fewer subspecies, thereby acknowledging that some characters are variable within populations and that some features may exhibit clinal variation (i.e., gradual change in an anatomical character through a species' geographic distribution). In pristine times, before intermediate populations had been extirpated, it would have been difficult to draw a boundary between Rocky Mountain and desert bighorn geographic ranges because they undoubtedly exhibited clinal variation in all phenotypic attributes. The same difficulty exists in determining the boundary between Mexican, Nelson's, peninsular, and Weem's desert bighorn. Because of the arbitrary nature of desert bighorn subspecific boundaries, we have not designated boundaries for desert subspecies (fig. 1.1).

Color intergradation between Dall's and Stone's sheep populations has been documented. Sheldon (1919) carefully mapped the color variations in Dall's and Stone's sheep, showing a complete intergradation from pure white Dall's types to dark Stone's types. In fact, no localities exist in the intervening areas where sheep populations have uniform color characters (Osgood 1909). Fannin's sheep, or the saddle-backed sheep, is a color variant among other color phases within Stone's populations; for this reason, Fannin's sheep has been invalidated as a subspecies. Coloration is an especially weak diagnostic feature in delineating Stone's sheep populations.

In some instances, subspecies were named on the basis of specimens exhibiting slight differences from the nearest recognized subspecies or on the basis of an atypical specimen such as Sheldon's bighorn. In 1916, Charles Sheldon collected a single, immature ram that was below normal size in the Sierra del Rosario, Sonora, Mexico. Merriam (1916) described the ram as a new species based on its small size. Later, detailed studies revealed that these size differences were within the variation exhibited by individuals in populations of previously described subspecies; therefore, the new species was invalidated.

Table 1.4. Classification of North American mountain sheep.

Upper Taxa
    Class Mammalia; Order Artiodactyla (even-toed ungulates); Family Bovidae (sheep, goats, bovines, antelopes); Tribe Caprini (sheep, goats); Genus *Ovis* (true sheep).

Species
Thinhorn Sheep: *Ovis dalli*
  Ovis dalli dalli Nelson, 1884 (Dall's sheep)
    Type locality: Mountains south of Fort Yukon on west bank of Yukon River, Alaska, probably Tanana Hills.
    Synonym: *Ovis dalli kenaiensis* J. A. Allen, 1902. Type from head of Sheep Creek, Kenai Peninsula, Alaska.

  Ovis dalli stonei J. A. Allen, 1897 (Stone's sheep)
    Type locality: Headwaters of Stikine River, British Columbia.
    Synonym: *Ovis dalli fannini* Hornaday, 1901. Type from Dawson City, Yukon.

Bighorn Sheep: *Ovis canadensis*
  Ovis canadensis canadensis Shaw, 1804 (Rocky Mountain bighorn)
    Type locality: Mountains on Bow River, Alberta.

  Ovis canadensis californiana Douglas, 1829 (California bighorn)
    Type locality: Uncertain, either near Mount Adams, Yakima County, Washington, or falls of the Columbia River, near mouth of the Deschutes River.

  Ovis canadensis auduboni Merriam, 1901 (Audubon or Badlands bighorn)
    Type locality: Upper Missouri, probably Badlands between Cheyenne and White Rivers, South Dakota.

  Ovis canadensis nelsoni Merriam, 1897 (Nelson's desert bighorn)
    Type locality: Grapevine Mountains on boundary between Inyo County, California, and Esmeralda County, Nevada.

  Ovis canadensis mexicana Merriam, 1901 (Mexican desert bighorn)
    Type locality: Lago de Santa Maria, Chihuahua.
    Synonyms: *Ovis canadensis texianus* V. Bailey, 1912. Type from Guadalupe Mountains, El Paso County, Texas.
      Ovis canadensis sheldoni Merriam, 1916. Type from El Rosario, northern Sonora.

Table 1.4. *Continued*

Ovis canadensis cremnobates Elliott, 1904 (peninsular desert bighorn)
   Type locality: Matomi, Sierra San Pedro Martir, Baja California.

Ovis canadensis weemsi Goldman, 1937 (Weem's desert bighorn)
   Type locality: Cajón de Tecomaja, Sierra de la Giganta, Baja California.

Note: Refer to Hall (1981) for a complete synonymy.

For the same reason, zoologists now synonymize the Texas bighorn and the Mexican bighorn. The validity of the California bighorn also is questionable because there are no phenotypic differences that unequivocally differentiate it from Rocky Mountain bighorn. Valdez (1982) recognized only two subspecies of desert bighorn *(nelsoni* and *cremnobates)* and one extant northern bighorn subspecies *(canadensis)*. Wehausen and Ramey (1993) synonymized *nelsoni* and *cremnobates* based on Ramey's (1993) mitochondrial DNA analysis and a reanalysis of Cowan's (1940) taxonomy of North American wild sheep.

Evolution

Asia is the ancestral home of wild sheep. The fossil record indicates that a large argaliform sheep existed in Asia during the early Pleistocene about 2,000,000 B.P., suggesting that argaliform sheep evolved in Asia during the early Pleistocene. Large sheep are found in the Pleistocene of Europe, but only moufloniform sheep existed in western Asia and Europe in the late Pleistocene. The oldest known fossils of wild sheep in North America are from Yukon deposits (Kurten and Anderson 1980). Fossil remains of mountain sheep in North America are pachyceriform types (Korobitsyna et al. 1974).

Since 1960, concentrated studies have been conducted on the chromosome numbers of wild sheep to analyze relationships. These studies revealed considerable variation. Chromosome numbers vary from 52 in Siberian snow sheep to 58 in urials. Numbers are most variable within the various populations of moufloniforms; however, there is uniformity in some groups. All North American forms possess 54 chromosomes, although Bottrell et al. (1978) reported one abnormal specimen with 55 chromosomes. All argaliforms so far sampled possess 56 chromosomes, whereas Asiatic mouflons possess from 54 to 58 chromosomes. European mouflons and most domestic breeds, with few exceptions, possess 54 chromosomes. Bighorn mated with

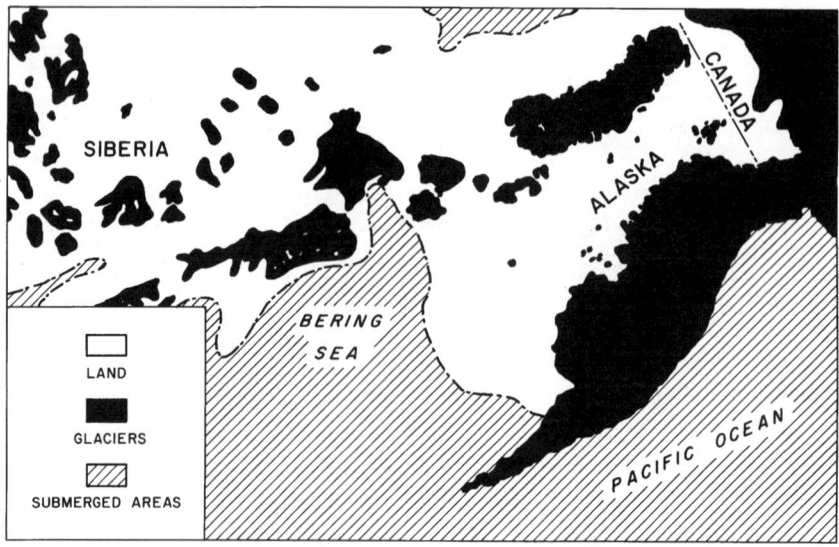

Figure 1.5. The Bering land bridge, where pachyceriform sheep evolved their distinctive characteristics during the last (Wisconsin) glaciation (after Hopkins 1967).

domestic sheep produce fertile hybrids (Young and Manville 1960). A detailed treatment of chromosomes, transferrins, and hemoglobins of North American mountain sheep is presented by Bunch (this volume).

Zoogeographers have postulated that the ancestors of North American wild sheep became isolated from their Asiatic ancestors during the Pleistocene, which was a turbulent period. A massive, continental ice sheet, in some places up to 3,000-m thick, covered approximately 8,000,000 km^2 of North America. The ice sheet's southern margin extended coast to coast from what is now Washington state to New Jersey. In North America, there were four recognized major glaciations followed by warm periods of interglacials. Because of the large quantities of water tied up in the ice sheets, sea level fell at least 150 m. The lowered sea level resulted in the creation of the Bering land bridge in the area between eastern Siberia and western Alaska. Even during glacial maxima, there was an ice-free area in this region, which became a refugium (i.e., an area with suitable habitat) for animals (Hopkins et al. 1982, Kontrimavichus 1986). During the interglacials, when the glacial masses melted and sea levels rose, Asia and North America became separated along with their faunas. North American mountain sheep evolved their unique characteristics during this isolation from Asiatic ancestors (fig. 1.5).

There are two principal hypotheses regarding the evolution and origin of

North American wild sheep. The reversed migration hypothesis, first proposed by Severtzov (1873, cited in Sushkin 1925), postulates that the first mountain sheep to reach North America across the Bering land bridge were argaliforms that, after becoming isolated south of the continental glaciers, evolved into pachyceriforms. After a postglacial period, these southern pachyceriforms reached Beringia and then gave rise to the snow sheep of Siberia and Dall's sheep in Alaska (Nadler et al. 1973*b*).

Cowan (1940) proposed the second principal hypothesis, that pachyceriforms evolved in the Beringian region rather than strictly in North America south of the continental glaciers. All fossil North American sheep so far described are pachyceriforms, which adds support to Cowan's hypothesis. Based on reexamination of fossil sheep originally described by Stokes and Condie (1961) from Pleistocene deposits of Lake Bonneville near Salt Lake City, Stock and Stokes (1969) concluded that specimens were more closely related to *O. canadensis* then *O. ammon*. Guthrie (1968) also assigned fossil sheep material from Alaska dated from the Illinoian glacial period to pachyceriforms. Based on paleontological, biogeographical, and chromosomal data, Korobitsyna et al. (1974) concluded that (1) pachyceriforms evolved their distinctive characteristics while isolated in the ice-free Beringian glacial period; (2) pachyceriforms migrated southward into the western United States when the continental and cordilleran glaciers blocking this route melted, probably in the Sangamon interglacial period; (3) isolation of pachyceriforms in eastern Siberia from those in Alaska by waters of the Bering Strait, probably in the Sangamon interglacial period but possibly as late as the Holocene (10,000 B.P.), led to differentiation and speciation of those two populations, whose modern descendants are the Siberian snow sheep and Dall's sheep, respectively; and (4) isolation of pachyceriforms in the western United States during the Wisconsin glaciation period resulted in the development of modern Rocky Mountain and desert bighorn. K. L. Dixon (1979), Geist (1985 *a,b*; 1987 *c*), Pielou (1991), and Geist (this volume) provide further details of wild sheep evolution in North America and faunal interrelationships during glacial and postglacial periods.

Geist (1971) hypothesized that during pachyceriform evolution, male sheep underwent selection for neoteny, that is, the presence of juvenile characters in adults. One consequence of neoteny is delayed maturation of rams; in terms of body size and social behavior, maturity is not reached until the fifth or sixth year. By assuming a juvenile appearance until physically and socially able to participate in the rut, younger rams are able to associate with other males with a minimum of antagonistic interactions. Hence, neoteny favors selection for sociality. Females are pedogenic, that is, they retain a juvenile appearance throughout life (Geist, this volume).

The melting of the extensive glaciers covering North America some 10,000 years ago created extensive sheep habitat. The grassy, montane habitats favored by bighorn were more extensive than today. Mountain sheep eventually invaded all favorable habitats. The extensive, high-quality forage afforded by the receding glaciers resulted in high-quality populations characterized by a high reproductive rate and large, vigorous, early-maturing individuals. Males in such populations interacted often, and selection favored those with large horns.

The increase in sheep habitat following the initial glacial retreat was later reversed. Sheep habitat became patchy due to the eventual encroachment of forests up the mountain slopes, which had greater amounts of precipitation. In this more limited environment, selection favored less-active, longer-lived, smaller individuals. Indeed, a decrease in bighorn body size at the close of the Pleistocene has been documented by A. H. Harris and Mundel (1974). Patchy environments also favored selection against dispersing juveniles because all available habitats would likely be colonized already. The preceding discussion is based on Geist's (1971) dispersal theory.

Deming (1962) first proposed the hypothesis that desert bighorn were relict Pleistocene populations relegated to an environment to which they were poorly adapted. Supposedly, desert bighorn are poorly adapted to the high temperatures, aridity, and unpredictable precipitation of the American Southwest (J. A. Bailey 1980, 1984). Geist (1985b) postulated that North American mountain sheep were best adapted wherever they had access to ancestral alpine landscapes and poorest adapted where they existed on native American flora and that desert bighorn had not evolved adequate adaptations to a desert environment. McCutchen (1981), M. C. Hansen (1982b), and Wehausen (1984) refuted the hypothesis that bighorn are poorly adapted. Any evaluation of the adaptability of desert bighorn to desert environments and their present precarious existence should carefully consider Europeans' drastic changes to their fragile habitats.

Numbers

Seton (1929) estimated that in pristine times there were approximately 2,000,000 mountain sheep in the contiguous United States and another 2,000,000 in Canada and Alaska combined. Seton's (1929) estimate of 4,000,000 sheep in pristine North America often is cited as a reliable approximation. Valdez (1988) doubted that sheep numbers ever exceeded 500,000 for all of North America. Mountain sheep are highly selective in their habitat preferences, and it is a misconception that sheep were uniformly distributed

Table 1.5. Estimated number of mountain sheep in North America in 1991.

LOCATION	DALL'S SHEEP	STONE'S SHEEP	ROCKY MOUNTAIN BIGHORN	CALIFORNIA BIGHORN	DESERT BIGHORN	TOTAL
Canada	26,500	14,500	12,700	4,700	0	58,400
Alaska	73,250	0	0	0	0	73,250
Contiguous United States	0	0	25,269	5,116	18,555	48,940
Mexico	0	0	0	0	4,500	4,500
Total	99,750	14,500	37,969	9,816	23,055	185,090

Note: Estimates were provided by biologists from each state or province.

throughout the montane terrain of western North America. In Alaska, where most wild sheep populations still occupy vast areas of pristine habitat, numbers do not exceed 74,000 (Nichols 1975).

Buechner (1960) reviewed historical and contemporary sheep distribution in the United States and estimated there were 15,000 to 20,000 wild sheep in the contiguous United States. Sheep numbers probably approached the upper estimate of 20,000 because many populations remained unsurveyed at the time of his estimate. Table 1.5 gives the estimated total numbers of mountain sheep in North America in 1991. Alaska supported the largest number (73,250), followed by Canada (58,400) and the contiguous United States (48,940), whereas Mexico had an estimated 4,500. Relative to subspecies, Dall's sheep far outnumbered all others (99,750), followed by Rocky Mountain and California bighorn combined (47,785), desert bighorn (23,055), and Stone's sheep (14,500). An estimated 185,090 wild sheep presently inhabit North America (tables 1.5, 1.6, 1.7, 1.8).

Conservation

Dall's and Stone's sheep populations have remained relatively unimpacted by humans and have retained their historical distribution and numbers. Mountain sheep populations of southwestern Canada, the western United States, and northern Mexico have declined due to human impacts. Mountain sheep have made the transition from relative abundance to one of the rarest

Table 1.6. Estimated number of Dall's and Stone's sheep in North America in 1991.

LOCATION	DALL'S SHEEP	STONE'S SHEEP	TOTAL
Canada			
British Columbia	500	11,500	12,000
Northwest Territory	7,000	0	7,000
Yukon Territory	19,000	3,000	22,000
United States			
Alaska	73,250	0	73,250
Total	99,750	14,500	114,250

Note: Estimates were provided by biologists from each state or province.

Table 1.7. Estimated number of Rocky Mountain and California bighorn in North America in 1991.

LOCATION	ROCKY MOUNTAIN BIGHORN	CALIFORNIA BIGHORN	TOTAL
Canada			
Alberta	9,700	0	9,700
British Columbia	3,000	4,700	7,700
Contiguous United States			
Arizona	500	0	500
California	0	200	200
Colorado	6,300	330	6,630
Idaho	4,000	1,500	5,500
Montana	5,250	0	5,250
Nevada	300	500	800
New Mexico	610	0	610
North Dakota	313	0	313
Oregon	400	2,000	2,400
South Dakota	380	0	380
Utah	400	0	400
Washington	266	586	852
Wyoming	6,550	0	6,550
Total	37,969	9,816	47,785

Note: Estimates were provided by biologists from each state or province.

Table 1.8. Estimated number of desert bighorn in North America in 1991.

LOCATION	TOTAL
United States	
Arizona	6,500
California	3,400
Colorado	330
Nevada	5,500
New Mexico	295
Texas	230
Utah	2,300
Mexico	
Baja California	2,000
Sonora	2,500
Total	23,055

Note: Estimates were provided by biologists from each state or province.

ungulates in North America. An entire subspecies, Audubon's (or Badlands) bighorn, which originally inhabited the badlands of the Yellowstone and Missouri Rivers in eastern Montana, eastern Wyoming, western North and South Dakota, and northwestern Nebraska, was extirpated. Also eliminated were Rocky Mountain and/or California bighorn in Washington, Oregon, northern California, Nevada, and New Mexico and desert bighorn in Texas and the Mexican states of Chihuahua and Coahuila. Transplant programs have been successful in reestablishing extirpated populations in localized areas in all American states where they were extirpated (Buechner 1960, Trefethen 1975). By 1990, 14 state wildlife agencies had transplanted sheep onto more than 200 historic ranges (J. A. Bailey and Klein 1997). They have increased significantly in several areas within the contiguous United States. Private landowners in Sonora, Mexico, are actively involved in successful wild sheep conservation programs that include monitoring and protection from poaching. In Sonora and Baja California Sur, desert sheep populations have successfully been established on offshore islands (Valdez 1997).

The major decline of mountain sheep populations occurred during the latter half of the nineteenth century. It was during this period and the early 1900s that cattle and domestic sheep overgrazed much of the northwestern United States and probably southwestern Canada. Heavy grazing of northwestern Mexico and the southwestern United States occurred in the early

1800s (Holechek et al. 1995). Diseases of domestic sheep also were an important decimating factor. Additional factors that contributed locally or collectively to mountain sheep reduction were habitat loss and disturbance resulting from dam, canal, fence, and road construction. Other factors included mining, logging, urban expansion, extensive off-road vehicle use, unregulated hunting during the 1800s and early 1900s, unregulated outdoor recreation, introduction of exotic ungulates and their diseases, competition with mule deer and elk, mineral extraction, oil and gas exploration, and the usurpation of water resources. The spread of the aoudad, or Barbary *(Ammotragus lervia)*, an exotic caprine from North Africa, poses one of the greatest future threats to desert bighorn because aoudad can readily displace bighorn (Simpson 1980). Simpson and Krysl (1981) estimated there were 6,500 aoudad in the southwestern United States.

John C. Phillips (1928) best described the radical changes to which North American wild sheep have been subjected when he wrote of the big ram of Rockwell Mountain, a ram he unsuccessfully hunted in the region of Glacier National Park, Montana:

> I have often wondered what became of the Giant of Rockwell. . . . Did he die of old age or was he carried to his last resting place on the wings of a spring avalanche. And now that all his former domain is park and sanctuary and the stall-fell tourist gazes at his summer pastures from the veranda of a great hotel, what excitement is there left for his nimble progeny? Decoyed, perhaps to a man-made salt-lick close to a motor road, shorn of all their alertness, indifferent alike to the clang of passing vehicles or the alien chatter, do they, I wonder, serve a higher purpose than they did of old when they played with life and death on the highest peaks of the Great Divide.

Mountain sheep, like all other native fauna and flora, are part of the structure and heritage of North America. Despite all of the efforts exerted toward their conservation, wild sheep face a precarious future. They are an ecologically fragile species, adapted to limited habitats that are increasingly fragmented. Future conservation efforts will only be successful if land managers are able to minimize fragmentation (Bleich et al. 1990*b*, 1996). According mountain sheep their rightful share of North America and allowing them to inhabit the wilderness regions they require is a responsibility all Americans must shoulder. It is our moral and ethical obligation never to relent in the struggle to ensure their survival.

two

NATURAL HISTORY OF THINHORN SHEEP

Lyman Nichols and Fred L. Bunnell

Introduction

The thinhorn group of mountain sheep is composed of one species *(Ovis dalli)* and two subspecies, the white Dall's sheep of Alaska and northwestern Canada (*O. dalli dalli;* fig. 2.1) and the dark Stone's sheep of northwestern Canada (*O. dalli stonei;* fig. 2.2). A third subspecies has been considered in the past (*O. dalli kenaiensis;* J. A. Allen 1902), but comparisons of blood chemistry suggest it is no different than *O. d. dalli* (Heimer 1973). North American thinhorn are closely related to the dark snow sheep *(O. nivicola)* of eastern Siberia (Cowan 1940, Rausch 1961, Cherniavski 1962). A detailed description of the species was provided by Bowyer and Leslie (1992).

Habitat

DALL'S SHEEP

Dall's sheep are largely animals of the alpine. They are dependent upon steep, rugged cliffs and rock outcrops that provide escape terrain from predators; nearby open grass and sedge meadows for feeding; and most importantly, winter habitat with light snowfall and strong winds to remove snow and expose forage. The climate affecting sheep habitat varies considerably by locale and latitude, but is typically cold and dry during winter in the interior and northern mountains, with temperatures rarely reaching above the freezing point. In 1983, mean January temperatures in the western Brooks Range,

Figure 2.1. A full-curl Dall's sheep ram in summer pelage. (Photograph by Tom Walker.)

Alaska, were reported at -20.6°c, with extreme lows reaching -60°c (Ayers 1986). Snow usually is relatively shallow and soft, offering little restriction to sheep movement or to pawing for forage (Simmons 1982).

Toward the southern limits of Dall's sheep range, particularly where the range may be affected by maritime conditions, more precipitation and warmer temperatures prevail in winter. Occasional thaws with subsequent crusting and deeper snow may prevent free movement and restrict sheep bands to relatively small areas where forage is available. Summers generally are cool and moist in the south and warm and dry in the interior. High winds are common in fall and winter in interior and southern habitats; sustained winds greater than 160 km/hr have been recorded (Nichols 1973, Hoefs and Cowan 1979).

Ranges are commonly vegetated with sedges (*Carex* spp.); rushes (*Juncus* spp.); bunchgrasses such as fescues (*Festuca* spp.) and bluegrasses (*Poa* spp.); species of forbs; low shrubs including dwarf willows (*Salix* spp.), huckleberry (*Vaccinium* spp.), mountain avens (*Dryas* spp.), and crowberry (*Empetrum* spp.); heather *(Cassiope tetragona);* lichens; and mosses. On the lower slopes, which often are used in winter, dwarf birch (*Betula* spp.), alder (*Alnus* spp.), and mountain hemlock *(Tsuga mertensiana)* may form dense stands inter-

Figure 2.2. A nearly ⅞-curl Stone's sheep ram in summer pelage. (Photograph by John P. Boone.)

spersed with larger willows, cottonwood *(Populus balsamifera),* aspen *(Populus tremuloides),* and occasionally white spruce *(Picea glauca).* During summer, forage on these relatively dry mountain meadows is abundant and varied, but in winter forage is reduced in availability and variety.

STONE'S SHEEP

Stone's sheep also use precipitous escape terrain with adjacent grass-sedge feeding areas and snow-free winter feeding sites. Stone's sheep are not as much restricted to the alpine as are Dall's sheep and often are seen in subalpine brushlands and even lower forested areas. They may be found on glaciers or in isolated areas far from the nearest mountains. The largest concentrations, however, are found in lower mountainous areas to the northeast of tall mountain chains. These areas are most favorable because of their fairly high precipitation but windy regime that favors graminoid production, winter snow removal, and summer drying. A significant feature of this environment is the chinook, a high-velocity wind that is warmed by moisture as the air rises over the high mountains then descends on the lee side, where the pressure increases. Wind speed on these ranges can exceed 160 km/hr (Luckhurst 1973).

Overall, vegetation composition on Stone's sheep ranges differs little from that of Dall's sheep ranges. Graminoids, usually species of wheatgrass (*Agropyron* spp.), bluejoint (*Calamagrostis* spp.), sedges, ryegrass (*Elymus* spp.), fescue, and bluegrass, are common on alpine and subalpine ranges. A wide variety of forbs are present. Major shrub species include bearberry (*Arctostaphylos uva-ursi*), juniper (*Juniperus communis*), soapberry (*Shepherdia canadensis*), and wild rose (*Rosa acicularis*) on drier sites and several willows, shrub birch (*Betula glandulosa*), blueberry (*Vaccinium uliginosum*), and lingonberry (*V. vitis-idaea*) on moist sites. Unaltered subalpine forests are dominated by white spruce, black spruce (*Picea mariana*), alpine fir (*Abies lasiocarpa*), and lodgepole pine (*Pinus contorta*). Unless altered by fire or other disturbance, which encourages growth of deciduous species such as cottonwood or aspen and willows (*Salix* spp.), these forests provide much less forage than alpine communities. In southern portions of their range, similar conditions hold for the brush zone (i.e., willow and birch), which provides the transition between alpine and forest. Unless opened by disturbance, these areas may be too dense to permit access to sheep or to allow significant graminoid growth.

Diet

FORAGE PREFERENCES

Thinhorn sheep inhabit an extensive area and encounter a great variety of potential forage species. The more detailed analyses of diet reveal that between 50 and 120 species are eaten by any specific population (Viereck 1963, Egorov 1967, Hoefs and Cowan 1979). A few generalities emerge. First, significantly fewer species are consumed in winter than in summer and only 10 to 15 species are taken throughout the year (Egorov 1967, Hoefs 1971, Hoefs and Cowan 1979, Seip 1983). Second, most thinhorn are primarily grazers that utilize grasses and sedges. Murie (1944) examined 75 rumens of Dall's sheep collected year-round and found graminoids in all, with an average volume of 81.5%. Luckhurst (1973) examined five rumens of Stone's sheep during winter. Among the major forage classes, the mean values for percent volume were as follows: graminoids, 87.6%; forbs, 3.0%; and shrubs, 8.5%. Using fecal fragment analyses, Seip and Bunnell (1985*b*) determined that graminoids constituted 57% of the summer diet of Stone's sheep. Third, thinhorn sheep ingest more terrestrial lichens and mosses than other sheep species (Murie 1944, Egorov 1967, Hoefs and Cowan 1979, Simmons 1982, Seip and Bunnell 1985*b*, Ayers 1986).

Specific studies indicate the variability associated with these generalities.

In the Kenai Mountains, Dall's sheep fed mostly on graminoids: alpine fescue *(Festuca altaica)*, sedges *(Carex* spp.), red fescue *(Festuca rubra)*, holy grass *(Hierochloe alpina)*, and lesser amounts of other species, all of which made up more than 66% of the average annual diet. Shrubs and forbs, primarily willows with smaller amounts of mountain avens *(Dryas* spp.), fleabane *(Erigeron* spp.), blueberry, crowberry *(Empetrum nigrum)*, and others, made up about 17% of the annual diet. Lichens, mostly reindeer moss *(Cladonia* spp.), and mosses *(Selaginella* spp.) comprised the remaining 17% (L. Nichols and R. M. Hansen, Colorado State Univ., unpubl. data).

During winter, when forage was greatly reduced in quantity, quality, and number of species available, sheep consumed fewer sedges, more bunchgrasses that remain green and more nutritious at their bases, and markedly more lichens and mosses than at other seasons. Lichens and mosses are consumed more in winter than in other seasons because other species are unavailable (Cherniavski 1967, Hoefs and Brink 1978). However, Hoefs and Cowan (1979) found that in the Yukon, lichens and mosses were consumed more in midsummer when choice was greatest. Ayers (1986) reported that lichens, which are high in carbohydrates, made up 74–80% of Dall's sheep's diet during fall in a Brooks Range study, whereas willows, which are high in protein, were the forage of preference in spring. Seip (1983) also noted use of lichens by Stone's sheep during summer (5.4% occurrence in fecal pellets). Egorov (1967) reported that snow sheep in Siberia ingested large amounts of terrestrial lichens: "They are eaten intensively in autumn, spring and winter. They constitute more than 50% of the stomach contents examined during these seasons." Fedosenko (1986) also reported snow sheep feeding heavily on lichens in May and June.

In the Kenai study, Dall's sheep fed largely on sedges during summer when the choice of species was at its best, suggesting that sedges were a highly preferred food. Sedges were not as available during late winter as were bunchgrass or alpine fescue. Mountain avens was readily available but generally avoided by Dall's and Stone's sheep. However, in the Alaska and Brooks Ranges mountain avens was used heavily by Dall's sheep (Whitten 1975, Winters 1980, Ayers 1986), again illustrating the variability in food selection by area.

In the Yukon's Sheep Mountain herd, graminoids (i.e., bluejoint and bluegrass) and other less important species made up 46% of the annual diet; browse and forbs, primarily willows, wormwood *(Artemisia* spp.), vetch *(Astragalus williamsii)*, oxytrope *(Oxytropis* spp.), yarrow *(Achillea lanulosa)*, and mountain avens, made up 53%; and mosses and lichens made up less than 1% (Hoefs and Cowan 1979). Only 13 species made up more than 75% of the annual diet, with the 4 most important being sedges, wormwood, bluejoint, and willows. In the Kenai Mountains herd, alpine fescue, sedges, willows, and lichens made up more than 75% of the annual diet. Interior Alaska studies also demonstrated the importance of graminoids, especially alpine fescue, sedges, and willows, which made up a large percentage of the winter diet of Dall's sheep (Whitten 1975; Heimer 1980*a;* Gasaway and Heimer, n.d.).

In winter, Stone's sheep feed principally on grasses, especially fescue *(Festuca scabrella* and *F. ovina)* and bluegrass, and sedges, although on many ranges wild rye *(Elymus innovatus),* wheatgrass *(Agropyron subsecundum),* and holy grass are important. Mountain avens also provide some winter forage, as does wormwood on those ranges (usually burned) where it is found (Seip and Bunnell 1985*a*).

In spring and early summer, Stone's sheep feed on grasses, sedges, forbs, and browse, especially willow; in summer, they seek forbs, including vetch and oxytrope, plus some browse, namely willows and poplars *(Populus* spp.; Seip and Bunnell 1985*b*). Luckhurst (1973) determined that on one range in northern British Columbia, wild rye, wheatgrass, bluegrass, oxytrope, willow, and huckleberry *(Vaccinium* spp.) were the most important forage species used by Stone's sheep. Although diets vary by area and forage availability, for the most part thinhorn sheep depend on relatively few plant species; grasses, sedges, willows, and in some areas wormwood or mountain avens make up the greatest portion of their annual diet.

FORAGING BEHAVIOR

Thinhorn sheep optimize their summer diets by seeking the most nutritious forage available. In spring, they move down from their alpine winter ranges to feed on newly emergent green vegetation, especially willows. As the snowline retreats upward, allowing new forage to emerge, sheep also move upslope to continue making use of this lush, high-quality forage in preference to the more abundant but maturing, and thus lower-quality, vegetation at lower altitudes. As summer progresses, thinhorn sheep may even move onto cooler, moister northern slopes, where late season grasses and forbs are just emerging (Whitten 1975, Winters 1980, Curby 1981, Jakimchuk et al. 1984, Seip and Bunnell 1985*b*, Ayers 1986).

Thus, sheep are able behaviorally to extend their all-important season of high protein intake throughout the summer, enabling them to accumulate the growth and body reserves needed to survive the following winter. Areas with a high degree of topographic diversity (i.e., steep and higher terrain) have a greater spread in time and space for early growth forage, thus extending the growing season and allowing sheep to obtain higher-quality forage over a longer period. Such habitats usually are associated with better forage and, thus, higher-quality sheep (Whitten 1975, Winters 1980).

By early fall, when most forage has matured, dried out, and passed its more nutritious stages, some thinhorn sheep may move back down to alpine or subalpine valley floors and lower slopes, again feeding on willows (Jakimchuk et al. 1984) or switching from diets high in protein (i.e., willows, graminoids, and forbs) to those high in carbohydrates (i.e., lichens; Ayers 1986).

As winter progresses, snowpack forces sheep onto windblown, alpine, and subalpine ridges, where forage has been exposed by wind action and/or snow is shallow and can easily be pawed through. Here sheep seek out the still-green bases of bunchgrasses, such as alpine fescue, but also consume lichens and other desiccated plants of poor nutritional value.

FORAGE RELATIONSHIPS

Density is often used in judging the condition of a herd in relation to its forage base. For example, Hoefs and Cowan (1979) estimated that there were about 6.8 Dall's sheep/km^2 on winter range in Kluane Park, a stable herd presumably maintaining itself at range carrying capacity. Winter densities were only about 1.0–1.3 sheep/km^2 in the Mackenzie Mountains, Northwest Territories (Simmons et al. 1984), and approximately 1.1 sheep/km^2 in a Brooks Range herd (Jakimchuk et al. 1984). All three of these herds were considered stable and probably at range carrying capacity.

Heimer (1980*a*) estimated a density of 5.3 sheep/km^2 on winter range of the Alaska Range Dry Creek herd that had been slowly declining, that was undoubtedly at or above the carrying capacity of its winter range, and that he considered to be a low-quality herd. On the nearby Sheep Creek winter range, Heimer estimated 1.8 sheep/km^2. He considered this a high-quality herd, primarily on the basis of horn size and herd density.

Nichols (1976) estimated winter densities of 10.6–24.7 Dall's sheep/km^2 on three Kenai Peninsula herds presumed to be at or near winter range carrying capacity. Thus, a considerable variation in densities was encountered on winter ranges of herds all thought to be at a nearly similar relative status, whereas comparatively low density was encountered in a herd judged by Heimer (1980*a*) to be high quality partly on the basis of its density. However, he noted that forage on the low-quality herd's range was abundant, whereas that on the high-quality herd's range was sparse.

The density of sheep relative to the amount of forage produced on their range more accurately describes their status. Heimer's (1980*a*) low-quality herd on abundant forage (i.e., large number of animals per unit of area) might be no denser in relation to its food base than his high-quality herd on sparse forage (i.e., low numbers of animals per unit of area). He did not determine the amount of forage per animal-unit on either area. A more detailed comparison of animal nutritional and reproductive condition suggested that this high-quality herd was, indeed, existing at a higher nutritional level, strongly implying that plane of nutrition, rather than density alone, is an important determinant of sheep quality (Gasaway and Heimer, n.d.).

An accurate assessment can be made of the relationship between sheep numbers and their range by comparing animal numbers with the actual biomass of usable forage produced on their summer or winter range. Little research has been undertaken on this subject for thinhorn sheep. Nichols (1976) and Hoefs and Brink (1978) have attempted the calculations for Dall's sheep. Hoefs and Brink (1978) found that primary production of 81.2 g/m^2 (i.e., 15.5 kg/sheep-day, our calculation) was associated with good lamb survival (i.e., 71%). Lower levels of primary production, $65-70 \text{ g/m}^2$, were associated with poorer lamb survival (i.e., 60%). Bunnell (1978) corroborated their findings and reported that horn growth and lamb survival were associated with favorable growing conditions for forage.

Nichols (1976) compared three herds using only the production of graminoids, the most important forage class. He estimated that about 570–1,031 kg (air dry weight) of grasses and sedges were available per sheep before winter. By the second fall of his study, two of the herds had increased, whereas forage production had decreased (primarily due to a dry summer), resulting in about 425 kg of forage being available per sheep on both areas. The third herd had been reduced by controlled hunting in the interim, and production of grasses and sedges on its range had increased to about 960 kg/sheep, despite the drier summer. Reproduction in this herd improved significantly, whereas that in the other two herds did not. Furthermore, the other two herds declined in numbers a year later.

More study is needed on the relationship between quantity and quality of usable forage available and the numbers and health of sheep using the forage. The preliminary work suggests a method of assessing carrying capacity by estimating the weight of primary forage species per animal per unit area of winter range.

With much larger areas of habitat available, summer densities have been found to be considerably lower on average than densities in winter. Examples of recorded summer densities have ranged from fewer than 0.4 sheep/km^2 in the western Brooks Range to 2.0 sheep/km^2 in the central Alaska Range (F. J. Singer et al. 1983).

FORAGE VALUE AND DIETARY EFFECTS

During late winter, forage on the exposed ridges becomes dried and wind shattered. However, the protected, lower stems of dense bunchgrasses remain green and are heavily grazed all winter. Other plants that have been protected by snow cover also may remain green throughout most of the winter and therefore provide some nutritious forage when they can be reached.

The general quality of forage is lowered in winter, however. In the Kenai diet study (L. Nichols and R. M. Hansen, Colorado State Univ., unpubl. data), crude protein (CP) was measured from summer- and winter-collected forage. During summer, CP constituted 11–14% of the sheep's estimated diet, exceeding the minimum level of about 8% recommended by the National Research Council (1968) for domestic sheep. In winter, the CP level dropped to less than 5% of the diet. Protein levels on other Dall's and Stone's sheep ranges were similar (Whitten 1975, Seip and Bunnell 1985*b*).

For all plant communities combined, Luckhurst (1973) reported fall CP levels of 9.1% (range = 7.5–10.9%). Levels in the weathered, early spring vegetation declined over winter to an average of 6.9% (range = 4.0–7.1%). J. P. Elliott (1978) found a key grass on Stone's sheep range (wild rye) to contain 15–18% CP in summer, declining to 4–6% in winter.

If one assumes that dietary requirements of thinhorn sheep and domestic sheep are similar, it is apparent that the former's late winter diet is below that required for maintenance and that sheep are then in negative protein, and likely energy, balance (i.e., sheep use more protein and energy than they consume). Whereas in summer the sheep are able to gain weight and store fat, in winter fat is catabolized for energy and weight is lost.

Fat stored in the bone marrow is one of the last sources to be used by the body, and femur marrow fat is commonly used as an indicator of body condition. In a series of sheep collected throughout one winter in Alaska's Kenai Mountains, the percentage of fat in the femur marrow declined from more than 90% in early winter to about 30% in April. Marrow taken from winter-killed sheep suggested that when the percentage of marrow fat approached 10%, death would almost certainly result (Nichols 1971, 1978*b*). Thus, a particularly severe or long winter can accelerate or extend the period of body fat loss in Dall's sheep and result in death from malnutrition.

RANGE ENHANCEMENT

Although lack of natural habitat generally has not been a problem for thinhorn sheep in the vast northern areas they occupy, some accessible or intensively managed herds might benefit from enhancement of their range under certain circumstances. For example, herds that have stabilized at their range carrying capacity might be increased or improved in quality (i.e., larger horns, higher birth rates) by improving the quantity or quality of their forage. Unsuitable areas might be altered so that they could support sheep where none now exist. Bighorn sheep established themselves in reclaimed open-pit mines in Alberta, using the pit walls as escape terrain and feeding on the revegetated grasses and legumes (MacCallum 1988, MacCallum and Geist 1992). Similarly, Dall's sheep moved into revegetated coal strip mines in Alaska that provided suitable new habitat (C. L. Elliott and McKendrick 1984), and they grazed on nonnative, revegetated grasses in the fall along parts of the Trans-Alaska Pipeline in the Brooks Range (Jakimchuk et al. 1984).

Range burning has been used in Stone's sheep habitat in British Columbia. This practice reversed a decline in one herd by increasing the lamb:ewe ratios, improving the horn size of rams, and lowering spring lungworm loads in sheep wintering in burned areas (J. P. Elliott 1985, Seip and Bunnell 1985a). Earlier spring forage growth in the burned areas was believed to give these animals a head start during their critical period of spring growth. Additional winter forage also was made available. Similar results were achieved in burned-over bighorn sheep ranges, where larger carrying capacities and better nutrition were obtained (Hobbs and Spowart 1984, Hobbs and Swift 1985, McWhirter et al. 1992).

The effects of a 1974 wildfire in Dall's sheep habitat on Round Mountain, Kenai Peninsula, Alaska, were studied by the U.S. Forest Service (Culbertson et al. 1980, Culbertson and Walker 1981). This very hot burn altered the habitat from a shrub-dominated subalpine type to one dominated by forbs and mosses but slightly lowered the production of grasses and sedges. Nevertheless, the researchers found that sheep used the burned area over twice as much during summer as the adjacent unburned slopes. This herd increased steadily from 60 sheep in 1974 to a peak of 151 sheep in 1991 (L. Nichols, unpubl. data). Prescribed burning is probably the most promising and economical technique available for large-scale use in sheep range enhancement.

The U.S. Forest Service began experimenting with fertilizer enhancement of Dall's sheep winter range. Preliminary results have demonstrated a significant increase in biomass of preferred forage (S. Howell, U.S. Forest Service, Chugach National Forest, unpubl. data).

WATER

Lack of drinking water has never been documented as a problem for thinhorn sheep. In winter they obtain water from snow; in spring and summer water is available from late-melting snowbanks or numerous springs or creeks. Certainly, some alpine sheep ranges become dry in late summer, probably forcing sheep to seek water in valley-bottom creeks. In most sheep habitat, surface water appears to be plentiful throughout the year.

MINERAL LICKS

Natural mineral licks are a conspicuous and important part of thinhorn sheep habitat. Mineral licks are found in most Alaskan and Canadian sheep ranges. Lick sites may develop at seep springs high on alpine slopes, in gullies, or in timbered valley floors. They often are characterized by moist, claylike soils that have been disturbed by the sheep's use and churned into mud holes. These sites generally form the focus of a spiderweb pattern of sheep trails, which converge from many directions and make the lick obvious to an observer. Sheep may spend hours at a time eating the soil in a lick to the extent that their droppings appear to be formed entirely of clay. The soil in a lick is further muddied and mineralized by urine and feces deposited over many years of use.

The importance of mineral licks may be illustrated by a site in the Alaska Range that served a sheep population of approximately 1,500 animals, some of which traveled as far as 19.3 km to the lick. Heimer (1974) found the site to be so important to these sheep that all ewes and more than 80% of rams would return each year. Lactating ewes spent more time at the lick than other classes of sheep, with approximately 45% of a ewe's time being spent there during early summer. Simmons (1982) reported that few rams older than three years of age used licks in the Mackenzie Mountains, Northwest Territories.

In Alaska's Watana Hills, Tankersley (1984) reported that sheep traveled through forested areas for approximately 8 km to reach a relatively low-elevation lick located in a forested bluff above a creek. Rams used this lick exclusively from 11 to 28 May; their use diminished by mid-June. Ewes and young began use on 29 May and continued licking until early August, with a peak of use in mid-June.

Mineral licks generally are used most intensively from late May through mid-June, but some use occurs throughout the summer. Maximum daily use occurs from early morning through midday, and intensity of use has been correlated directly with air temperature (Heimer 1974, Klingel et al. 1974, Summerfield 1974, Jakimchuk et al. 1984). Linderman (1972) stated that the early summer distribution of all sheep in the Brooks Range was influenced by their use of licks, whereas Simmons (1982) felt that the entire annual range of Dall's sheep in the Mackenzie Mountains, Northwest Territories, was heavily influenced by mineral lick locations.

Stone's sheep use of licks is similar. The most intensive use occurs in late spring and early fall, but sheep may descend from alpine ranges throughout the summer to use low-elevation licks. The apparent earlier use of licks in summer (i.e., April and May) may be a result of Stone's sheep occupying more southerly ranges than Dall's sheep and licks being located where earlier plant growth flourishes. Like Dall's sheep, Stone's sheep may travel more than 16 km to licks (Luckhurst 1973).

The benefits gained by eating at licks are not yet completely resolved. Geist (1971) hypothesized that licks were used to replace skeletal minerals lost during winter. By placing different chemicals in several types of choice boxes, Stockstad et al. (1953) determined that other species of big game were seeking sodium. Heimer (1988) hypothesized that magnesium was the paramount mineral sought at licks, particularly in wetter springs. Hebert and Cowan (1971) believed that mountain goats *(Oreamnos americanus)* were using licks to curb the increased metabolic loss of sodium in spring, which occurs during the change in diet from dry to succulent forage. New, succulent, spring forage eaten by ungulates is typically high in potassium and low in sodium, calcium, and magnesium, thus causing a potassium imbalance. Ingesting sodium, magnesium, and possibly calcium at licks helps to correct this dietary problem (R. L. Jones and Hanson 1985).

A side benefit from lick use occurs when subpopulations of sheep intermingle at licks. Because of their propensity for following older rams, young rams frequently are exchanged between herds (Luckhurst 1973, Heimer 1974, Simmons 1982). Such a shuffling of individuals would serve to enhance genetic exchange between sheep populations.

Mineral licks are scarce in Alaska's Kenai Mountains. The few that have been located are not used as regularly or by as large a portion of the nearby sheep herds as are those in interior Alaska. It is possible that the herds' forage contains adequate trace elements for metabolic needs, or that suitable lick sites simply do not exist at enough locations to service the main sheep population. Whitten (1975) noted that in Mount McKinley National Park no mineral licks were being used intensively during his study, despite the fact that more than one had been used heavily in the past. Thus, the need for trace elements, as indicated by lick use, may vary over time for as yet unknown reasons. Sheep and mountain goats on the Kenai Peninsula use salt blocks (i.e., sodium chloride), suggesting either a need or a preference for salt (L. Nichols, unpubl. data).

Bighorn sheep were attracted to left-over construction minerals around abandoned well sites in Alberta and were probably seeking sodium (Morgantini and Bruns 1988). Dall's sheep similarly used construction by-products as licks at abandoned construction sites along the Trans-Alaska Pipeline in the Brooks Range (Jakimchuk et al. 1984).

An interesting use of a nonnatural lick by Dall's sheep began in June 1986 along the Seward Highway just south of Anchorage, Alaska. A series of large, south-facing cliffs effectively extends sheep habitat down from the alpine in the Chugach Mountains to a point on the highway 5 m above sea level. Sheep come down to lick soil in the ditch adjacent to this busy highway, possibly obtaining sodium and/or calcium from gravel that is spread on the road during winter and is later washed into the ditch. The attraction is so great that ewes, lambs, and young rams will remain licking despite noisy groups of photographers, stopped automobiles and tour busses, and heavy traffic passing or stopped within 10 m of the animals. This may be the only place where thinhorn sheep range extends from sea level to the high alpine.

FALSE MINERAL LICKS

In Alaska's Kenai Mountains, Dall's sheep use the false hellebore plant *(Veratrum escholtzii)* in a similar way to mineral licks used elsewhere (Nichols 1973). Sheep use a few select plants at a given location, usually near the upper altitudinal limit of the plant's range, and often in association with old marmot *(Marmota caligata)* burrows. Sheep seek primarily the lower stems and rootstocks of these plants. Possibly, such sites are first selected where the marmot's digging has exposed the roots. The sheep spend many hours digging and scraping at these exposed rootstocks, apparently consuming only small amounts of the plant and leaving surrounding plants of the same species untouched. Eventually, such use creates many small pits, surrounding areas trampled and grazed bare of other vegetation, and the typical mineral-lick spiderweb of trails radiating from the areas. The sites and the sheep's behavior suggest that these are true mineral licks. The main difference in use is seasonal; sites are most heavily utilized in late summer and early fall, with some use even after the ground has frozen. No studies have been undertaken to learn the intensity of use by individual sheep or the substance being sought from the plant.

False hellebore is poisonous to domestic sheep. Ingestion in sufficient quantities causes death within six to eight hours. Sublethal doses cause excess salivation and urination, weakness, irregular gait, rapid respiration, and irregular heartbeat. Very small amounts over a long period seem to have no effect on domestic sheep; however, if consumed at about the fourteenth day of gestation, false hellebore causes congenital, cyclopian-type malformations in lambs. Continued ingestion causes spontaneous abortion in many domestic ewes (Binns et al. 1963). The toxic agent in false hellebore is cyclopamine. It was found to be highly concentrated in the roots, particularly after the middle of August in Utah specimens (Keeler and Binns 1971).

Further investigation is needed to determine the significance of these "licks" to Dall's sheep. They may be a substitute in the absence of true mineral licks and supply sheep with required trace elements or nutrients. It is possible that late season use in December could lead to reduced success in lambing. Use of these false licks has not been documented elsewhere.

Growth

BODY GROWTH

From a birth weight of 3.2 to 4.1 kg in late May and early June, Dall's lambs reach a weight of about 29.5–30.4 kg by early winter (Bunnell and Olsen 1976, Nichols 1978a). Stone's sheep weigh about the same at birth; female and male yearlings in the following July average 23 and 28 kg, respectively (Seip and Bunnell 1984). During one winter in Alaska's Kenai Mountains, lambs lost approximately 40% of their fall weight and by spring weighed about 17.7 kg. During their second summer, yearlings in that herd reached an average weight of 44.9 kg, but again lost more than 40% of their late fall weight during their second winter, weighing only 24.0 kg by age 23 months (Nichols 1978a). Because much of their second summer's feeding goes into growth (without the benefit of mother's milk), yearlings probably are unable to store fat reserves proportionally as large as adults. Lack of such reserves may contribute to their winter weight loss, failure to give birth, and to mortality. Growth of both sexes was largely complete by age 6 in the Yukon's Sheep Mountain herd (Hoefs 1984a).

In late summer, adult ewes in good condition averaged 56.3 kg in Alaska. By spring, average weight of ewes was 47.2 kg, a loss of 16% during winter (fig. 2.3). Of course, winter weight loss for all age classes depends upon winter severity; the reported weight losses were those following a fairly severe winter (Nichols 1972, 1978b). Weight loss over winter was greater in pregnant bighorn ewes compared to those that were not pregnant, whereas summer growth rates were more rapid in nonlactating versus lactating ewes (Jorgenson and Wishart 1984), demonstrating that the costs of reproduction affect ewes and rams. Similar relationships undoubtedly hold true for thinhorn ewes.

Bunnell and Olsen (1976) found similar weights for Dall's ewes in the Yukon Territory in late winter (48.8 kg). Adult rams weighed an average of

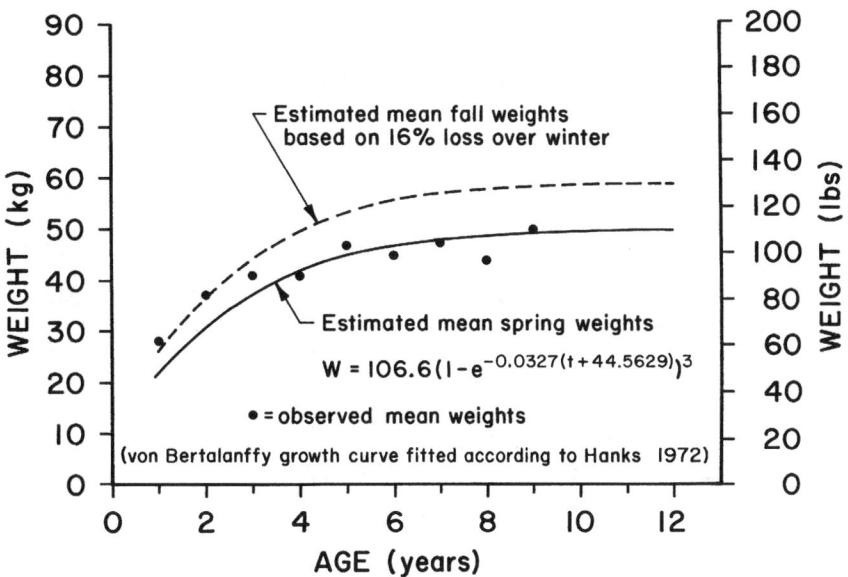

Figure 2.3. Estimated mean spring and fall weights of Dall's sheep ewes (data from Heimer 1972).

74.6 kg during winter. This average may be somewhat biased on the high side for spring weights, because several of these rams were weighed in early winter when they were in prime condition. The heaviest ram weighed 82.1 kg. Heimer (1972) weighed a number of adult Dall's rams in spring in Alaska; their average weight was 69.6 kg (fig. 2.4). Few actual weights of adult rams in Alaska in fall are available, but if one assumes a winter weight loss similar to ewes, the average weight of those rams in fall would be expected to have been about 83 kg. Large rams exceed 91 kg, and a few exceed 113 kg. Geist (1971) reported one ram from the Yukon that weighed 102.5 kg in the late fall, and J. P. Bone (B.C. Fish and Wildlife Branch, Victoria, pers. comm.) recorded ram weights of 120 and 129 kg from the same area. The average maximum weight of Dall's rams in the Sheep Mountain, Yukon, herd was 104 kg (Hoefs and Barichello 1984).

Shoulder height of a sample of Dall's rams more than six years old averaged 93.3 cm (fig. 2.5), and ewes averaged 84.4 cm (Heimer 1972, Nichols 1972; fig. 2.6). Dall's rams at Sheep Mountain, Yukon, were less than 102.6 cm shoulder height (Bunnell and Olsen 1976) and heavier than those reported by Heimer (1972) and Nichols (1972). Data on weights and measurements of free-ranging Stone's sheep are sparse. Seip and Bunnell (1984) reported that adult ewes from northeastern British Columbia weighed an average of 49.6

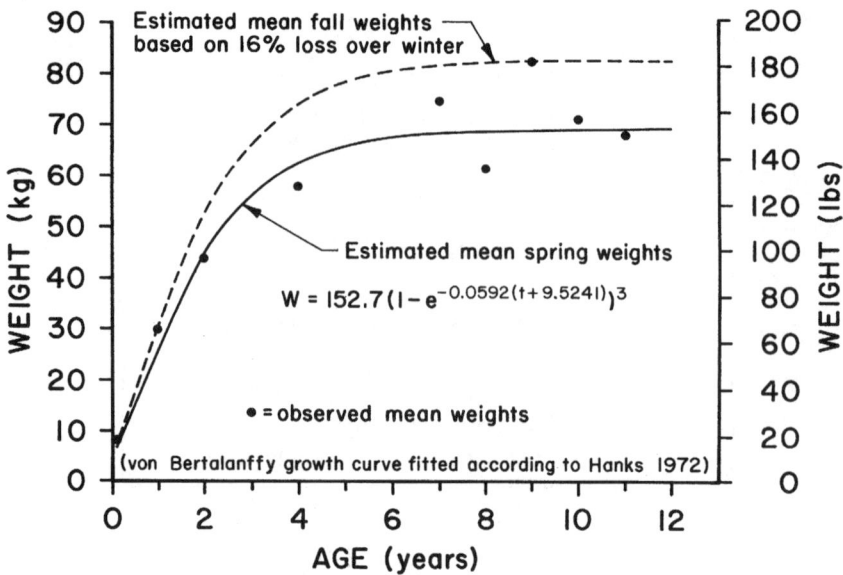

Figure 2.4. Estimated mean spring and fall weights of Dall's sheep rams (data from Bunnell and Olsen 1976; W. E. Heimer, unpubl. data; L. Nichols, unpubl. data).

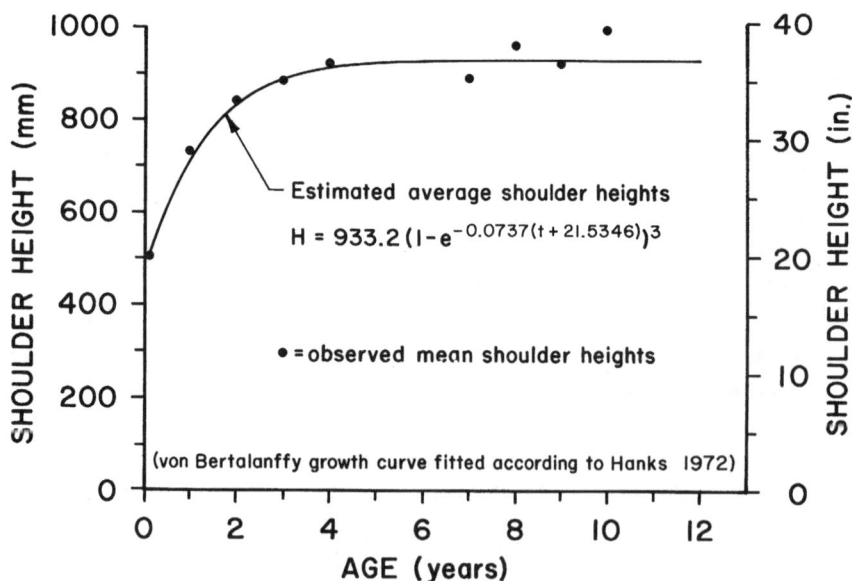

Figure 2.5. Estimated mean shoulder heights of Dall's sheep rams (data from Heimer 1972).

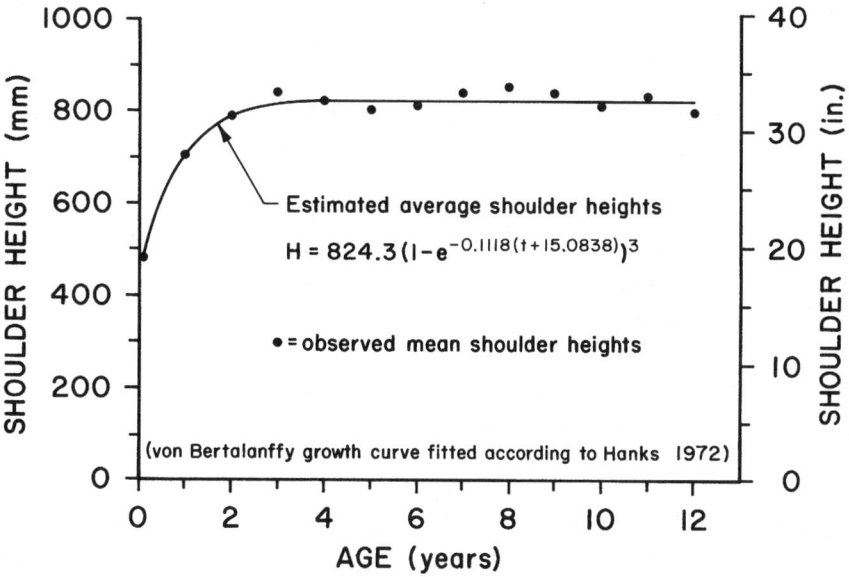

Figure 2.6. Estimated mean shoulder heights of Dall's sheep ewes (data from Heimer 1972).

kg in July. Average shoulder height was 88.3 cm. Geist (1971) reported the body weight and shoulder height for a 6.5-year-old, emaciated ram (77.3 kg, 101.6 cm) and a ewe more than 16 years old (61.4 kg, 88.9 cm). Data available indicate no differences in gross morphology between Stone's and Dall's sheep. Weights and sizes of both subspecies are variable depending on season, plane of nutrition, and genetics.

Bunnell (1980b) reported correlations between body weight, body length, chest girth, and horn length in Dall's sheep. He developed methods of estimating weight by correlating known weights with other body parameters from sample sheep. Body length and chest girth provided the most accurate estimates of weight in his sample.

HORN GROWTH

Both sexes of thinhorn sheep have horns. Those of the females are relatively small, straight, thin, and roughly oval in cross section. A sample of 54 Dall's sheep ewe horns older than six years averaged 24.5 cm in length. Mean horn length of adult Stone's ewes is about 26 cm (Seip and Bunnell 1984).

Ram's horns are more massive and longer than those of ewes. They are roughly triangular in cross section and grow in a flaring helix. In Alaska, the average length of a full-curl ram's horns is about 90.1 cm. "Full-curl" means the horn tips have attained a 360° curl about their axis. When viewed into the axis of the horn curl, the outside surface of the horn follows a circular path. Measurements of the diameter of this curl in adult Alaskan rams averaged 27.0 cm and in the Mackenzie Mountains, Northwest Territories, 28.8 cm. Horn basal circumference averaged about 36 cm (J. A. Erickson 1970, Heimer and Smith 1975, Nichols 1978*b*, Simmons et al. 1981, Bayer and Simmons 1984). Of course, rams with larger horns may be found; the longest Dall's horn ever recorded was 123.5 cm (Reneau and Reneau 1993).

Stone's sheep rams are similar in horn size to Dall's sheep. No statistical differences ($P < 0.001$) could be detected between subspecies in average horn score when comparing the top-scoring 50 heads of each from the Boone and Crockett Club's record book (Reneau and Reneau 1993). The longest Stone's sheep horns recorded were 131.6 cm in length.

Horns begin to grow in both sexes by the second week in July during their first summer and continue growing until late December, reaching a length of about 4 to 6 cm by their first winter. Thereafter, rams' horns grow faster than ewes' horns. Maximum growth in length takes place in the second summer and decreases each summer thereafter (Hemming 1961, 1969; Hoefs and Nette 1982). Horns of both sexes continue to grow throughout life (fig. 2.7).

Horn growth ceases each fall from late September to early November, probably as a result of food shortage and hormonal changes during rut. By the time rams reach maturity at five to six years of age, the annual horn growth period has decreased to only about six months, with 75–80% of annual growth completed by 1 August. Growth begins again in late February to late April (Hemming 1961, Bunnell 1978, Hoefs and Nette 1982).

Except for the first winter, when horn growth may not completely stop, a growth ring (or annulus) is deposited during each dormant winter period, enabling the observer to estimate accurately the animal's age by counting these rings (Geist 1966*a*, Hemming 1969). Both sexes may be so aged, but care must be taken to count only true growth annuli, not the many ridges (or false annuli) deposited during active horn growth. Old ewes are more difficult to age than rams because growth rings may be closely spaced and hard to detect among the numerous false annuli (Hoefs and Konig 1984). However, careful brushing with a stiff bristle brush and examination through a lens will nearly always make apparent the grayish white lines of the true annuli (L. Nichols, pers. obs.). A strong correlation between tooth cementum layers and horn annuli confirmed the value of the latter in aging thinhorn sheep (Hemming 1969, Simmons et al. 1981).

Horns grow only at the interface of the horn material and the bony cone-

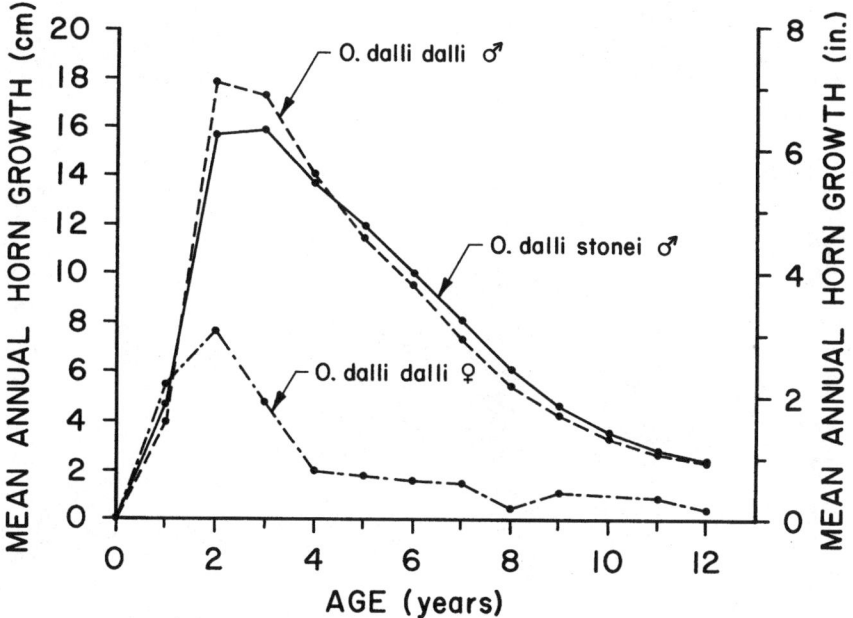

Figure 2.7. Mean annual horn growth of Dall's and Stone's sheep rams and Dall's sheep ewes (from W. E. Heimer, unpubl. data; J. P. Elliott; unpubl. data).

shaped horn core that is part of the skull. The resultant, cross-sectional form of horn growth is similar to a curling stack of cones, each overlapping the one nearer the skull. The annulus is formed where the proximal (inner) end of the year's "cone" reaches the horn's surface. The distal (outer) cones are each smaller in diameter because the tips represent the horns earliest growth stages (fig. 2.8).

Cumulative and annual growth rates of Dall's and Stone's rams are similar (fig. 2.9). Heimer and Smith (1974, 1975) discussed Dall's sheep ram horn growth in detail by volume, age, curl, and mountain range.

Konig and Hoefs (1984) found that the volume of horn material in rams' horns increased more uniformly with age than did length or circumference, with the greatest growth occurring in the fourth through sixth years, rather than the second as with length. Bunnell (1978) found that Dall's rams' horns showed maximum growth during early summer and that early forage production, as influenced by spring precipitation, was of primary importance in annual horn growth. Rates of horn growth were not uniform each year for individual animals, but varied directly with spring weather and, hence, nutrition. Bunnell (1978) also reported that horn growth was influenced for

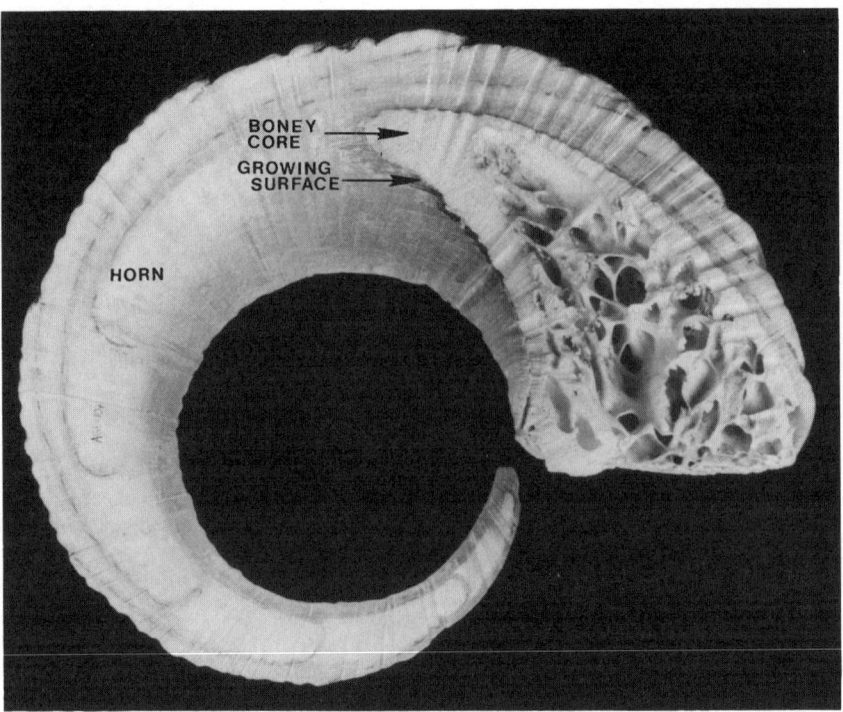

Figure 2.8. Ten-year-old Dall's sheep ram horn sectioned along the midline showing annual growth segments and the heavy, bony core adapted for the shock of clashes. Growth occurs only at the interface of bone and horn sheath. (Photograph by Lyman Nichols.)

approximately five years following birth by the condition of the ram's mother during pregnancy and lactation. Data for Stone's sheep rams relating horn growth to range fire history also suggest a nutritional effect on growth (J. P. Elliott 1978). The effect is most pronounced in the yearling increment and may disappear in older animals (Seip and Bunnell 1984). Mean early horn growth showed an increasing cline when comparing sheep from the cold, dry Brooks Range in northern Alaska through the warmer, moist Kenai Mountains near the southern end of Dall's sheep range, further supporting the relationship between nutrition and horn growth (Bayer and Simmons 1984).

Geist (1966a) related individual patterns of horn growth to survival in Rocky Mountain bighorn *(O. canadensis canadensis)*. Animals dying younger tended to exhibit more rapid early horn growth, thus earlier sexual maturity. Implicit in Geist's (1966a) concept is the assumption that individuals showing relatively rapid early growth of horn show relatively slower later growth.

Bunnell (1978) evaluated the assumption for Dall's sheep and found it to be valid.

Hunting regulations usually are based on the degree of rams' horn curl. Most Dall's rams reach ¾-curl (270°) by age 5, ⅞-curl (315°) 1 or 2 years later, and full-curl (360°) by age 9, although this varies by location, nutrition, and possibly genetics. For example, in one Brooks Range study, the mean age at reaching ¾-curl was 6 years, ⅞-curl was 10 years, and full-curl was 10.5 years (Klingel et al. 1974).

Not all rams will attain full-curl horns, regardless of their age. On average in Alaska, approximately 15% of rams will never reach full-curl in their lifetime because of the growth rate or growth form of their horns. This potential to reach full-curl varied by mountain range, with 90–100% of rams from the Wrangell, Kenai, and Chugach Ranges having this potential, whereas only 63% of rams from the Brooks Range could be expected to reach full-curl in their lifetime (Nichols 1984). In addition, although growth continues throughout life at the horn bases, wear of rams' horns at the tips begins to equal or exceed growth by age 8 or 9, so little if any length is added to most horns after that age (Hoefs and Nette 1982, Hoefs 1984a). "Brooming," or breakage of the horn tips during fights, also reduces total horn length in many rams.

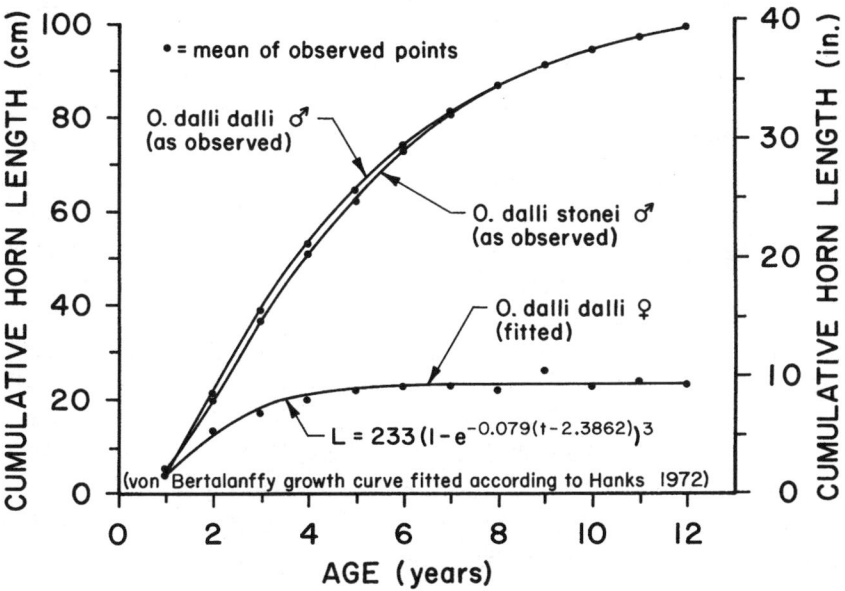

Figure 2.9. Cumulative horn growth of Dall's and Stone's sheep rams and Dall's sheep ewes (from W. E. Heimer, unpubl. data; J. P. Elliott, unpubl. data).

In the Sheep Mountain herd, Hoefs (1984b) estimated that approximately 121 "nursery sheep" were required (i.e., ewes, lambs, yearlings, and some two-year-old rams that form these groups) to produce 10 ¾-curl rams and 7 ⅞-curl rams per year. Working in the Mackenzie Mountains, Simmons et al. (1981) estimated that only 7.3% of rams born would ever reach full-curl.

There were approximately 27% less full-curl rams than ¾-curl rams in the Sheep Mountain herd (Hoefs 1984a) due to natural mortality in the intervening years. In the Alaska Range, Heimer and Smith (1975) found this reduction to be 37%, whereas in the arctic Brooks Range, Summerfield (1974) estimated the reduction to be 61–69%. It is possible that these investigators included those rams not reaching full-curl because of horn growth form and horn tip wear in their "natural mortality" estimates. Nevertheless, a substantial percentage of rams whose horns reach ¾-curl will not reach full-curl status in their lifetime, and thus would be lost to hunters under a full-curl law versus a ¾-curl law.

Activity Patterns

DAILY ACTIVITY PATTERNS

Activity patterns of thinhorn sheep vary with the seasons and possibly latitude, primarily as a function of day length, but also with energy demands. At Sheep Mountain, Yukon Territory, day length on 1 January is 5 hours 42 minutes, increasing to 19 hours 11 minutes by 30 June. When day length was shortest, Dall's sheep foraged only 5 hours daily; 74% of foraging occurred between sunrise and sunset. By 15 April, average time of foraging had increased to 9.4 hours daily and reached 13.6 hours daily by the peak of lambing on 18 May (F. L. Bunnell and A.C.M. Farr, Univ. British Columbia, unpubl. data). Once day length reached 14 hours, 90 to 97% of the foraging occurred during daylight hours.

Stone's sheep show similar patterns. They are active for 5–6 hours daily (72% during daylight) during winter, but increase their active period to 11–13 hours (90 to 95% during daylight) by the lambing period and early summer (Geist 1971, Seip and Bunnell 1985b).

The role of energy demand is most obvious during the May–June lambing period when some females bear the additional cost of lactation (table 2.1). Lactating females forage for 55 to 70% of reported observation periods, whereas adult rams forage during only 40 to 55% of the same periods. At Sheep Mountain, lactating Dall's ewes foraged for 11.9 hours daily, nonlactating ewes for 10.9 hours (some of these were pregnant), and adult rams for 9.1

Table 2.1. Activity budgets of thinhorn and Rocky Mountain bighorn sheep during lambing season (percent of observation period).

SPECIES	ANIMAL CLASS	ACTIVITY			
		FEEDING	LYING	STANDING	TRAVELING
Dall's sheep[a]	Adult males	55	41	2	2
	Lactating females	62	28	4	6
	Lambs	20	52	7	21
Stone's sheep[b]	Lactating females	54	33	6	7
	Lambs	20	55	14	11
Stone's sheep[c]	Adult males	49	42	6	3
	Lactating females	62	29	5	4
Rocky Mountain bighorn[d]	Lactating females	72	15	8	5
	Lambs	12	42	30	16

Sources:
[a] Bunnell and Farr (unpubl. data).
[b] Geist (1971:306).
[c] Seip and Bunnell (unpubl. data).
[d] Geist (1971:306).

hours daily (F. L. Bunnell and A.C.M. Farr, Univ. British Columbia, unpubl. data). The lactating ewes utilized 82% of available daylight hours in foraging versus 75 and 63% for nonlactating ewes and rams, respectively. This behavior indicates the increased energy demands of lactation.

Lengths of foraging bouts clearly are related to body size and probably to rumen size. Adult Dall's rams exhibit only 3.6 ± 0.4 foraging bouts/day (foraging bouts are periods when the animals are up, feeding, or searching for forage); comparable values for adult ewes are 5.2 ± 0.3 foraging bouts/day. Time spent feeding generally is greatest in the evening, prior to darkness. At that time, mean bout lengths are 179.1 ± 19.4 minutes for adult rams, 106 ± 7.6 minutes for lactating ewes, and 68.3 ± 6.4 minutes for ewes without lambs (F. L. Bunnell and A.C.M. Farr, Univ. British Columbia, unpubl. data). Observations available for Stone's sheep are not as detailed, but suggest similar patterns (Geist 1971, Seip 1983).

Generally, during the long summer days in the arctic and subarctic regions, thinhorn sheep are most active early in the morning and late in the evening, with shorter midday feeding periods. However, some sheep may be seen feeding at any daylight hour. Between feeding periods, they spend their time resting, ruminating, and sometimes moving about. Rams spend time in dominance-related activities. In the arctic Brooks Range, sheep were active during all of the 24 hours of summer daylight. After darkness began again in late summer, observed activity was condensed to daylight hours (Curby 1981).

Thinhorn sheep usually prefer to rest in or very close to escape terrain at night, moving away during the day to preferred feeding sites. Thus, the daily activity pattern usually shows a drifting away from rocky cliffs and outcrops during the day, with a return as darkness approaches. This behavior is most pronounced when nights are dark and particularly so in winter when the only available feeding sites might be some distance from protective escape terrain. The extent to which thinhorn sheep feed during the hours of darkness is poorly documented. In the arctic, where the sun does not rise in midwinter, Dall's sheep have little choice but to feed in the dark or, at best, in very dim twilight.

MOVEMENTS

Sheep of all sexes and ages congregate during the late fall rut, often on areas that become winter range. Winter ranges for thinhorn sheep are determined primarily by snow cover, or lack thereof, and escape terrain. These may be in portions of the summer range or in entirely different areas, but must offer snow-free feeding sites, preferably near rugged cliffs. Much of the habitat used in summer is covered with deep snow and is unavailable to sheep in winter. Thus, the overall annual range of a sheep herd may consist of relatively unlimited summer habitat, restricted only by accessibility to escape terrain and perhaps bounded by major topographic features such as river valleys or snowfields, that shrinks to a much smaller usable size in winter because of snow cover. The maximum size of the herd is limited by the number of animals that can be supported by the restricted winter range.

In winter, Dall's sheep usually remain in the alpine, depending for food on areas blown free of snow or where snow is soft and shallow and can be pawed to expose forage. On some ranges where conditions permit, sheep may use suitable parts of the upper boreal zone. Stone's sheep winter primarily in alpine habitat; however, some herds make considerable use of low-elevation, open grassland sites.

After the November–December rut, deepening snow restricts movement and both sexes generally occupy the same winter range. Older rams often segregate themselves from the ewe groups, even though both groups may feed on the same windblown ridges and rest in the same cliffs. In some thinhorn sheep herds, following the rut the older rams move to ranges distinct from those occupied by ewes and young during winter. Such separation would be of benefit in distributing the grazing pressure to the advantage of gestating ewes, but to the potential disadvantage of those rams using poorer quality ranges. It is not unusual, however, to see rams and ewes in mixed groups during winter.

With approaching spring, Dall's sheep frequently move downward to feed on the first emergent vegetation of the warmer, lower slopes. In some areas, they can be seen feeding on open or brushy slopes below treeline in early spring, often where avalanches have swept the slopes clear of trees and brush. Stone's sheep also leave winter range prior to lambing, and their spring feeding is done on lower portions of their summer range (Seip and Bunnell 1985*b*). As the snowline retreats upward, followed by greening forage, sheep again move upslope. By following the receding snowline as summer progresses, sheep are able to continue grazing on the most succulent, newly emergent forage for much of this season.

Lambing season for Dall's ewes coincides with this upward movement in mid-May, and parturient females leave for the isolation of higher and steeper sanctuaries to give birth. At the same time, rams begin to gather into bands in areas peripheral to those occupied by ewes, yearlings, and new lambs. After a period on this late spring range, the rams begin to disperse onto higher summer ranges, sometimes leaving in a single day, but usually in scattered bands. En route to summer feeding grounds, rams may pause to congregate on other intermediate ranges, spend time at mineral licks, or merely follow retreating snows to preferred summer habitat. Stone's sheep rams also move either toward rugged, high-elevation mountains in late spring, or to brush-covered, lower-elevation mountains.

With the end of lambing, most Dall's nursery groups (i.e., ewes, yearlings, and lambs) leave winter-spring ranges in mid to late June for summer habitat, often, but not always, following the rams by a week or two (L. Nichols, unpubl. data). Hoefs and Cowan (1979) found that nursery groups often preceded the rams to summer range. These movements may be related to spring weather conditions, with earlier movements occurring during unusually warm and dry circumstances that adversely affect forage condition. Unlike Dall's sheep movements, Stone's ewes normally leave their winter ranges prior to parturition, and lambing occurs on early summer areas.

Interestingly, this same general tendency is followed by Dall's sheep at about the same time period from their southernmost range on the Kenai Peninsula to their northernmost range in the Brooks Range, Alaska, which are some 960 km apart (Klingel et al. 1974, Summerfield 1974, Whitten 1975, Winters 1980, Jakimchuk et al. 1984, Ayers 1986). This suggests a similar seasonal pattern of snow melt and plant growth despite the large difference in latitude.

Although in many areas the change in habitat use, especially by ewes, from winter to summer ranges may be merely an expanding and contracting movement relating to forage availability and succulence, some true migrations do occur where part or all of a population moves many kilometers between ranges. Sometimes, these migrations cross forested valleys normally avoided by sheep. Such movements are probably traditional in nature, with routes being passed on from generation to generation by the habit of younger animals following older ones (Geist 1971, 1975). Sheep are able, therefore, to exploit isolated patches of suitable habitat separated by normally unsuitable areas.

Such migrations may be quite fragile and the routes easily lost to succeeding generations. Thinhorn sheep evolved in postglacial alpine and subalpine grasslands that probably were contiguous for many generations, allowing unrestricted movement among suitable ranges. As forests invaded valleys and separated mountain ranges into isolated areas of alpine grassland, sheep maintained migration routes between these separated habitats. However, catastrophes (i.e., heavy hunting, epizootics, excessive predation) could remove the older animals that led and perpetuated migration routes, thus destroying the herd's memory of traditional routes. Many areas of apparently suitable habitat are now unoccupied, possibly for this reason (Geist 1975).

Nevertheless, sheep occasionally do tend to pioneer new ranges, possibly reoccupying traditional ranges previously lost. Simmons (1982) found that young rams and even nursery groups sometimes dispersed to new habitats in the Mackenzie Mountains, Northwest Territories, becoming established and not returning to their former ranges. Two ½-curl rams were found in habitat occupied only by mountain goats about 13 km and across an icefield from the nearest sheep habitat (L. Nichols, unpubl. data). An Alaska Department of Fish and Game report (unpubl.) mentioned two young rams that established themselves on a river bluff some 48 km across forests and rivers from the nearest known sheep habitat. In 1971, F. J. Singer et al. (1983) found sheep began inhabiting areas in the western arctic where they had not been seen during this century and that were some 48 km from the nearest known populations. Although such movements may succeed in establishing new populations, many fail, and mortality is high among sheep venturing into unsuitable surroundings.

During summer, most rams remain segregated from the ewe-young bands. F. J. Singer (1984) found that rams on summer range mixed more with nursery groups in high density, McKinley Park herds than in low density,

Brooks Range herds. More young rams and few older rams mixed with nursery groups in the low-density herds, whereas equal numbers of full-curl and ¼-curl rams were found with ewe groups in McKinley Park. Ram and nursery groups mingled only 6% of the time on winter range and 4% on summer range in the central Brooks Range (Jakimchuk et al. 1984).

Occasionally, large bands of rams may be seen, but usually rams are scattered in small groups and as singletons, frequently in extremely rugged habitat. Although they may feed on gentler slopes, some rams may be found bedded during the day on the highest peaks and pinnacles, probably as much for heat and insect relief as for security. Hoefs and Cowan (1979) found that rams used the highest elevations of their range from August until the latter half of October, whereas ewe bands grazed at their uppermost average elevation in July, descending again in August while drifting back toward wintering areas. Many rams do not return to wintering grounds until just prior to the rut.

Social Organization

NURSERY GROUPS

The social organization of female groups centers around the ewe-lamb, or ewe-yearling-lamb units. These units may form bands with other family groups, possibly tied loosely together by the lambs' propensity to associate with their peers. However, female groups often separate freely into smaller family units. Interactions between lambs are common, and group play occurs frequently. Groups of ewes with lambs tend to remain separate from ewes without lambs during early summer, but later, both nursery and nonnursery groups become mixed. Interactions between adult females are less frequent, although some dominance conflicts arise over feeding or bedding sites. The older animal usually is the winner by virtue of a more aggressive horn threat. Ewes occasionally butt aggressively, but do not horn clash as do rams.

A behavior often mentioned is baby-sitting. Groups of lambs may be seen in the company of one or two adult ewes during the day while their dams feed at some distance. Ostensibly, this appears to be a deliberate act by adult ewes; however, perhaps lambs merely prefer the company of their own age class and so remain for a time with other lambs, not following their own mothers entirely. This behavior allows the ewes freedom to feed while still offering the lambs the protection and leadership of more than one alert, adult female. Occasionally, groups of lambs have been found bedded more than 1 km from the nearest feeding adults (Curby 1981).

RAM SOCIAL ORGANIZATION

Social organization among rams is more complex, consisting for the most part of establishing and maintaining dominance rank (Geist 1971). In general, rams establish a dominance hierarchy based upon horn size. By frequent horn display and clashing, rams learn to associate the force of a horn clash with the size and curl of the horns delivering the force. In this manner, a ram can determine his relative rank by visual observation of the opponent's displayed horns. Subdominant rams behave like females toward dominants, nuzzling them and allowing the dominant ram to mount them, presumably in a display of dominance, not sexual behavior.

Serious conflicts appear to occur only between rams of approximately equal horn size that are otherwise unable to determine their relative status. Nichols (1971) witnessed several battles between rams with equal-sized horns, one of which lasted over 90 minutes. During these conflicts, the rams' clashes were interspersed with short display and rest periods and short displacement-activity grazing periods. At the end, winners could not always be determined, and usually both rams would merely begin grazing together. Certainly, these battles were not over possession of ewes, because no ewes were present. Such battles for dominance are fought throughout the summer, mostly between younger rams whose dominance status has not yet been established, become more frequent in the fall, and usually end by the rut. Established dominance status enables rams to live in proximity with one another in a relatively ordered society with rare serious physical injury (Geist 1971). Established dominance status also enables rams to conserve energy by determining social standing before the prewinter rut, a period when any energy surplus will be needed to survive winter.

Conflicts over ewes in estrus do occur, but are less formal and more rough-and-tumble. Conflicts usually are settled in short order by the largest ram present who takes possession of the estrous ewe and guards her against all other rams. Smaller rams may remain near the guarding ram, interacting with each other and occasionally attempting to approach the ewe. A horn threat, rush, or merely a horn display usually discourages them. Approaches by other rams appear to increase the libido of the guarding ram, who often renews his reproductive efforts following such an approach.

After leaving their mothers during parturition, yearling rams often begin following older rams and thereby join ram groups for the summer. Simmons (1982) found that young rams frequently follow new ram groups from mineral licks in early summer, thus changing their ranges. Presumably, this also occurs during chance meetings of ram groups, thus benefiting the population by constantly mixing the genetic pool among different subpopulations.

Reproduction

MATING SEASON

The reproductive cycle begins with the rut that occurs in early winter. The rutting season of Dall's sheep in the Kenai Mountains, Alaska, begins about mid-November and extends to mid-December, with a peak of activity on 30 November (Nichols 1978*a*). No information is available on the timing or length of the rut in the northern parts of Dall's sheep range, but because the lambing period is similar, it is assumed to be about the same as that in the south.

The rut of Stone's sheep peaks and extends later than the rut of Dall's sheep. Rutting activity is induced by onset of estrus in ewes, which appears strongly controlled by photoperiod in all but desert populations of North American wild sheep (Bunnell 1980*a*, 1982). However, photoperiod varies considerably with latitude from the southernmost to the northernmost thinhorn sheep ranges. There are more than six hours of daylight when the rut begins on the Kenai Peninsula, whereas there may be no true daylight at all in the northern Brooks Range. Therefore, if photoperiod does regulate the beginning of the rut, it must be related to the number of days after the autumnal equinox, which is the same at all latitudes. More research is needed on this subject.

Rutting behavior has been described for Stone's sheep by Geist (1971) and for Dall's sheep by Nichols (1971). Rams, whose dominance hierarchy has been discussed above, begin to move restlessly between ewe groups, testing the anestrous ewes by scent and behavior. When a ewe approaches estrus, rams gather about her, continually testing her status. As she comes into heat, there may be a chase by all rams present with the ewe apparently trying to evade them in the cliffs. During the early stage of the rut, when few ewes are in estrus at a given time and rams are greatly excited, younger, ¾-curl males have been observed to outrun the older rams and copulate with the ewe on the run (Nichols 1971). Eventually, the ewe stops and usually is claimed by the largest ram present, who then defends her by threat displays, rushes, and occasional clashes against other rams. He remains with her during her estrous period. Subordinate rams rarely threaten the dominant ram, although they do attempt to chase the ewe from him. These attempts sometimes are successful; the ewe, however, appears to favor dominant rams and tries to elude subdominants. Copulations appear to be stimulated by the attempts of subordinates to reach the ewe. If the guarding ram is not sexually responsive, the ewe may adopt subordinate, ramlike behavior to encourage mounting.

Although Murie (1944) mentioned "harem gathering" by large rams during the rut, no such behavior was observed by Nichols (1972) during two seasons each of detailed observations on two herds of Dall's sheep. He did notice weak attempts to gather harems by young rams in a herd where large rams were not present, but thought this probably was abnormal behavior.

Estrus lasts one to three days and includes many copulations. Time of service relative to ovulation affects the pregnancy rate in domestic sheep (Blockey et al. 1975). Ewes bred 11–15 hours before ovulation were more likely to conceive than those bred during any other period while the ewe was in estrus. If this holds true for Dall's sheep, natural selection would favor a ram (i.e., usually the large, guarding ram) copulating frequently with a ewe during her entire estrus period, rather than the quicker, young rams that might breed her only during the initial onset of estrus.

Estrous periods in Dall's sheep are separated by 12 to 14 days (Bunnell 1980a), similar to that of feral Soay sheep (Jewell and Grubb 1974) and the mean of 16.5 days generally recorded for domestic sheep (Asdell 1964). Bunnell (1982) reported a mean lambing period of 29.8 ± 2.4 days for 18 populations of North American mountain sheep (almost precisely two interestrous periods in length). Dall's ewes have two and possibly three estrous cycles during the rut if conception does not occur; however, most ewes conceive during the first two cycles (Bunnell 1980a, 1982).

As the rut progresses, more ewes are in estrus simultaneously and the rams become increasingly fatigued, which results in fewer rams about each female. There is reputed to be considerable weight loss by rams during the rut, although specific data are not available. This loss may be especially true of mature, actively mating rams and may increase overwinter mortality of this age group (Geist 1971). Because ewes expend comparatively little energy during the rut, significant weight loss at this time would not be expected.

Age of first reproduction among ewes appears governed by nutrition. Reproduction for lambs has not been reported, but three of four yearlings from the Kenai Mountains examined after the rut were pregnant (Nichols 1978a). In that population, all ewes older than two years that were examined were pregnant ($n = 18$). In the Northwest Territories, Simmons et al. (1981) found 78% of Dall's sheep ewes older than two years were pregnant after the rut ($n = 94$); three of seven yearlings also were pregnant. It is unknown how common yearling pregnancy is in general. Hoefs and Cowan (1979) suggested that most Dall's sheep from Sheep Mountain, Yukon Territory, first gave birth at three years of age. Using individually marked sheep, Bunnell and Olsen (1982) reported that wild ewes of that population first gave birth at four years, and the birth rate did not exceed 50% until ewes reached five years of age. Two of four ewes captured as lambs from that same population and provided abundant feed, lambed at two years of age or were pregnant at 18 months. It appears that thinhorn ewes usually become reproductive upon reaching their adult weight, rather than at a specific age.

This relationship may be general. Asdell (1964) found that puberty in domestic sheep was related to weight, rather than age alone. Frisch (1980) concluded that onset of menarche in human females and animals is related to body weight, rather than age, and in particular to the percentage of fat in the tissue. Well-fed, fast-growing animals reached reproductive maturity earlier than slow-growing individuals. Frisch (1980) also found that the ratios of weight and height at menarche were about 85 and 96%, respectively, of fully adult weight and height. The yearling ewes collected in the Kenai Mountains that were pregnant had reached about 80 and 92% of the mature weight and shoulder height of ewes from that area. These observations strongly suggest that ewes (and rams) on nutritious ranges mature earlier than those on poor ranges.

Some yearling Dall's males may be physiologically capable of reproduction by 18 months (Nichols 1978a). However, because of the dominance hierarchy among males in populations containing large rams, most are prevented from mating until reaching adult size with horns larger than ¾-curl (i.e., approximately five to seven years old for Dall's rams, but younger in Stone's sheep on good range). Where most or all large males were removed by hunting under the ¾-curl law, mating was performed by three- and four-year-old rams (Nichols 1972). More harassment of anestrous ewes by these young rams was noted, including some herding and attempted harem gathering. This atypical behavior in the absence of socially dominant, older rams is possibly detrimental in that it unnecessarily stresses both sexes prior to the rigors of winter. However, Stewart (1980) found no relationship between dominance and mortality in a study of bighorn rams.

Both rams and ewes appear capable of reproduction throughout their adult lives. Whereas most Dall's sheep do not live beyond 12 years of age (Murie 1944, Bunnell 1978, Hoefs and Cowan 1979), at least some exceed that limit. Ewes from 13 to 16 years old have been found to be pregnant or with a lamb at heel (Nichols 1978a, Bunnell and Olsen 1982). No senile, nonmating adult rams have been recorded for either subspecies.

Survival of lambs conceived too early or too late may be poor. In average years, those conceived too early are born before spring plant growth provides sufficient nutrition for lactation; those conceived much later are born too late for adequate growth prior to their first winter. Although the breeding season may be timed directly by photoperiod, it appears to be controlled via a fixed gestation period and indirectly by the onset and length of the average summer's growing season, which in turn controls lamb survival and hence the timing of the lambing season (Bunnell 1982, Rachlow and Bowyer 1991).

LAMBING

The lambing period for Stone's sheep extends from 23 May to 16 June (Bunnell 1982). Comparable dates for Dall's sheep are variable, with first lambs observed as early as 3 May on the Kenai Peninsula, 4 May in the northern Brooks Range (Jakimchuk et al. 1984; L. Nichols, pers. obs.), and as late as 20 May in the Brooks Range (Roseneau and Stern 1974). Although there may be considerable variation between years, the average median date of lambing probably lies between 17 and 24 May. Lambing is usually completed by mid-June.

Singleton births are the rule among Dall's and Stone's sheep. Only one set of twins has been reported for a Dall's ewe in captivity and three sets in the wild (Bunnell 1980a, Hoefs 1978, Nichols, 1978a).

Lambs, which weigh 3.2 to 4.1 kg at birth, are born in late May and early June after a gestation period of about 171 days (Bunnell and Olsen 1976, Hoefs 1978, Nichols 1978a). Considerable differences in day length exist between the southernmost and northernmost thinhorn sheep populations, thus differences in lambing period are expected. Bunnell (1982) reviewed lambing periods for 28 populations of North American mountain sheep, but did not treat thinhorn separately. We examined seven Dall's and Stone's populations ranging from 56°20' to 63°58'N latitude. There were no significant correlations between latitude and onset, termination, or duration of lambing. However, lambing was shorter in duration and terminated earlier in areas of least favorable plant growth (table 2.2). There is a significant tendency for populations in harsh environments to have shorter lambing periods and to terminate lambing earlier (Bunnell 1982). The tendency to begin later

Table 2.2. Tests of hypotheses regarding lambing period in North American thinhorn sheep (H_0: no difference; P for one-tailed tests).

ALTERNATIVE HYPOTHESES		r_s[a]	P
H_1:	Onset of lambing is positively correlated with phenology	0.22	> 0.10
H_2:	Termination of lambing is negatively correlated with phenology	−0.73	< 0.05
H_3:	Duration of lambing is negatively correlated with phenology	−0.67	< 0.05

[a] Spearman's rank correlation coefficient.

Table 2.3. Mean dates (± SE in days) for onset, end, duration, and peak of lambing period for mountain sheep taxa.

	LAMBING PERIOD			
TAXA	ONSET (DATE)	END (DATE)	DURATION (DAYS)	PEAK (DATE)
Desert bighorn	n = 9	n = 9	n = 9	n = 4
	14 Jan	23 June	160.2	12 April
	± 14.6	± 15.5	± 24.4	± 14.3
Rocky Mountain and	n = 11	n = 11	n = 11	n = 5
California bighorn	15 May	15 June	30.9	31 May
	± 3.1	± 4.5	± 3.2	± 3.0
Dall's and Stone's sheep	n = 7	n = 7	n = 7	n = 6
	17 May	12 June	25.6	27 May
	± 4.3	± 4.0	± 5.2	± 3.6

in more harsh environments was insignificant for the populations tested. Lambing periods of thinhorn sheep do differ dramatically from desert populations (Krausman et al., this volume), but less so from Rocky Mountain and California bighorn (Shackleton et al., this volume; table 2.3).

Features other than the bioenergetic regime can influence lambing period. The peak of lambing occurred on 18 and 19 May in the unhunted Yukon Sheep Mountain herd and on 20 May in a protected herd on the Kenai Peninsula, Alaska. Lambing was later in two adjacent Kenai herds in which hunting had removed most older rams, suggesting insufficient rams to impregnate all ewes during their first estrous periods.

As parturition approaches, thinhorn sheep ewes seek isolation in the most rugged terrain available, usually still on their winter-spring ranges. Ewes abandon and tend to drive off any yearlings that may have been accompanying them. Small groups of these yearlings may be seen at this time, occasionally following a dry ewe, often appearing nervous and "lost." After the ewes with new lambs regroup with other nursery groups of sheep, the yearling females may rejoin their mothers. However, the yearling males, for the most part, follow older rams thereafter, although a few may be found with nursery groups throughout the summer. In contrast, the majority of Stone's sheep yearling males remain with the ewe groups.

In one witnessed birth, the lamb was born 3 minutes after the ewe was first observed lying in labor with the lamb's head just beginning to protrude. The

Figure 2.10. Dall's sheep ewe nursing her lamb in early summer. During nursing, the ewe licks and cleans the lamb's anal region. (Photograph by Tom Walker.)

ewe immediately stood up, ate the birth membranes, and began pawing and licking the lamb. In 20 minutes the lamb was on its feet and in 65 minutes, after several unsuccessful attempts, it first nursed (fig. 2.10). The ewe was attentive to her lamb during this period, but did graze intermittently in the near vicinity. Two days later, the same ewe and lamb were seen in company with another ewe and lamb in precipitous terrain some 150 m from the birth site. Both lambs were playing actively (L. Nichols, unpubl. data).

Pitzman (1970) reported that most ewe-lamb pairs left their isolated birthsites and joined other ewes with lambs within one to two days after parturition. The period of strict seclusion is often less than 24 hours (F. L. Bunnell, unpubl. data). Groups of ewes with young lambs tend to remain separated from barren females and yearlings for several weeks after lambing, but gradually integrate as summer progresses.

Although nearly all mature ewes may become pregnant during the rut, not all fetuses remain viable, nor do all newborn lambs survive parturition. Mortality over the first two weeks may be approximately 28% (F. L. Bunnell, unpubl. data). The midsummer lamb:ewe ratio varies greatly from year to year and ranged from 8:100 to 81:100 in Alaska, with an average of 37:100 (Nichols 1978a).

Adequate data are not available for Stone's sheep on age of sexual maturity or fertility rates. Lambing occurs during the latter part of May and the first 10 days of June. At that time, ewes are already on summer ranges. Lambing behavior is probably very similar to that of Dall's sheep. Lambs of both subspecies normally nurse until fall, but probably are capable of subsisting on forage alone by the end of their second month. Lambs usually remain with their mothers until the following spring, at which time they are displaced by the arrival of new lambs.

Population Dynamics

LAMBING RATES AND SURVIVAL

The lambing rate, the number of viable lambs observed after parturition, (usually expressed as the ratio of lambs to 100 ewes over two years of age) is a useful first indicator of population status, even though survival of these lambs to reproductive age may be variable due to such factors as harsh winters or predation. Lambing rates as observed are not necessarily representative of the actual birth rate. During two lambing seasons studied, Hoefs and Cowan (1979) found in their Kluane Dall's sheep study that 15 and 20% of newborn lambs died in their first month. F. L. Bunnell (unpubl. data) noted that rates were higher in the first two weeks (24 to 28%), as compared to the next two weeks (10 to 17%). In the Mackenzie Mountains of Canada's Northwest Territories, Simmons et al. (1984) estimated about 20% of lambs died between birth and midsummer. Depending upon the nutritional level of the dam and weather conditions prevailing at birth, mortality can be as high as 100% (L. Nichols, unpubl. data). Some of that early mortality is omitted from observed lambing rates, which are usually obtained in midsummer. These are conservative estimates of birth rates.

The observed lambing rate, however, is indicative of effective reproduction. Rates can be extremely variable depending upon habitat conditions; observed rates have ranged from zero to more than 80 lambs per 100 adult ewes (older than two years). The average Dall's sheep lambing rate in a number of populations and over a number of years in Alaska was 37 lambs per 100 ewes (SE = 2.10, n = 57; Nichols 1978a). If pregnancy rates of 75 to 100% are typical, this suggests an average mortality rate of about 50 to 60% at and just after birth. Preliminary results in an Alaska Range study showed a mortality rate of 37% over the first year of life of 19 lambs radiocollared just after birth (Scotton 1996).

In one summer survey of 521 Stone's sheep ewes, J. P. Elliott (B.C. Fish and Game Dep., unpubl. data) observed a lamb:ewe ratio of 47:100. Ratios greater than 80:100 have been reported from good ranges (Luckhurst 1973). As opposed to summer surveys of the Dall's sheep, most data have been collected in winter for Stone's sheep. Thus, average lamb:ewe ratios for Stone's sheep during summer and over a number of years and areas are not available. For nine-month-old lambs, an average lamb:ewe ratio of 35:100 was obtained during winter surveys (J. P. Elliott, unpubl. data).

Survival of lambs to yearling age also is variable and depends on winter conditions, plane of nutrition throughout the summer growing season and their first winter, and predation. Based on average survival rates over time in increasing, stable, and declining Dall's sheep herds (Pitzman 1970, Murphy 1974, Nichols 1976, Hoefs and Cowan 1979, Heimer 1980b), the overall average survival of lambs to yearlings (11–14 months old) was approximately 60%. Here again, short-term survival rates vary widely in relation to winter conditions and spring-summer nutrition and have been observed to range from only 7% in a sharply declining population to 100% in a rapidly increasing herd. The long-term average yearling:ewe ratio over many Dall's sheep herds and under varying conditions was 21:100 (SE = 0.0154, n = 59; L. Nichols, unpubl. data). The long-term average yearling:ewe ratio for Stone's sheep at 21 months old was 15:100 (J. P. Elliott, unpubl. data).

RECRUITMENT

Regardless of how many lambs are born, they do not contribute to a herd's productivity until they reach reproductive age. Recruitment is the number of sheep of breeding age added to a herd each year. In combination with adult mortality and including any immigration or emigration, recruitment determines whether a herd will increase, remain stable, or decrease. Recruitment is difficult to assess unless the average minimum breeding age of ewes is known for a given herd. This may be 18 months for sheep under good nutritional conditions, 30 months in herds under less favorable conditions, and 4 to 5 years in still other herds, all apparently dependent upon growth rate.

Where conditions are good and first breeding is achieved at 18 months, adult recruitment is equal to the number of yearlings in the herd. Assessment is relatively easy because this age is recognizable even from aircraft in early summer. Where sheep mature more slowly, the accurate assessment of adult recruitment may be much more difficult, requiring the marking and close observation of known-age animals, or the sampling of populations by capture or collection. It is difficult at this point to generalize about recruitment in thinhorn sheep; each herd must be assessed according to its own condition.

SEX RATIOS

Although the sex ratio at birth is probably equal, the observed adult sex ratios are variable by year and herd. In unhunted populations, the observed ram:ewe ratio averaged 60–67:100 in Mount McKinley (Denali) and Gates of the Arctic National Parks. In hunted portions of Noatak, Gates of the Arctic, and Lake Clark Preserves the ram:ewe ratio averaged 59:100 (Murphy and Whitten 1976, F. J. Singer et al. 1983). The ram:ewe ratio averaged 93:100 in unhunted Kluane Park, Yukon Territory, over a five-year period (Hoefs and Cowan 1979). In a lightly hunted Brooks Range herd in arctic Alaska, a ram:ewe ratio of 87:100 was noted during a one-year study (Klingel et al. 1974), whereas that in another lightly hunted Brooks Range herd averaged 54:100 over two years (Jakimchuk et al. 1984).

One Kenai Mountains herd represents an extreme example. This Surprise Mountain herd has been hunted heavily and every legal ram has been removed almost every year. The average adult ram:ewe ratio over 23 years was only 33:100. However, the nearby Crescent Mountain herd, also hunted heavily, had an average ram:ewe ratio of 69:100, whereas that for the adjacent, protected Cooper Landing Closed Area herd averaged 56:100 over this same period. Reproductive success was similar in all three herds over this period ($P < 0.10$), with a mean lamb:ewe ratio of 38:100 (L. Nichols, unpubl. data). In a survey of 2,036 Stone's sheep from an extensive area open to hunting in British Columbia, J. P. Elliott (unpubl. data) calculated a ram:ewe ratio of 35:100 in those sheep over two years old.

These wide variations in adult sex ratios can result from differences in hunting pressure and/or differences in survival of cohorts from birth to adulthood. That ratios of rams to ewes are almost invariably less than equal, even in unhunted populations, suggests differential mortality during adult life, with rams dying at a greater rate than ewes. Geist (1971) attributed this to a greater stress on rams than ewes during mating, resulting in a shorter average life span. Heimer et al. (1984) further hypothesized that removal of older rams through hunting allows young rams to become active breeders by altering the dominance hierarchy. This, in turn, should place more stress on these young rams and so reduce their life expectancy. This unproven hypothesis also predicts disruption of the rut with the result that lamb production should suffer.

However, Hoefs and Barichello (1984), compared hunted and unhunted populations of Dall's sheep and concluded that the annual removal over a decade of up to 80% of the mature rams in the hunted herd resulted in no lowered life expectancy of rams nor did age-specific mortality differ between herds. Lamb production was higher in the hunted herd and trophy quality improved. In a study of breeding behavior, Nichols (1972) observed that young, ½-curl Dall's rams were very active, albeit not very efficient, during the rut despite the presence of large, dominant rams. This implies young rams suffer some degree of breeding stress whether or not large rams are present. No average difference in lambing success could be found when comparing three herds over two decades: one herd with all adult rams removed by hunting almost every year, the second herd with almost all removed annually, and the third herd protected from hunting and containing numerous large rams (L. Nichols, unpubl. data). No conclusive evidence has yet been presented to support the hypothesis of harvest-related effects on survival of sublegal rams (Murphy et al. 1990, F. J. Singer and Nichols 1992).

Limiting Mortality Factors

WEATHER AND SNOW

The ultimate factor limiting population size and growth in thinhorn sheep is the quantity and quality of forage available to them. This, in turn, is affected by summer and winter weather, topography, density of sheep, and predation. Weather during the short summers in these northern sheep ranges determines forage production that, in turn, affects animal growth rates and survival over the winter. Weather delineates the period of successful parturition that, because of a fixed gestation period, specifies the breeding season.

Availability of usable, low-elevation habitat, which includes open grassland and adjacent escape terrain, can be significant to the annual range carrying capacity of Stone's sheep populations. Early growth of green forage on low-elevation ranges can precede that on alpine ranges by at least one month in spring (Seip 1983). Likewise, low-elevation ranges retain succulent vegetation later in the fall than do alpine ranges. This allows increased fall fat deposition by sheep and a shorter period of dependence on dry winter forage, thus increasing the potential carrying capacity of the annual range.

Summer weather affects the growth of forage on summer and winter ranges of thinhorn sheep. Good growing conditions promote the quantity and quality of plant growth, not only allowing sheep to reach good physical condition and to store surplus energy in the form of fat reserves by fall, but also giving them abundant and more nutritious food when on winter range.

Thus, sheep are able to overwinter in better condition, whereas pregnant ewes are better able to devote more energy to producing larger and healthier fetuses. Additionally, healthy ewes produce more milk, lambs grow faster, and are better able to face their first winter. Bunnell (1978) and Hoefs and Cowan (1979) suggested that, in the relatively cold and dry Kluane area of Canada and probably other interior sheep ranges, Dall's sheep populations were at least partially limited in size by summer rainfall that directly affected nutrition of the ewes during pregnancy in the following winter and that then determined lamb production, early survival, and ultimately recruitment into the herd as reproductive adults. Years of high population numbers, good overwinter survival of lambs, and good lamb production were correlated with wet summers and good growing conditions the previous summer (Burles and Hoefs 1984, Hoefs 1984*b*).

Sheep use virtually all of their body fat reserves to survive the winter food shortages. The relatively short summer growing season in these northern climes is critical to their ability to replenish these reserves. A late spring or dry summer can reduce their chances of surviving the following winter or successfully reproducing by shortening the time period wherein they regain lost weight and store body fat or by decreasing the amount of forage available in summer and the next winter.

A late spring storm in the central Alaska Range, which deposited 25 cm of snow in mid-May 1989, delayed successful parturition so that the first live lamb was not observed until 22 May, as compared to 9 May the previous year. Such late snows and/or wet, windy conditions kill early born lambs by hypothermia (Rachlow and Bowyer 1991).

Winter weather, primarily snow cover, determines the extent of winter range by affecting the sheep's ability to reach forage regardless of its abundance. In the relatively cold and dry interior, snowfall usually is limited and snow normally remains powdery and is easily removed from open ridges and slopes by strong winter winds, thus exposing forage. Available sheep winter range in such areas is limited by snow depth as affected by wind action. Winds usually remove snow from exposed ridges and benches and deposit it in draws, gullies, and other areas that are protected. Heimer et al. (1994) hypothesized that the best sheep ranges are on the slopes of long, relatively straight valleys lying perpendicular to the general winter wind flow. The alpine, windward slopes of these valleys thus are swept more free of snow, making forage available.

Stone's and Dall's sheep rarely attempt to dig for forage in snow depths greater than 30 cm and usually remain feeding on snow-free sites (Hoefs and Cowan 1979, Seip and Bunnell 1985*b*, Nichols 1988). Thus, under conditions of soft snow and dry cold, sheep are effectively limited to winter feeding areas with less than 30 cm of snow cover, depths that in turn are affected by wind direction and force. In the Mackenzie Mountains of Canada's Northwest Territories, where snowfall was light and winters cold, snow did not inhibit sheep movement significantly (Simmons 1982).

In winters of heavy snow, feeding areas become restricted in size and sheep are less able to obtain nutritional needs. Both Murphy (1974), in McKinley National Park, and Nichols (1978a), on the Kenai Peninsula, found an inverse relationship between winter snow depth and lambing success the following spring, suggesting the inability of gestating ewes to obtain adequate nutrition and thus to produce and support viable lambs.

In the Kenai Mountains, which are affected by maritime weather conditions and frequently are warmer and wetter in winter than interior ranges, snow depth and snow hardness can adversely affect survival and lamb production. Nichols (1976) found that during warm, wet winters, snowfall was heavier and often partly melted, refroze, and formed a hard crust through which sheep were unable to dig. Under these conditions, winds could not remove the snow and further hardened the crust by packing. Sheep were then restricted to even less winter range than normal, being able to feed only on the few ridges left exposed. Nichols (1988) measured the depth and hardness of snow around sheep feeding craters, some of which had been successfully pawed through the snow to forage beneath and some of which had not. He also found no attempts to dig in soft snow deeper than 30 cm and that there was a measurable limit in combined depth and hardness through which sheep apparently were unable to dig (fig. 2.11). There was a significant negative correlation between his depth-hardness index and lambing success in three herds over five years. Nichols (1988) also measured percent snow cover on ranges occupied by sheep in midwinter and found that sheep distribution was defined by the amount of ground covered by snow.

During particularly severe winters in the Kenai Mountains, large numbers of sheep died from starvation. Over 12 years that included several such winters, herds under study, which totaled almost 900 sheep at their peak, declined 66%. Long-term resident wildlife photographer Cecil Rhode mentioned a major decline in the early 1940s associated with severe winters. The sheep herds on the Kenai Peninsula with which he was familiar contained only dozens of animals where he had observed hundreds before winter. After that crash, herds increased during a series of mild winters until they again declined in the early 1970s (Nichols 1976).

Such population fluctuations seem to be the rule in these southern Alaskan mountains where winter conditions vary between cold and dry, which allow sheep herds to increase, and warm and wet, which cause severe mortality. In

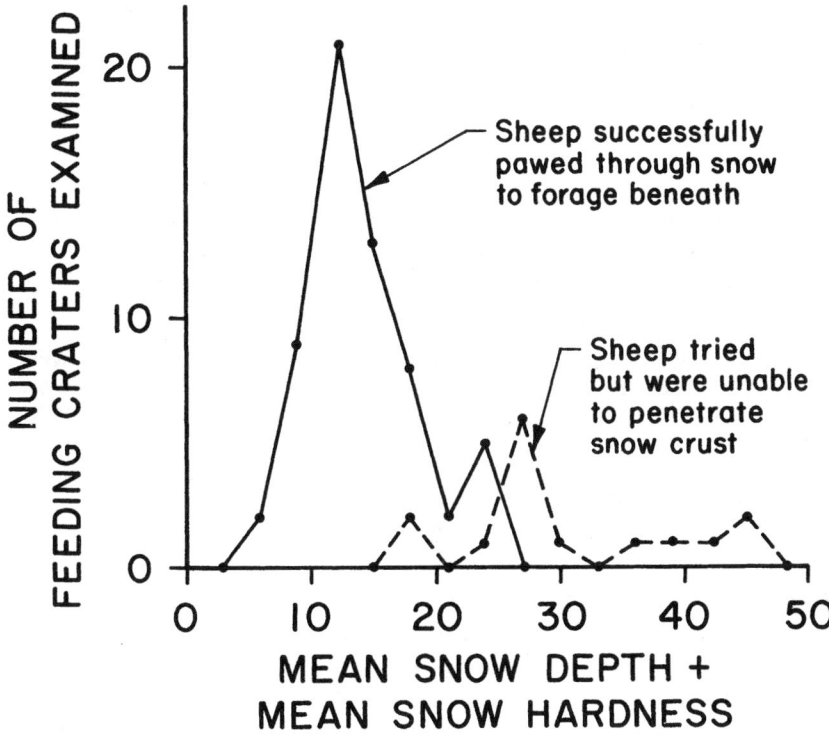

Figure 2.11. Sheep's ability and inability to dig through snow as indicated by measurements of depth and hardness at feeding crater sites (Nichols 1988).

this case, severe winter weather for sheep is a combination of heavy snowfall and warm temperatures that form deep, hard-crusted snow cover. Mild winters are those with light snow, continual cold, and strong winds to remove the snow, winters that usually are typical of interior winter ranges.

Although most interior Dall's sheep herds appear to have been relatively stable over recent years, primarily because winters have been consistently cold with comparatively light snowfall and no cycles of freezing and thawing (Simmons 1982, Hoefs 1984b), similar population crashes have been documented in the past and can be expected at infrequent intervals (J. S. Dixon 1938, L. J. Palmer 1941, Murie 1944). In fact, several interior Dall's sheep herds, after a long period of relative stability, were reduced by 20–25% in the exceptionally severe winter of 1981–1982, which followed a dry summer of poor forage production (Burles and Hoefs 1984, Burles et al. 1984, S. M. Watson and Heimer 1984). The following winter also was one of deep snow and severe cold in the central Yukon, causing a 40% decline in sheep numbers (Barichello and Carey 1988a). In this case, almost the entire 1982 and 1983 lamb crops were lost, and many older animals died. In the early 1960s, the McKinley Park herd suffered a single-year decrease of 67% because of severe winter conditions (Whitten 1975). Dall's sheep numbers in portions of the Alaska and Brooks Ranges and the Wrangell Mountains apparently declined by as much as 60% during the harsh winters of 1991, 1992, and 1993 (Alaska Dep. Fish and Game, unpubl. memos), further demonstrating that interior herds are not immune to drastic reductions brought on by occasional severe winters.

Dall's sheep seem to be little affected by cold or wind alone and may be seen feeding or bedded on exposed ridges on cold or windy days. However, they do respond to combined wind and cold (i.e., wind chill) by seeking protected areas out of the wind. Hoefs and Cowan (1979) found that sheep occupied lower elevations on colder winter days than on warmer days and that rams usually grazed lower than ewe-lamb groups. In Alaska's Kenai Mountains, L. Nichols (pers. obs.) detected little difference in elevation use related to temperature alone, but observed use of higher elevations on colder days with little or no wind. Most sheep in this area winter above the elevation of temperature inversions (indicated visually by the frostline in valleys) that occur on cold, calm days and, therefore, were subjected to relatively warmer temperatures. Although no specific measurements were made, in winter rams appeared to occupy higher elevations than ewes. Even so, sexes frequently were mixed during this season. Thus, it appears that Dall's sheep reaction to temperature in winter may vary between areas or be related to temperature extremes. Most Stone's sheep also remain at high elevations on cold, calm days, taking advantage of the temperature inversion.

Because thinhorn sheep are best adapted to a winter habitat of continued cold and windy weather, one would expect them to be physically well adapted to the cold. Scholander et al. (1950) found that Dall's sheep and arctic foxes *(Alopex lagopus)* were the best insulated of northern mammals, better than caribou or polar bears *(Ursus maritimus)*. In the late 1960s, L. Nichols (unpubl. data), in attempting to census Dall's sheep by thermal infrared scanning, found that so little body heat was lost on cold days that it could not be detected by that method. The winter coat of Dall's sheep is composed of a dense layer of hollow, crinkly, guard hairs, thicker than 5.1 cm on their backs, with a thinner layer of fine wool close to the skin. These hairs individually are white and shiny and collectively reduce heat loss from radiation and from convection. In late spring, these heavy winter coats are molted, being replaced by a short summer coat that enables sheep to tolerate the relative heat and strong radiation of the alpine. The animals then appear sleek and trim.

PREDATION

Murie (1944) wrote that predation by wolves *(Canis lupus)* was the main factor limiting Dall's sheep numbers in Mount McKinley (now Denali) National Park. Heimer (1980a) and Heimer and Stephenson (1982) believed that wolves were depressing the herd in Dry Creek, Alaska Range, which responded favorably following wolf control. However, Gasaway et al. (1983) found no differences in this same herd in response to wolf control when compared to nearby herds, where wolves were abundant and uncontrolled. Other investigators have found predation to have little direct impact in controlling Dall's sheep population size (Nichols 1978b, Hoefs and Cowan 1979). The sheep herd in Mount McKinley National Park increased between 1947 and 1961 at an average annual rate of about 11%, despite the presence of wolves that were protected from hunting and presumed to be abundant (Murphy 1974). On the Kenai Peninsula, where wolves were essentially absent during the period, three herds under study increased at average annual rates from 11 to 14% from 1949 through 1968 (Nichols 1976). Thus, when habitat conditions (i.e., winter weather, food, and escape terrain) were suitable there was little difference in rate of increase between Dall's sheep herds free of wolf predation and those where wolves were present. Barichello and Carey (1988b) also compared areas of low and higher wolf density, but could find no differences in sheep demography between the two populations. However, J. P. Elliott (1985) believed that Stone's sheep were declining in British Columbia partially because of an increase in wolf numbers.

The importance of escape terrain to thinhorn sheep has been mentioned. The basis for this requirement is, of course, predation (other than by eagles). For example, L. Nichols (unpubl. data) observed a pack of 12 wolves chasing a band of Dall's sheep down an alpine ridge in winter. The sheep reached several rocky outcrops, whereupon the wolves immediately ended their chase, making no attempt to follow into the cliffs. A similar ending from a chase by a wolf of a single ewe was documented by Hoefs et al. (1986). There is good reason for wolves to be cautious in pursuing sheep into rough terrain. Child et al. (1978) witnessed a Stone's sheep and a wolf fall to their deaths during one such episode.

The restriction of range utilization to areas adjacent to escape terrain reduces the potentially available habitat and thus the size of the sheep population. The degree to which sheep would utilize range further from escape terrain if there were no predators is an important way predation could restrict an otherwise food-limited population.

When sheep seek forage further from their sheltering cliffs or in timbered areas where predators can approach undetected, wolves can inflict considerable mortality, especially among animals weakened from malnutrition. Such was the case during Murie's (1944) studies. Similar effects were observed during an especially severe winter near Twin Lakes in the Alaska Range (R. Proenneke, pers. comm.) and in the Kenai Mountains (L. Nichols, unpubl. data). In Kluane Park, Yukon Territory, the winter of 1981–1982 was especially severe, leaving sheep in poor condition by spring. Several successful wolf attacks in which wolves surprised and killed sheep on open slopes were documented by Hoefs et al. (1986). Sheep carcasses showed the animals to be suffering from severe malnutrition at the time (Burles and Hoefs 1984). Later that summer, after the sheep had recovered from their weakened state, other chases failed when the sheep outran their attackers into escape terrain. However, wolves have been observed occasionally to catch and kill adult sheep in midsummer when sheep were in the open and away from escape terrain (Heimer and Smith 1972). Huggard (1993) found that wolf predation on elk *(Cervus elaphus)* increased with increasing winter snow depth, which made traveling, and thus escaping, more difficult for the prey but easier for the wolves. Such conditions probably affect predation on Dall's sheep in winters of deep snow when they have been forced to feed away from escape terrain. Thus, although wolf predation alone normally might not limit populations, adverse conditions could enhance the effectiveness of predators and thus seriously affect sheep population status.

Other predators that have been observed to kill thinhorn sheep include golden eagles *(Aquila chrysaetos)*, coyotes *(Canis latrans)*, wolverines *(Gulo gulo)*, black bears *(Ursus americanus)*, and lynx *(Lynx canadensis)*. Hoefs and Cowan (1979) wrote that eagles may have killed a number of newborn lambs in the Kluane area. Nette et al. (1984) documented several successful golden eagle attacks on newborn lambs in that area following the bad winter of 1981–1982, when sheep were weak and alternate prey species were scarce. Eagles are seen commonly flying among the cliffs used by Dall's sheep as lambing grounds and have been seen diving at adults and lambs. It appears that their aggressiveness and success in capturing lambs varies between areas and perhaps years, depending upon the abundance of their normal prey and the terrain occupied by sheep. It has not been established, however, that eagles are a significant factor in limiting thinhorn sheep populations.

Next to wolves, coyotes probably are the most successful predators of thinhorn sheep. Several actual kills have been witnessed. L. Nichols (unpubl. data) saw an unusually large pack of 13 coyotes that appeared to have killed a mature Dall's ram on an open alpine slope in early winter and observed another adult ram that had been killed by more than two coyotes in late winter. Other witnessed kills have been of ewes and young by lone or small groups of coyotes (Hoefs and Cowan 1979). Hoefs (1984*b*) considered coyotes to be the main predator of Dall's sheep in Kluane Park.

Stephenson et al. (1991) reported several instances of lynx successfully killing Dall's sheep in winter. In one case, the lynx chased a ram down a frozen stream bed until the ram broke through the overflow and was slowed down. The lynx jumped on his back and killed him by a bite through the back of the neck. The ram was nine years old and appeared to have been in good condition. Sheldon (1930) provided a vivid description of lynx killing sheep, also in winter.

A. C. Smith (Alaska Dep. Fish and Game, unpubl. data) observed a wolverine kill a sheep, F. D. Priester (camper, pers. comm.) saw a large black bear kill a sheep on the Kenai Peninsula, and Hoefs and Cowan (1979) suggested red foxes *(Vulpes vulpes)* as possible predators of Dall's sheep. They also listed ravens *(Corvus corax)* and magpies *(Pica pica)* as contributing to the deaths of already wounded or crippled sheep. However, none of these predators, with the exceptions of wolves and coyotes, are suspected of exerting enough pressure to influence the size of sheep populations.

ACCIDENTS

Accidents, particularly those involving falls, account for some mortalities. L. Nichols (unpubl. data) examined a number of Dall's sheep that died naturally in the Kenai Mountains. Some, if not most, of these deaths were the result of falling off icy cliffs in winter; broken bones were not uncommon. During their falls, several sheep had their horns caught in brush in such a manner that they could not extricate themselves. During one particularly severe winter, Alaska Fish and Wildlife Protection Officer J. Taylor (pers. comm.) observed a yearling in a long fall down an icy avalanche chute. When he approached to retrieve the carcass, he found three other dead sheep at the same site. Since then, other sheep have been found there. Apparently, a sheep trail traverses the head of the chute. When the trail is icy, animals lose their footing and fall.

Nearly all of these sheep were in poor physical condition, as indicated by bone marrow fat level (usually less than 20% and often less than 10%; L. Nichols, unpubl. data). Consequently, the sheep were probably in weakened condition, which made them less sure of foot. However, one healthy adult ram that had fallen to his death was found in early autumn. Possibly this and occasional similar accidents result from carelessness during ram clashes. Another healthy ram was found drowned in a mountain lake in midsummer (L. Nichols, pers. obs.). The banks were extremely steep and the ram may have been unable to climb out.

Avalanches also take their toll, but the number of sheep killed by them presumably is small. Altogether, accidents probably are not a significant direct cause of thinhorn sheep mortality.

DISEASE AND PARASITES

Neither parasites nor diseases, which will be covered elsewhere (Bunch et al., this volume), have yet been documented as primary causative factors in controlling thinhorn sheep populations.

INTERSPECIFIC COMPETITION

Competition between thinhorn sheep and other species of wildlife appears to be minimal, mostly because few other animals use sheep wintering areas. Where mountain goats and Dall's sheep shared winter range in Alaska's Kenai Mountains, the goats' winter food consisted of more than 37% grasses and willows (Culbertson et al. 1982). Klein (1953) found goats to be actively feeding on ferns and grasses during winter on a mountain used by goats and sheep; both studies suggested possible competition with sheep that depend largely on graminoids. On a nearby area used only by goats, Hjeljord (1971) determined their winter diet to consist primarily of ferns and shrubs with lesser amounts of grasses. It is not known whether goats and sheep use the same portions of the shared winter range and thus compete directly. In any case, overlapping winter range by goats and Dall's sheep is not extensive. Although both goats and sheep share summer ranges in some areas, the abundant forage should make competition of little importance during the growing season.

Marmots are common in Dall's sheep ranges and also feed extensively on grasses and sedges (R. M. Hansen 1975). However, most of their feeding seems limited to plant communities exposed only in summer and not used by sheep in winter. Hoefs and Cowan (1979) suggest the arctic ground squirrel *(Citellus parryi)* as the only possible competitor in their Yukon study area, but consider competition negligible. Of course, both of these rodents hibernate during winter, so they could only compete significantly by using forage from sheep wintering sites during summer.

Competition for forage between Dall's sheep and caribou *(Rangifer tarandus),* which occasionally occupy sheep habitat, has not yet been documented, but may occur. Henshaw (1970) reported two surprising instances of interaction between Dall's sheep and caribou in the Alaska Range wherein rams deliberately rushed at small bands of caribou, which ran from them.

Hoefs and Brink (1978) found that free-ranging domestic horses were important competitors of Dall's sheep at Kluane in Yukon Territory. Fortunately, livestock use of thinhorn sheep range is limited and cannot at this time be considered of general importance as a limiting factor. It may become so in the future with increasing human encroachment into sheep habitat.

Elk inhabit about 25% of Stone's sheep range. These species have considerable overlap in diet; however, the preference for adjacent timber cover by elk and escape terrain by sheep reduces the potential competition, with the result that there is little mutual use of the critical ranges. Bison *(Bison bison)* also have similar diet; however, they are restricted to shallower slopes, whereas the need for escape terrain tends to keep sheep on steeper slopes. The preference of caribou for more lush, summer vegetation and a preference for feeding in forests in winter reduces competition with Stone's sheep. Competition is minimal between Stone's sheep and moose *(Alces alces)*. The preference of mountain goats for a denser but poorer quality forage than sheep's also tends to minimize competition. Mule deer *(Odocoileus hemionus)* and white-tailed deer *(O. virginianus)* are present in Stone's sheep areas in low numbers and only in the low-elevation portions of their range.

TOOTH WEAR AND LONGEVITY

An important factor in limiting an individual animal's life span is tooth wear (Hoefs and Cowan 1979). As sheep age, their teeth become progressively worn and/or lost, until adequate forage intake and mastication become impossible. Wear is accelerated in areas where dust and silt collect in the forage.

Thus, life span may be shortened in areas where winds carry heavy loads of silt, such as near large glacial valleys like that adjacent to Kluane Park's Sheep Mountain herd. Bunnell and Olsen (1982) reported that, of the 26 examined on the Kluane Range, no sheep older than nine years of age had its full compliment of incisors. No sheep older than 13 years was collected (Bunnell 1978). The life expectancy of rams in that area, which receives frequent, silt-laden winds, was found to be 12–13 years (Hoefs and Bayer 1983).

On other Yukon ranges receiving much less glacial dust and silt, life expectancy rose to 14–15 years (Hoefs and Barichello 1984), and a few Dall's rams 17 years of age have been harvested. In Alaska, of more than 500 sets of rams' horns examined, only two had reached age 13 (Nichols 1984). In a collection of horns from Alaska's Brooks Range, the oldest male was estimated at 14 years and the oldest female at 17 years (Hemming 1961).

In addition to wear, thinhorn sheep suffer from tooth loss brought about by disease and/or breakage (Hoefs 1974). In the Mackenzie Mountains, 46% of male and 27% of female Dall's sheep had dental anomalies, including bone-cell proliferation, uneven wear, and missing or drifting teeth (Simmons et al. 1984). In general, then, longevity appears to be determined by loss of functional teeth, both by wear and by loss; most rams live to only 12–14 years of age and ewes average 1–2 years older.

ADAPTATIONS TO HUMAN DISTURBANCE

Dall's sheep appear able to habituate readily to human disturbance, especially if that disturbance occurs frequently. In the early stages of construction of the Trans-Alaska Pipeline, aircraft traffic nearly always caused sheep to react, either mildly or severely, and there was concern that construction activities might provoke the animals to abandon part of their range (Linderman 1972). However, as construction progressed sheep in the heavily traveled Atigun Canyon soon became accustomed to truck and aircraft traffic, the most common forms of disturbance. Although sheep still showed severe response to low passes by helicopters and fixed-wing aircraft, flights beyond 200 m from the animals elicited little reaction. Humans on foot within 500 m were less common and caused stronger responses. Sheep commonly crossed the haul road and were observed to cross under overhead sections of the pipeline (Price 1972:23–38, Jakimchuk et al. 1984).

In a more remote section of the Brooks Range, Curby (1981) noted that light, fixed-wing aircraft caused little reaction by sheep, but sheep reacted strongly to humans on foot. Sheep in Denali National Park have become so accustomed to humans on foot that photographers and others have little trouble in approaching them.

In the Cooper Landing Closed Area, which lies along a heavily traveled highway and just above the village of Cooper Landing on the Kenai Peninsula, Dall's sheep, here protected from hunting, show no reaction to noisy vehicles, chainsaws, human voices, or occasional gunshots directly below their habitat. Dall's sheep do, however, respond negatively to the less common occurrence of humans approaching on foot at or above their level. During more than 25 years of annually surveying this and other nearby sheep herds from a light aircraft, L. Nichols (pers. obs.) has observed little or no reaction to low passes in most instances. Despite repeated low passes (< 50 m), some sheep refuse to move or even stand up. The only animals that commonly react with fright to these flights are groups of yearlings that often gather together during lambing, having been driven off by their mothers. Desert bighorn sheep *(O. c. mexicana)* showed little reaction to light aircraft as long as they were flown greater than 50 m above the sheep (Krausman and Hervert 1983), but reacted severely when approached by helicopters (Bleich et al. 1990a). Low passes by helicopters almost invariably cause strong reactions by all classes of Dall's sheep.

In spring 1991, a forest fire burned just below and partially up into sheep habitat on Surprise Mountain, not far from Cooper Landing. This was in late May, just as lambing began. It was thought that the heavy smoke and activity of fire-fighting crews, including almost constant helicopter flights (mostly below the alpine areas occupied by sheep), noisy crews of people on foot, and even four-engined borate bombers would negatively impact lambing here. However, lambing apparently was normal despite the disturbance, with a midsummer lamb:ewe ratio of 59:100, which is somewhat better than average (L. Nichols, unpubl. data).

Thinhorn sheep seem able to adapt to various harmless human disturbances that become sufficiently routine for the animals to get used to them. Unusual or unexpected activities are more likely to trigger flight reactions. Of course, being hunted teaches sheep to flee from human approach. Too much hunting pressure may even force sheep to abandon their normal range permanently (Nichols 1976).

MORTALITY RATES

Mortality curves and life-expectancy tables for Dall's sheep have been published by several investigators (Deevey 1947, Murphy and Whitten 1976, Hoefs and Cowan 1979). In general, mortality is high during the first year of life, moderate to low during the second, low from age 2 to 8, and then increases sharply to 100% by about age 13 (although in some herds, a few individuals may live longer; fig. 2.12). Data for ewes was constructed from life-table data from a stable, relatively unhunted population in the Mackenzie Mountains (Simmons et al. 1984), whereas that for rams came from the protected Sheep Mountain herd, which was also considered "stable" at that time (Hoefs and Bayer 1983). Data from other populations and times show similarly shaped curves, but vary by the herds' condition and even by the method of data collection.

Hoefs and Cowan (1979) calculated an average annual mortality rate of 20% for adult Dall's ewes in Kluane National Park, whereas Simmons et al. (1984) estimated the average annual mortality rate of Dall's ewes older than one year to be about 15% both in the Mackenzie Mountains of northwestern Canada and in McKinley National Park, Alaska. Murphy and Whitten (1976) showed that no single curve or time span could be applied to all sheep

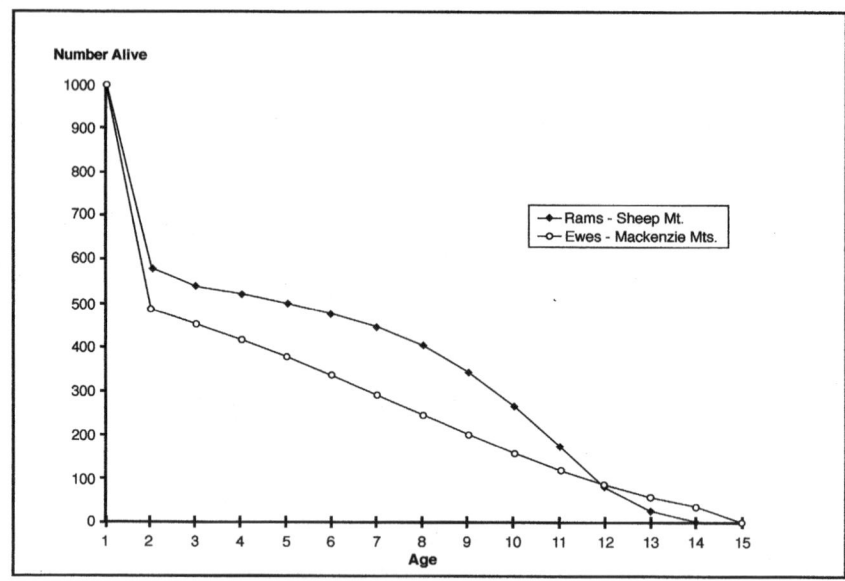

Figure 2.12. Examples of mortality curves for both sexes of Dall's sheep (data from different populations at different times; ewes: Simmons et al. 1984; rams: Hoefs and Bayer 1983). Curve forms are generally similar for all herds but vary in degree by herd, cohort, time frame, and method of data collection.

populations because mortality rates vary for each cohort by existing conditions and stocking rates.

POPULATION CHANGES AND QUALITY

Despite differences in periodicity, most thinhorn sheep herds fluctuate greatly in numbers (J. S. Dixon 1938, Murie 1944, R. V. Scott et al. 1950, Murphy and Whitten 1976, Nichols 1976). These fluctuations probably were caused indirectly by changes in winter weather that, in turn, affected the animals' nutrition or predator vulnerability. Some herds in interior Alaska and Canada may have reached density-limited stability under conditions of relatively unchanged weather patterns over time (Hoefs and Cowan 1979, Heimer 1980b). However, past and recent history suggests that this was a temporary stability and that abrupt declines followed by increases may be expected at infrequent intervals.

Dall's sheep herds on Alaska's Kenai Peninsula are at the southern extremity of this species' range and are affected and apparently limited in distribution by maritime climatic conditions. Severe (i.e., warm, wet) winters occur with more regularity than in the interior, and sheep herds have shown large fluctuations. Fortunately, the buildup of several of these herds was documented by periodic surveys (Pitzman 1970, Nichols 1976; fig. 2.13). These herds increased from their low points in the 1940s to maximum sizes in the late 1960s and early 1970s, then decreased sharply again following another series of harsh winters.

These sheep herds and the interior Alaskan herd in McKinley National Park have been documented as following the classic ungulate pattern of population irruption described by Caughley (1970, 1976). Curves were mathematically fitted to the populations to illustrate the sigmoidal shape of the increase stage, an important point to the game manager (fig. 2.13). During this increase stage, these herds grew at an average annual rate of between 11 and 14%, showing the capability of thinhorn sheep for fairly rapid population growth. As herds approached their maximum size, presumably limited by their winter food supply, their growth rate progressively declined to zero, which was followed by a negative growth rate as herds decreased. On the Kenai Peninsula, the negative growth rate appears to have ended and the herds presently have begun recovering from their lowest level of this cycle.

In his quality hypothesis, Geist (1971) proposed that sheep in expanding populations exhibited certain characteristics that were common to high-quality herds. These included rapid body and horn growth, early sexual maturity, high lambing rates, high juvenile survival to adulthood, and shorter life expectancy. Because these characteristics are undoubtedly related to plane of nutrition (Geist 1971; Murphy 1974; Bunnell 1978, 1980a; Winters 1980), one would expect available forage, which is especially important on winter range, to be abundant and sheep density in relation to forage production to be low. With forage abundant, competition would be light.

Geist's (1971) quality hypothesis further proposed that nonexpanding, or stable, populations would exhibit low-quality characteristics, including slower body and horn growth, delayed sexual maturity, lower lambing rates, higher juvenile mortality, and longer life expectancy. Winter range should be stressed, sheep density should be high in relation to forage availability, and competition should be strong. In declining populations, these low-quality characteristics would be somewhat magnified; in particular, juvenile and adult mortality should be excessive.

Examination of the limited data available for increasing, stable, and decreasing thinhorn sheep herds confirms at least some of these expected characteristics (i.e., increasing herds on Surprise Mountain [Pitzman 1970, Nichols 1976], McKinley National Park [Murphy 1974], and Nevis Mountain, British Columbia [Luckhurst 1973]; relatively stable herds in Kluane Park [Hoefs and Cowan 1979] and McKinley National Park [Murphy 1974]; a

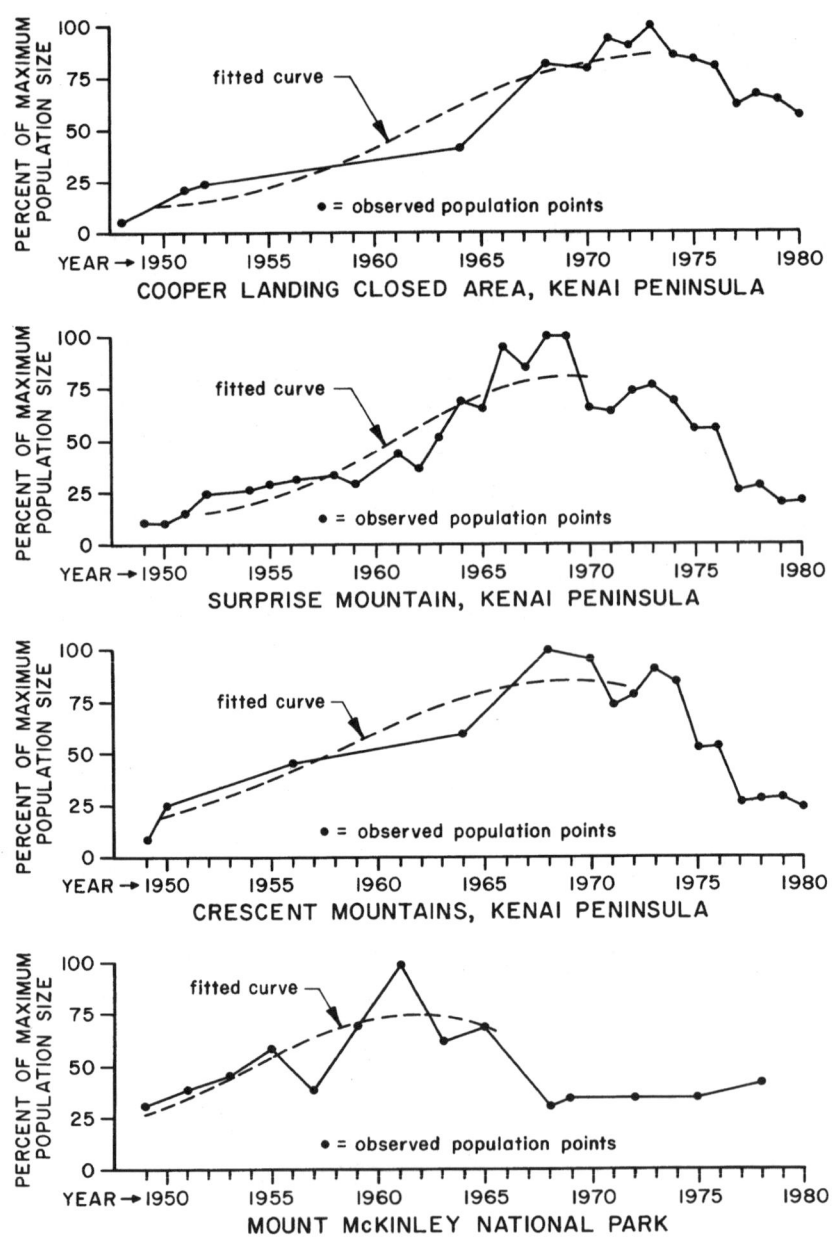

Figure 2.13. Four Dall's sheep populations in Alaska with mathematically fitted curves to illustrate the sigmoid growth pattern. All populations are reduced to the same scale for comparative purposes.

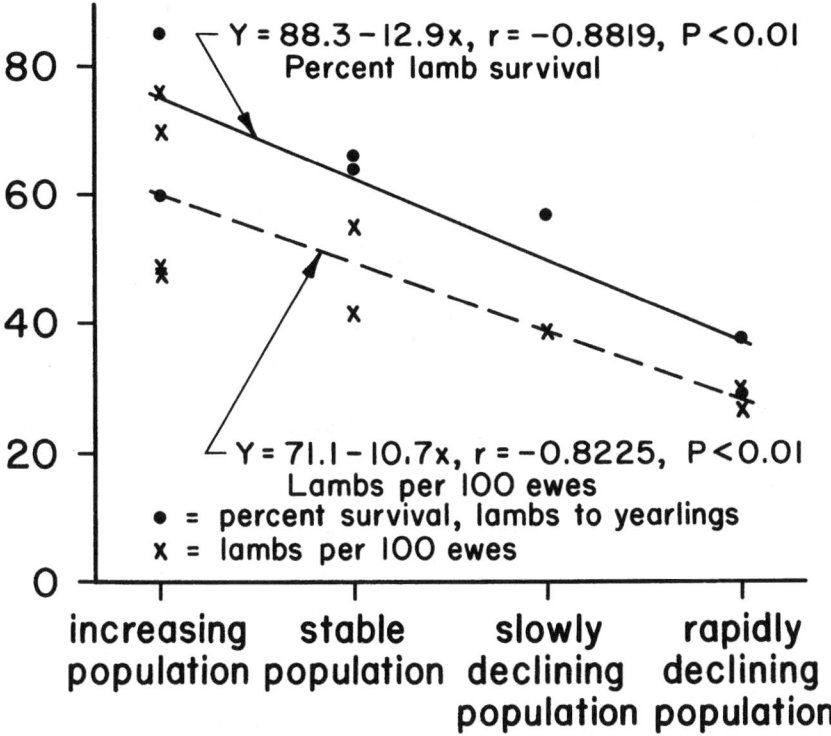

Figure 2.14. Relationship between lamb production, lamb survival, and population status.

slowly declining herd in Dry Creek, Alaska Range [Heimer 1980b]; and rapidly declining herds on Surprise Mountain [Nichols 1976] and McKinley National Park [Murphy 1974]). The average ratios of lambs per 100 adult ewes declined significantly ($P < 0.02$) from increasing through stable, slowly decreasing, and rapidly decreasing herds. Survival of lambs to yearlings also declined significantly ($P < 0.01$; fig. 2.14). Thus, as the status of these herds changed from expanding through stable to declining, lamb production changed from above average to below average, and survival of lambs through their first year fell from high to low.

Other indicators of herd quality in relation to herd status were observed in various studies. In increasing herds, neonatal mortality was low; yearlings were large and sexually mature; adult ewes bred annually; density per unit of winter range was low; and rams showed rapid horn growth, reaching ¾-curl by age 4 (Pitzman 1970, Murphy 1974, Nichols 1976). In stable or slowly declining herds, neonatal mortality was higher; yearlings were small and did not mate, or did mate in some herds but not others; adult ewes bred either annually (but began breeding later in life) or only in alternate years; ram horn growth was slower, reaching ¾-curl by age 5.5 or 6; and density was high (Bunnell and Olsen 1976, Hoefs and Cowan 1979, Heimer 1980*a*). Thus, as sheep herds attain the point of maximum range carrying capacity and density per unit of forage increases, lowered nutrition, particularly during winter, affects the animals' growth rates, survival, and ability to reproduce.

That the level of nutrition is of major importance to population quality was further quantified when four female Dall's sheep lambs were removed just after birth from the late-maturing, stable Kluane herd, and were fed a better diet in a game farm. Two of them reached a sufficient size to mate successfully as yearlings (Bunnell and Olsen 1976). Captured sheep from the low-quality Dry Creek herd exhibited exceptional horn growth when fed a high-quality diet in a zoo, as did a male lamb taken from the Kluane herd (Bunnell 1978, Gasaway and Heimer, n.d.). Following an artificial herd reduction by either-sex hunting in a Kenai herd, winter range forage production increased significantly, as did lamb production and survival (Nichols 1976).

The classical population irruption pattern (Caughley 1970) includes the original sigmoidal increase stage, followed by progressively smaller oscillating decreases and increases as the population approaches stability with its food supply. This type of pattern is found in a density-dependent population that is limited solely by its food supply, which, in turn, results in progressively lower productivity and carrying capacity until a balance between foragers and forage is reached. Although the illustrated declines in the Dall's sheep herds previously described apparently resulted from more than one severe winter, these herds and those in the Alaska Range and Kluane Park exhibited at their maximum points the characteristics of stable, or environmentally limited, populations. This indicates that thinhorn sheep, like other ungulates, are density dependent. In other words, the size of a herd (barring abnormal weather, excessive predation, or disease) is self-limiting through its pressure on its own food supply in winter as the herd's numbers increase to a point where no more animals can be supported on a given range. If a herd remains fairly stable while at this stage of high density relative to available forage, its quality declines. Of particular significance to hunters is that the annual surplus of sheep available for harvest is greatly decreased and body and horn size of rams decreases.

Of course, catastrophic winters, disease outbreaks, or excessive predation can disrupt the pattern of density dependence, causing a population to decline regardless of its relative food base. The previously mentioned population declines in several "stable" central Alaska and western Yukon herds are good examples of winter-weather-related setbacks, whereas heavy predation by wolves apparently lowered Stone's sheep populations despite their having adequate forage (J. P. Elliott 1985). Thus, herds that appear to have reached stability at the carrying capacity of their ranges may be subjected to catastrophic "crashes" at irregular intervals because of factors other than those imposed by their forage base.

three

NATURAL HISTORY OF ROCKY MOUNTAIN AND CALIFORNIA BIGHORN SHEEP

David M. Shackleton, Christopher C. Shank,
and Brian M. Wikeem

Introduction

In this chapter we review the scientific literature on the ecology, behavior, growth, bioenergetics, and population biology of Rocky Mountain *(Ovis canadensis canadensis)* and California *(O. c. californiana)* bighorn sheep. We hope it provides the reader with a synthesis of material that can be used as source material and as a guide for future research.

We maintained the distinction between California and Rocky Mountain bighorn even though the original criteria used to separate the two (Cowan 1940) may not be supported by contemporary taxonomy (e.g., Ramey 1995). Our reasons for recognizing both are allopatry (geographic separation), prior convention, and because no taxonomic revision has been made. Where comparative data are available, we have treated the two subspecies separately, otherwise we used the collective term "bighorn," though exclude desert sheep.

Habitat

PRESENT AND HISTORICAL DISTRIBUTION

Rocky Mountain and California bighorn sheep are widely distributed in localized populations throughout the drier, mountainous regions of western

North America (Cowan 1940, J. L. Clark 1978, Shackleton 1985, J. A. Bailey and Klein 1997, Shackleton et al. 1997). The distribution of Rocky Mountain bighorn extends from about 55°N latitude in Alberta and British Columbia through Montana, Idaho, Utah, Wyoming, Colorado and into northern New Mexico, at 36°N latitude. California bighorn historically ranged from the eastern slopes of the Coast Mountains in central British Columbia (51°N), south through Washington, Oregon, and Idaho to the Sierra Nevada in California (37°N). From the turn of the century until about 1954, California bighorn became extinct over much of their distribution. Since then, reintroductions have been made to California, Oregon, Washington, Idaho, Nevada, and North Dakota, primarily from British Columbian populations (D. A. Demarchi and Mitchell 1973).

Compared to their historical distributions, the present ranges of California and Rocky Mountain bighorn are greatly reduced (D. R. Smith 1954, Sugden 1961). Historical records also show that both races were well adapted to habitats far from the rugged mountain terrain now considered critical habitat (Cowan 1940, D. R. Smith 1954, Wishart 1958, Buechner 1960, Stelfox 1971, Shackleton 1985). For example, some populations of Rocky Mountain bighorn once inhabited the river valleys and surrounding prairies east of the Rocky Mountains in both Canada and the United States (Cowan 1940, D. R. Smith 1954, McCann 1956, Buechner 1960, Stelfox 1971). Also, many populations probably occupied low-elevation ranges for most, or all, of the year (D. R. Smith 1954, Sugden 1961), as several contemporary herds of California bighorn do today (Sugden 1961, Drewek 1970, Morrison 1972, P. Ebert 1978).

ELEVATION

Bighorn occupy ranges from as low as 450 m to over 3,300 m in elevation (Buechner 1960, D. A. Blood 1961, Sugden 1961, McCullough and Schneegas 1966, Berwick 1968, Stelfox and Taber 1969, J. K. Morgan 1970, Oldemeyer et al. 1971, Morrison 1972, Woodard et al. 1972, Riggs 1977, Becker et al. 1978). Individual populations, however, vary considerably in their use of elevational gradients, with herds that remain relatively sedentary (D. R. Smith 1954, Sugden 1961, Drewek 1970, Spalding and Bone 1970) and those that exhibit long, annual elevational migrations (D. R. Smith 1954, Wishart 1958, Berwick 1968, J. K. Morgan 1970).

TOPOGRAPHIC FEATURES

Bighorn sheep typically inhabit river canyons, foothills, and mountains (Geist 1971, Adams et al. 1982, Shackleton 1985). Their habitats are generally characterized by rugged terrain including canyons, gulches, talus cliffs, steep slopes, mountain tops, and river benches (D. R. Smith 1954, Beuchner 1960, Sugden 1961, Drewek 1970, Todd 1972a, Stelfox 1975, R. J. Hudson et al. 1976, Kornet 1978, Van Dyke 1978).

Proximity to escape terrain (i.e., security cover) is an important and common feature of most bighorn sheep habitats. These areas provide sheep with protection from terrestrial predators (Adams et al. 1982), especially during lambing (D. A. Blood 1961, Drewek 1970, Kornet 1978, Shank 1979, Akeson and Akeson 1992). In fact, in winter Rocky Mountain sheep can spend as much as 86% of their time within 100 m of rocky escape terrain and 75% of feeding time within this distance (Oldemeyer et al. 1971, G. L. Erickson 1972). Rocky Mountain bighorn generally remain within 800 m from escape terrain during all seasons (Pallister 1974).

Security cover may be more important to females than to males. For example, M. C. Hansen (1982a) found that females generally were located within 100 m of escape habitat, but moved farther away as spring progressed through summer. In contrast, males generally ranged more than 200 m from escape terrain. This difference might be because males are better able to defend themselves against predators than are females (Shank 1979) and because pregnant females may also make trade-offs between forage benefits and predation risk in their choice of habitats near or far from escape terrain (Festa-Bianchet 1989a).

CLIMATE

Usually, bighorn sheep live in areas that are relatively warm and arid, but with cold, dry winters (Cowan 1940, F. L. Jones 1950, D. R. Smith 1954). Indeed, temperatures vary from winter lows below -40°c to summer highs above 35°c (D. R. Smith 1954, Schallenberger 1966, Drewek 1970). In Oregon, California bighorn habitats have been characterized as drier and warmer than Rocky Mountain bighorn habitats in the same state (P. Ebert 1978). However, elsewhere annual precipitation on Rocky Mountain and California bighorn habitats varies from about 20 cm in British Columbia (Spalding and Bone 1970), Idaho (Drewek 1970), and Colorado (Todd 1972a) to over 40 cm in Colorado (Dale 1987), Montana (Schallenberger 1966), Oregon (Van Dyke 1978), and east-central British Columbia (Stelfox 1975).

Correlations of temperature and precipitation with long-term Rocky Mountain bighorn population data suggest that low annual precipitation over a 12-month period is most important to lamb survival (Picton 1984). This sensitivity to precipitation might also explain why bighorn sheep are absent from the wet coastal areas of western North America (Picton 1984).

Typically, most bighorn winter ranges are relatively snow-free because of light snowfall, aspect, or high winds. Snowfall on bighorn ranges is from less than 45 cm (D. R. Smith 1954, Sugden 1961) to over 120 cm (Stelfox 1975). Stelfox (1975) suggested that critical snow depth for Rocky Mountain bighorn is 30 to 44 cm for lambs, 32 to 48 cm for yearlings and females, and 36 to 54 cm for adult males. He concluded that mountain sheep generally avoid snow depths over 30 cm. Snow quality is probably another important factor on winter ranges because snow crusts, caused by freeze-thaw cycles, can constrain bighorn from digging for forage and impede their movements (Sudgen 1961).

VEGETATION

Recognizing that researchers define habitats in different ways, bighorn sheep do use a variety of habitats throughout their distribution and also within their home ranges (Schällenberger 1966, Pallister 1974, Stelfox 1975, Risenhoover and Bailey 1985, Dale 1987). Rocky Mountain bighorn in Montana can use as few as 4 habitats (G. L. Erickson 1972), and in Colorado as many as 19 habitats can be used between May and November (Goodson 1978). There are fewer habitat studies for California bighorn, but they have been found to use between 10 (M. C. Hansen 1982a) and 15 (Kornet 1978) different habitats annually in Nevada and Oregon, respectively.

Bighorn habitats include open grasslands, alpine, subalpine, shrub-steppes, rock outcrops, cliffs, meadows, moist draws, stream sides, talus slopes, plateaus, deciduous forests, clear-cut or burned forests, and conifer forests (D. A. Blood 1961, R. A. Demarchi 1965, G. L. Erickson 1972, Pallister 1974, Goodson 1978, Kornet 1978, Van Dyke 1978, M. C. Hansen 1982a, Risenhoover and Bailey 1985, Dale 1987). When available, these different habitats meet different needs, so the specific use bighorn make of them often varies daily and seasonally as their requirements for food, security cover, mating and lambing grounds, and thermal regulation change (Oldemeyer et al. 1971, Goodson 1978, Kornet 1978, Van Dyke 1978, M. C. Hansen 1982a, Risenhoover and Bailey 1985, Dale 1987).

Open grasslands are typically used for winter ranges by bighorn (D. A. Blood 1961, R. A. Demarchi 1965, G. L. Erickson 1972, Todd 1972a, D. A. Demarchi and Mitchell 1973, Pallister 1974, Dale 1987, Van Dyke 1978, M. C. Hansen 1982a). They are often covered by wheatgrasses (*Agropyron* spp.), fescues (*Festuca* spp.), bluegrasses (*Poa* spp.), mesquite-grasses (*Bouteloua* spp.), muhlys (*Muhlenbergia* spp.), needle grasses (*Stipa* spp.), and ricegrasses (*Oryzopsis* spp.) combined with mixed forbs and small shrubs. Naturally, the specific species composition varies considerably among geographic locations.

Alpine areas (fig. 3.1) often serve as summer range for foraging (D. A.

Figure 3.1. Members of a male group of Rocky Mountain bighorn sheep foraging in their alpine summer range. (Photograph by David Shackleton.)

Blood 1961, Sugden 1961, Woolf et al. 1970, Pallister 1974, J. D. Johnson 1975, Goodson 1978, Shank 1979). Dominant forage species may include kobresia (*Kobresia* spp.), sedges (*Carex* spp.), grasses, and forbs. Although mountain ranges are often contiguous with foothills and river systems, bighorn populations may spend the entire year on subalpine and alpine ranges. Rock cliffs providing security cover are especially common in alpine areas. Below timberline, fire may play an important role in producing and maintaining subclimax grassland or parklands habitats for bighorn sheep (Geist 1971, G. L. Erickson 1972) and therefore can be a useful tool for habitat management in some areas (McWhirter et al. 1992). Subalpine communities inhabited by bighorn may contain sedges, grasses, rushes (*Juncus* spp.), a variety of forbs, and a few shrubs (Pallister 1974).

Talus slopes, rock outcrops, and cliffs are often sparsely vegetated but provide habitat for resting, lambing, and security (G. L. Erickson 1972, Kornet 1978, Van Dyke 1978, Akeson and Akeson 1992). When these habitats are vegetated, especially with shrubs, they can be important areas for foraging (Pallister 1974, Van Dyke 1978). Depending on the location, shrubs may include gooseberry (*Ribes* spp.), cinquefoil (*Potentilla* spp.), sagebrush (*Artemisia* spp.), rose (*Rosa* spp.), buckbrush (*Symphoricarpos* spp.), maple (*Acer* spp.), serviceberry *(Amelanchier alnifolia),* kinnikinnick *(Arctostaphylos uva-ursi),* juniper (*Juniperus* spp.), and blueberry (*Vaccinium* spp.; Schallenberger 1966, G. L. Erickson 1972, Pallister 1974, Van Dyke 1978).

Generally, deciduous and coniferous forests are used sparingly by bighorn sheep (McCann 1956). In Colorado, Rocky Mountain bighorn inhabited open Douglas fir *(Pseudotsuga menziesii)* forest only when strong wind and cold temperatures forced them to seek shelter (Schallenberger 1966), whereas dense Douglas fir forest was never used (Dale 1987). Apparently, dense forest and other thick vegetation associations that restrict vision are avoided by bighorn (Kornet 1978, Van Dyke 1978, Risenhoover and Bailey 1985), probably because of higher predation risks. However, open forest stands can be important habitats for foraging and thermal cover (Sugden 1961, Spalding and Bone 1970, D. A. Demarchi and Mitchell 1973, Pallister 1974, Goodson 1978), especially in the summer when bighorn will bed under trees for shade (McCann 1956, Wikeem 1984).

Vegetation structure may be more important to bighorn sheep than the specific plant species composition on a range. For example, Risenhoover and Bailey (1985) and Wakelyn (1987) each reported that open habitats with high visibility were used most by Rocky Mountain bighorn, regardless of vegetation associations. However, even though open habitats provide opportunity for bighorn to detect predators (Wakelyn 1987), these habitats may not be used by bighorn if escape terrain is not readily available (McCann 1956).

SLOPE AND ASPECT

Most bighorn winter ranges are found on steep south, southwestern, or southeastern facing slopes (D. R. Smith 1954, McCann 1956, D. A. Blood 1961, Sugden 1961, McCullough and Schneegas 1966, Schallenberger 1966, Stelfox and Taber 1969, J. K. Morgan 1970, Geist 1971, Oldemeyer et al. 1971, G. L. Erickson 1972, Riggs 1977). Southerly exposures maximize heat gains from direct and indirect solar radiation, thus reducing cold stress on sheep while at the same time reducing snow cover and increasing forage availability in winter. The warmer temperatures also promote earlier spring green-up of forages on these slopes. During summer, however, when cold stress is not a problem, bighorn may move to north-, east-, and west-facing slopes to feed, particularly on alpine range (D. R. Smith 1954, McCullough and Schneegas 1966, Stelfox 1975, Goodson 1978).

Steep slopes appear to be a significant feature for most populations of

Rocky Mountain and California bighorn sheep, although percent slope varies considerably from habitat to habitat within and between populations. In Montana, a mean slope of 36% was used by Rocky Mountain bighorn in the Beartooth Mountains (Pallister 1974), as compared to a 60% mean slope used in Sun Valley (Frisina 1974). In Colorado, slopes between 61 and 80% were used most and slopes less than 20% were avoided (Fairbanks et al. 1987).

Slopes ranged from 6 to 100% among the 14 habitats used by California bighorn in Oregon (Van Dyke 1978). At Hell Creek, Nevada, bighorn used slopes of 34–50% most in spring and summer and 0–17% slopes in fall and winter (M. C. Hansen 1982a). In contrast, a second herd at Virgin Canyon mostly occupied habitats with 18–33% slopes in spring and summer, compared to 34–50% in winter. Seasonal use of different slope and aspect results in a mosaic of plant communities, phenological patterns, and hazards to predators that provide a variety of foraging and security opportunities for bighorn.

SALT AND WATER

There has been very little research on the specific needs of either subspecies of bighorn for trace minerals and water. Some studies suggest that salts may be nutritionally important for Rocky Mountain bighorn (Couey 1950, Geist 1971), and major declines in some mountain sheep populations have been attributed to mineral deficiencies (Packard 1946, L. O. Wilson 1968).

The significance of mineral licks for habitat use patterns has not been examined in detail. Habitat use by Rocky Mountain bighorn may be influenced by the presence of mineral licks in Colorado (Goodson 1978), whereas other populations seem to use mineral licks mostly in spring (D. R. Smith 1954, McCann 1956, R. J. Hudson et al. 1976), even though the licks are readily available throughout the year. No mineral licks have been reported on California bighorn ranges in British Columbia (Sudgen 1961, D. A. Demarchi and Mitchell 1973) or in Oregon (Kornet 1978, Van Dyke 1978), although F. L. Jones (1950) found rock fragments in bighorn sheep feces at a salt lick in California.

Mineral licks may be more important to Rocky Mountain bighorn than to California bighorn because of the low mineral content of granitic soils throughout the distribution of Rocky Mountain bighorn (Van Dyke 1978). D. R. Smith (1954) determined that soil mineral contents averaged only 250 ppm on Rocky Mountain bighorn range in Montana and speculated that forages grown on these soils may be low in mineral content. No similar work has been done on California bighorn range; however, soils in many of their habitats are derived from volcanic or sedimentary materials and probably have no significant mineral deficiencies (Sudgen 1961, Van Dyke 1978). The importance of mineral licks as a habitat component for bighorn sheep and for their nutrition requires further investigation.

Bighorn sheep requirements for free-standing water have not been clearly established. However, they appear well adapted to arid conditions and can subsist for long periods without free-standing water. Bighorn appear able to meet their water requirements from succulent vegetation in summer and from snow and ice in winter (McCann 1956, Kornet 1978, Van Dyke 1978).

Because he never saw them drinking from streams, McCann (1956) suggested that water availability does not limit bighorn. In Nevada, M. C. Hansen (1982a) observed that 70% of California bighorn were within 0.5 km of water and 90% were within 1.0 km. Nonetheless, he concluded that their proximity to water was likely more related to juxtaposition of other habitat features and not necessarily an attraction to water itself. On Poker Jim Ridge, Oregon, California bighorn were never seen drinking. The population at Hart Mountain only drank from streams between November and April, possibly to meet the higher demands for water during rutting and lambing (Kornet 1978).

A habitat model for desert-dwelling bighorn sheep based on vegetation, topography, and water availability was tested by Bleich et al. (1992b) using a geographic information system (GIS). The results indicated that while these three variables were important for the sheep, they could not be used to predict habitat use. This is the first published use of GIS for bighorn habitat assessment. While emphasizing the value of GIS for this type of work, the authors cautioned managers to exercise care when interpreting results derived from such an approach.

RANGES AND MOVEMENTS

Bighorn populations are divided into sexually segregated groups that migrate seasonally between different areas during their annual cycle. Although the number and types of seasonal ranges used in a year vary, there are usually at least two: winter and summer ranges. The Riske Creek population of California bighorn sheep may be an exception because they occupy the same general range year-round (Sugden 1961). The same may be true for other populations in similar habitats in south-central British Columbia, but their movements patterns are currently being studied (R. Lincoln, B.C. Wildlife Branch, pers. comm.). Most bighorn populations may have more than the two basic seasonal ranges and as many as five have been reported (Geist 1971). Besides summer and winter ranges, these may include a lambing or spring range, a salt lick range, and a fall or rutting range (Geist 1971).

Few studies have rigorously documented home ranges or seasonal movements of bighorn, but it is clear that mature individuals show a high degree of fidelity to specific, seasonal ranges, and return to them each year (J. K. Morgan 1970, Geist 1971, Becker et al. 1978). However, the degree of range fidelity may differ between the sexes, with females showing high annual fidelity to their winter range whereas some male bighorn do not (Festa-Bianchet 1986a,b).

The delineation and enumeration of seasonal ranges can sometimes reflect as much the interests of sheep biologists as they do the behavior of the sheep themselves. The preconceptions of researchers and how they define seasonal ranges may determine whether two, three, or more such units are recognized. Thus far, few published studies of seasonal movements and ranges of bighorn sheep have applied the same methods and statistical approaches used so successfully with other large mammals (e.g., following several radiocollared animals and calculating home-range size using minimum convex polygon or adaptive kernel techniques; Priede and Swift 1992).

SEASONAL MIGRATIONS

Seasonal migrations between ranges are generally interpreted from a functional and adaptive viewpoint. Seasonal altitudinal migrations of bighorn are found in most populations. An explanation for these migrations from low to high elevations during spring and summer (fig. 3.1) was first suggested by Klein (1965) for black-tailed deer *(Odocoileus hemionus sitkensis)* in Alaska. He argued that because the onset of plant phenology is retarded by increased elevations, a "wave" of fresh vegetation growth moves up the mountains, providing high-quality, readily digestible forage spread over time from spring through summer. Therefore, by moving upward in elevation during these seasons, herbivores could maintain or prolong a diet composed mainly of new, growing vegetation. Klein (1964, 1969) further proposed that differences in growth rates and other production parameters among deer populations were a reflection of differences in the altitudinal extent to which a population could move during its seasonal vertical migration. Consequently, the longer the vertical migration, the longer a population could feed on high-quality forage. Hebert (1973) examined this concept for Rocky Mountain sheep and concluded that they benefit from new growth and alpine vegetation in summer and that vertical migration plays an important role in their annual nutrition. D. A. Blood (1963) also suggested that the stimulus for spring migration in the California sheep in the Ashnola, British Columbia, is a response to vegetation condition at higher elevations. Not only may a highly nutritive diet be maintained by moving vertically following a green-up, but as Johnston et al. (1968)

showed, some forage plants growing in the alpine may have a higher nutritive value than the same species growing at lower elevations.

The consequences of seasonal movements and plant growth discussed by Klein (1964, 1965) for individual growth and population productivity of ungulates have important implications for theories of population variability (Geist 1971). Shackleton (1973) suggested that a major factor accounting for production, growth, and developmental differences between the expanding and stable populations that he studied was differences in vertical migrations made by bighorn. The expanding population comprised of fast-growing individuals made vertical migrations from winter to summer ranges that were twice as large as those made in the stable population. He calculated that this difference was equivalent to 30 extra days of foraging time on high-quality, growing forage.

Although temporal variation in vegetation growth seems to be a major factor influencing the upward vertical migrations of bighorn sheep in spring and summer, other motivating factors are probably involved. Where lambing grounds are separate from winter ranges, imminent parturition and the need for security cover probably stimulate pregnant females to migrate from winter ranges. Needs for security from predators can even override forage conditions. Festa-Bianchet (1988a) showed that, to give birth, pregnant females moved to areas that had lower forage quality but significantly greater security cover from predators. Security cover has also been suggested to be an important determinant of winter range use (Wishart 1958, Shannon et al. 1975).

Other movements by bighorn appear related, at least in part, to climatic factors. Migrations from alpine summer ranges down to low-elevation winter ranges seem motivated by snow accumulation that restricts movements on the high ranges in late summer and early fall (D. R. Smith 1954, Sugden 1961, D. A. Blood 1963). Noxious, biting insects may not be a factor in major movements of bighorn, because these pests can be equally common at high elevations (D. A. Blood 1963). In a number of populations, small groups of bighorn remain on the wintering areas year-round and apparently do not migrate with the rest of the herd (D. R. Smith 1954, McCann 1956, Sugden 1961, Oldemyer et al. 1971, Akeson and Akeson 1992). As in some herds in Idaho, this may be because sheep remain on winter range to access mineral licks found there (D. R. Smith 1954). However, in Kootenay National Park, British Columbia, although sheep made brief (one to two days), periodic visits to low elevations to use mineral licks, they always returned to distant, summer, alpine areas (D. M. Shackleton, unpubl. data). Overall, seasonal range movement appears to be an attempt by sheep to optimize constraints over an array of different conditions (Festa-Bianchet 1988a).

Timing of seasonal movements follows a similar pattern in most populations. Bighorn usually return to their winter ranges in October and November (D. R. Smith 1954, Sugden 1961, D. A. Blood 1963, Berwick 1968, J. K. Morgan 1970, Woolf et al. 1970, Geist 1971, Oldemeyer et al. 1971, Becker et al. 1978), with females invariably arriving before the males. Movements from the summer ranges may occur earlier in some areas (Berwick 1968, Geist 1971). Migration from the winter range, which is related to lambing and foraging requirements, occurs mainly in May and June, with some as early as April and as late as July; females may leave before or after males (D. R. Smith 1954, Sugden 1961, D. A. Blood 1963, Berwick 1968, Geist 1971, Oldemeyer et al. 1971, Becker et al. 1978).

As might be expected, linear distances of sheep migration between seasonal ranges vary, most probably in relation to the relative availability and distribution of suitable ranges. There appears to be no difference between the extent of movements between California bighorn (0–48 km; Sugden 1961, D. A. Blood 1963, McCullough and Schneegass 1966) and Rocky Mountain bighorn (5–51 km; D. R. Smith 1954, Wishart 1958, Berwick 1968, J. K. Morgan 1970, Festa-Bianchet 1986b, Akeson and Akeson 1992, Hengel et al. 1992).

Diet

Diet preferences of bighorn sheep have been studied since the 1920s (Seton 1929). Following is a review of available information on the diets of Rocky Mountain and California bighorn sheep that focuses on common factors in bighorn diet.

STUDYING BIGHORN SHEEP DIETS

Methods used to quantify bighorn diets have varied significantly, with sampling intensity ranging from examination of a few or even a single rumen (W. O. Hickey 1978) or a few day's field observation (Cowan 1947) to detailed monthly and seasonal diet analyses (Constan 1972; Todd 1972b; J. D. Johnson 1975; Stewart 1975; Wikeem and Pitt 1979, 1992; Goodson et al. 1991). Similarly, data collection methods have varied from stem counts, snow trailing, examination of feeding sites, observations from transects, direct observations of animals, to detailed rumen and fecal analyses. These varied approaches can be expected to produce different estimates of diet (L. O. Wilson 1976), making comparisons and generalizations difficult. In addition, certain plant species (e.g., pasture sage, *Artemisia frigida;* prairie sage, *A. ludoviciana;* and buckwheats, *Eriogonum* spp.) have been classified as forbs or shrubs by different researchers, thus hampering comparisons.

DIVERSITY OF BIGHORN SHEEP DIETS

Rocky Mountain sheep diet has been described as cosmopolitan (Todd 1972b) and they seem to eat almost every plant available to them at one time or another (Ellis 1941). Similarly, California bighorn sheep are also opportunistic feeders that adapt their diets to whatever forage is available (Sudgen 1961). From California to British Columbia, more than 267 species of plants have been recorded in California bighorn diets, including 45 species of grass and grasslike plants, 160 forbs, and 62 shrubs (Wikeem 1984). Diet breadth is extensive for individual populations of both Rocky Mountain and California bighorn sheep (table 3.1), with between 69 and 88 species reported to be consumed in different populations (J. D. Johnson 1975, Stewart 1975, M. C. Hansen 1982a, Wikeem and Pitt 1992). Forbs generally dominate diets for both subspecies, followed by grasses and browse, respectively (table 3.1).

COMPOSITION OF BIGHORN SHEEP DIETS

Although forbs often contribute the greatest number of species to bighorn diets, based on percent composition, grasses (including sedges and rushes) typically dominate both Rocky Mountain and California bighorn diets (table 3.2). Nonetheless, in some habitats the percent composition of forbs and shrubs in the diet can equal or surpass that of grasses (tables 3.2, 3.3). Additionally, the relative proportions that each of these three forage classes contribute to bighorn diets vary considerably among populations and even within a subspecies. Therefore, depending on the population, bighorn diets can be dominated ($> 66\%$) by either grasses, forbs, or shrubs (F. L. Jones 1950, Sugden 1961, D. A. Blood 1967, B. W. Brown 1974, Pallister 1974, Stelfox 1975, Bear 1978, W. O. Hickey 1978). In addition, large variation in dietary composition among forage classes has even been documented among individual sheep within a population (W. O. Hickey 1975).

Interpretation of seasonal variation in diet may be further complicated by variation among age-sex classes. Winter diets of Rocky Mountain bighorn females, lambs, and yearlings were more similar to each other than they were to the diet of adult males (Shank 1982). Shank (1982) attributed this difference to the spatial segregation of male and maternal groups onto ranges with different proportions of forage species available.

Variability in bighorn sheep diets over their entire distributional range is not unexpected. Plant species diversity varies considerably from habitat to habitat, thus altering forage availability. For California bighorn, although the vegetation physiognomy and the relative proportions of grasses, forbs, and shrubs were similar between the sites in Nevada (M. C. Hansen 1982a) and

Table 3.1. Number of grasses, forbs, shrubs, and total species in Rocky Mountain and California bighorn diets throughout their geographic distributions.

LOCATION SEASON	SOURCE	NO. SPECIES IN DIET			
		GRASSES	FORBS	SHRUBS	TOTAL
Rocky Mountain Bighorn					
Alberta	J. D. Johnson (1975)	10	38	21	69
Alberta	Stelfox (1975)	11	19	6	36
Colorado	Capp (1967)	4	6	3	13
Colorado	Todd (1972a)	14	16	12	42
Idaho	D. R. Smith (1954)	8	8	15	31
Montana	Schallenberger (1966)	8	14	15	37
Montana	Oldemeyer et al. (1971)	7	8	4	19
Montana	Constan (1972)	9	10	8	27
Montana	G. L. Erickson (1972)	6	28	8	42
Montana	Pallister (1974)				
West Rose Bud					
Summer		8	16	8	32
Fall		2	2	0	4
Winter		2	2	0	4
Stillwater					
Fall		8	11	3	22
Winter		4	4	2	10
Montana	Stewart (1975)				
West Rose Bud		18	53	13	84
Stillwater		21	35	15	71
California Bighorn					
British Columbia	Sugden (1961)	9	7	8	24
British Columbia	D. A. Blood (1967)				
Winter		6	2	2	10
Spring		6	3	3	12
British Columbia	Wikeem and Pitt (1992)				
Summer		10	19	11	40
Fall		11	18	13	42
Winter		10	14	11	35
Spring		10	28	11	49
Annual		14	47	18	79
California	F. L. Jones (1950)	5	18	6	29

Table 3.1. *Continued*

LOCATION SEASON	SOURCE	NO. SPECIES IN DIET			
		GRASSES	FORBS	SHRUBS	TOTAL
California	McCullough and Schneegas (1966)	8	0	18	26
Idaho	Drewek (1970)				
Winter		2	0	7	9
Spring		10	19	15	44
Nevada	M. C. Hansen (1982a)	20	57	11	88

British Columbia (Wikeem 1984), only about 45% of the genera available for grazing were the same and less than 15% of the species were common to both locations. Clearly, bighorn sheep are well able to adjust their diets to diverse habitat and forage conditions.

SEASONAL USE OF FORAGE

Seasonal changes in dietary diversity and diet composition appear common in Rocky Mountain and California bighorn populations (tables 3.1, 3.2). These shifts in diet are likely related to changes in the availability and palatability of forage species, as well as to the nutritional requirements of the animal. An increase in the number of forbs in both California (Drewek 1970, Wikeem and Pitt 1992) and Rocky Mountain (Pallister 1974) bighorn diets has been found between winter and spring (table 3.1). The greatest number of forbs in the diet are consumed in spring and summer, when these plants are most readily available, and the fewest forbs are eaten in fall and winter, when they become senescent and decay.

Based on the few available data, the pattern of grass and shrub use among populations of Rocky Mountain and California bighorn sheep is less clearly defined. For example, the greatest number of grasses and shrubs has been found in the Rocky Mountain diets in summer and spring, respectively (Drewek 1970, Pallister 1974). However, in California bighorn diets, the number of grasses and shrubs remained relatively constant in all seasons over two years, although the specific species in the diet changed between years (Wikeem and Pitt 1992).

Table 3.2. Geographic variation in percent composition of three main forage groups (grasses, forbs, and shrubs) in diets of Rocky Mountain and California bighorn sheep.

	GRASSES	FORBS	SHRUBS
Rocky Mountain Bighorn[a]			
Alberta			
Mean (± SEM)	66.2 (14.5)	20.5 (11.6)	12.2 (4.4)
n (range)	6 (0–92)	6 (2–78)	6 (0–28)
Colorado			
Mean (± SEM)	75.7 (15.2)	6.7 (1.9)	17.0 (14.0)
n (range)	3 (46–96)	3 (3–9)	3 (1–45)
Idaho			
Mean (± SEM)	31.4 (9.6)	9.3 (7.0)	59.0 (10.2)
n (range)	4 (12–57)	4 (1–30)	4 (40–86)
Montana			
Mean (± SEM)	41.2 (10.1)	39.4 (11.0)	18.9 (5.6)
n (range)	6 (10–72)	6 (17–78)	6 (6–43)
California Bighorn[b]			
British Columbia			
Mean (± SEM)	49.0 (19.7)	7.7 (5.0)	43.3 (18.8)
n (range)	3 (25–88)	3 (0–17)	3 (6–67)
California			
Mean (± SEM)	37.0 (31.0)	63.0 (31.0)	0
n (range)	2 (6–68)	2 (32–94)	2
Idaho			
Mean	50.7	10.5	38.8

Sources:

[a] Alberta: Cowan (1947), J. D. Johnson (1975), Stelfox (1975); Colorado: Todd (1972a), Goodson et al. (1991); Idaho: D. R. Smith (1954), Claar (1973), W. O. Hickey (1975, 1978); Montana: Schallenberger (1966), Constan (1972), G. L. Erickson (1972), Stewart (1975).
[b] British Columbia: Sugden (1961), R. A. Demarchi (1965); California: F. L. Jones (1950); Idaho: W. O. Hickey (1978).

Table 3.3. Seasonal variation in percent composition of three main forage groups (grasses, forbs, and shrubs) in diets of Rocky Mountain (RM) and California (CA) bighorn sheep.

	GRASSES		FORBS		SHRUBS	
	RM	CA	RM	CA	RM	CA
Winter						
Mean (± SEM)	51.9 (23.3)	66.3 (3.4)	13.6 (7.6)	11.0 (3.4)	34.8 (19.3)	21.7 (3.5)
n (range)	3 (22–98)	4 (64–76)	3 (2–28)	4 (4–15)	3 (0–67)	4 (17–32)
Spring						
Mean (± SEM)	72.5 (12.7)	62.8 (13.2)	11.0 (0.6)	24.7 (11.7)	16.5 (13.4)	12.0 (3.6)
n (range)	3 (57–88)	4 (30–94)	3 (0–12)	4 (3–58)	3 (0–33)	4 (3–20)
Summer						
Mean (± SEM)	30.6 (9.5)	60.5 (2.8)	19.6 (7.2)	31.1 (2.8)	49.7 (13.1)	7.9 (1.3)
n (range)	9 (3–86)	3 (55–63)	9 (0–55)	3 (27–36)	9 (0–92)	3 (6–10)
Fall						
Mean (± SEM)	70.0 (13.2)	68.5 (6.2)	5.2 (1.8)	16.7 (3.7)	24.7 (14.0)	13.1 (2.3)
n (range)	7 (6–98)	3 (58–80)	7 (0–12)	3 (11–24)	7 (0–92)	3 (9–17)

Sources: RM: D. R. Smith (1954), Todd (1972a), B. W. Brown (1974), Pallister (1974), Bear (1978), Rominger et al. (1988); CA: D. A. Blood (1967), M. C. Hansen (1982a), Wikeem (1984).

Gross analyses of seasonal changes in diet based on dietary diversity can, however, obscure more subtle changes in diet, as reflected by shifts in the percent composition by forage class and plant species eaten. Diet composition is not only influenced by the availability of forage species, but also by the productivity of available species. For example, even though Wikeem and Pitt (1992) found that the number of grass and shrub species remained the same among seasons, the percent composition of grasses varied from 55 to 80% between summer and fall 1978. Similarly, shrub composition in the diet ranged from 10% in summer to 20% in winter and spring 1977. Moreover, individual plant species, such as bluebunch wheatgrass *(Agropyron spicatum)* ranged from 30% of the spring diet to 16% of the summer diet. Variations in seasonal use of forage classes are also reported for many populations of Rocky Mountain bighorn (D. R. Smith 1954, Todd 1972*a*, Pallister 1974, J. D. Johnson 1975, Bear 1978, Irwin et al. 1993), and California bighorn (D. A. Blood 1967, M. C. Hansen 1982*a*; table 3.3).

FORAGE SELECTIVITY

Forage selection indices (sis) are calculated to indicate when use of forage species differs from what would be expected if animals grazed randomly. An index greater than 1.0 indicates selection for a plant species, an index less than 1.0 implies selection against that species, and an index equal to 1.0 suggests random grazing, with no selection for or against the plant species (van Dyne and Heady 1965).

However, few bighorn sheep studies have monitored simultaneously both forage availability and diet. Also, when studies have, they have used different methods to calculate sis. For example, for Rocky Mountain bighorn, Oldemeyer et al. (1971) calculated sis for 19 forage species by dividing the percent diet composition by the percent cover on the range. In contrast, Stewart (1975) produced sis for 31 plant species on spring and winter habitats by dividing the percent observed instances of use of plant species by the percent canopy coverage of the species in the habitat. In another approach, Wikeem and Pitt (1992) divided diet frequency by frequency of forage in the habitat to produce sis for 79 California bighorn forage species.

Generally, both Rocky Mountain and California bighorn eat grasses, forbs, and browse in proportions similar to their availability (Oldemeyer et al. 1971, Stelfox 1975, Stewart 1975, Wikeem and Pitt 1992). In fact, Wikeem and Pitt (1992) reported that most forage species on California bighorn range in British Columbia displayed sis between 0.7 and 2.5. Marked preferences for individual species, however, often are found within all forage classes among sites and seasons (Oldemeyer et al. 1971, Stelfox 1975, Stewart 1975, Wikeem and Pitt 1992). For example, sis for single species of grass used by Rocky Mountain bighorn have been found to vary between sites from less than 1.0 to 3.7 (Oldemeyer et al. 1971), with an si as high as 8.7 being estimated for some species such as reedgrass (*Calamagrostis* spp.) on some ranges (Stelfox 1975). Other grasses that have exhibited high si values with Rocky Mountain bighorn grazing include prairie sandgrass (*Calamovilfa longifolia,* si = 4.8–8.0), Idaho fescue (*Festuca idahoensis,* si = 6.8), and needle-and-thread (*Stipa comata,* si = 6.3; Stewart 1975).

In British Columbia, California bighorn ate rough fescue *(Festuca scabrella)* year-round in proportions exceeding availability (si = 23), but other grasses such as prairie Junegrass *(Koelaria cristata),* needle-and-thread, and Kentucky bluegrass *(Poa pratensis)* were only seasonally selected (Wikeem and Pitt 1992). Bluebunch wheatgrass was generally eaten in proportions less than available on the study site, despite its importance in the annual diet (> 20%).

Measures of bighorn feeding on forbs and shrubs have generally produced the highest sis, with values as high as 6.6 for aster (*Aster* spp.; Stelfox 1975) and 22.5–27.0 for lupine (*Lupinus* spp.; Oldemeyer et al. 1971) being reported in diet studies of Rocky Mountain bighorn. When data were averaged over all seasons for two years, sis calculated for forbs as a forage group indicated random grazing by California bighorn (Wikeem and Pitt 1992). Selection indices for all annual forbs generally fell below 0.7, probably because they produce very little forage compared to other plant species and have a short phenological period of availability. High sis were calculated for perennial forbs such as silky lupine (*Lupinus sericeus;* si = 9.2) and Thompson's paintbrush (*Castilleja thompsoniip;* si = 27) during summer, but these high values were attributed to a combination of factors including phenological patterns, plant morphology, environmental site characteristics, and grazing preference.

Less information is available for shrubs than grasses and forbs. Shrub sis exceeded 3.0 on three of eight sites (maximum si = 40.6) on a winter habitat dominated by bluebunch wheatgrass (Stewart 1975). Other Rocky Mountain bighorn browse species with sis over 10.0 were common chokecherry *(Prunus virginiana)* and prairie sage. In British Columbia, pasture sage averaged only 0.4% of California bighorn diet, with the highest percent composition in autumn (Wikeem and Pitt 1992). Selection indices (mean = 4.0) remained high, however, for pasture sage in all seasons but spring. Earlier studies from British Columbia also recorded pasture sage in California bighorn diets, especially in winter (D. A. Blood 1961, R. A. Demarchi 1965, Morrison 1972). Other small shrubs that also have high si values include Wyeth buckwheat *(Eriogonum heracleoides)* in winter (si = 6.1) and snow buckwheat *(E. niveum)* in spring (si = 4.0; Wikeem and Pitt 1992).

In Montana, Rocky Mountain bighorn diets contained 43% big sagebrush, with an SI value of 46.4 indicating preference (Stewart 1975). In contrast, other studies recorded very sparse use of this species on other bighorn ranges (M. C. Hansen 1982a, Oldemeyer et al. 1971, Wikeem and Pitt 1992). Indeed, in British Columbia, big sagebrush diet composition averaged only 1.0% and SIs ranged from 0.0 to 0.2, even though it was the most common shrub on the site (Wikeem and Pitt 1992). Such variable use of big sagebrush among bighorn populations may be associated with differential proportions of essential oils and palatability in sagebrush ecotypes (Plummer 1972).

Selection indices should be interpreted cautiously and values different from 1.0 should not be considered as synonymous with animal avoidance (<1.0) or preference (>1.0). Moreover, such indices focus on plant species as a whole and may obscure more subtle levels of selection within a plant. For example, California bighorn sheep selected for new growth over old growth in grasses, for flower heads of forbs and shrubs, and for leaves over stems on browse (Wikeem and Pitt 1987, 1992). Rocky Mountain bighorn in Colorado similarly selected leaves of true mountain mahogany (*Cercocarpus montanus;* Rominger et al. 1988). Plant selection by bighorn appears to be related to forage availability and the type of habitat used (Todd 1972b), although environmental factors such as weather, snow cover, topography, soil fertility, slope, aspect, and management practices may also influence selection. Thus, forage selection is determined by a variety of factors, only one of which is animal preference (Wikeem and Pitt 1979, 1992; Hanley 1982), so that some combination of both innate and learned factors are involved when sheep select what they eat (Wikeem and Pitt 1979, 1992).

FORAGE SELECTION IN RELATION TO FORAGE QUALITY

Forage quality of bighorn has been measured throughout both subspecies' distributions (McCullough and Schneegas 1966, R. A. Demarchi 1968, D. A. Demarchi 1970, Hebert 1973, J. D. Johnson 1975, Shannon et al. 1975, Stelfox 1975, Stewart 1975, Wikeem 1984, Irwin et al. 1993). Although sheep are reported to prefer succulent green material over old-growth forage (Capp 1967, Todd 1972b, Stelfox 1975, Goodson et al. 1991, Wikeem and Pitt 1992, Goodson and Stevens 1996), few quantitative data exist to support that they select individual plants based on nutritive quality alone. In fact, Shannon et al. (1975) and J. D. Johnson (1975) concluded that bighorn showed distinct preferences for various plant species, even when there was little apparent difference among them in either chemical composition or stage of growth. Similarly, in Montana, bluebunch wheatgrass, the most important winter forage for some Rocky Mountain populations, can have the lowest nitrogen content of any of the forage species sampled (Stewart 1975). However, during winter, sheep selected bluegrasses and fringed sagewort *(Artemisia frigida)*, which remained relatively green all winter, and also the terminal buds of chokecherry, suggesting that bighorn were selecting plant species on the basis of nutritional quality (Stewart 1975).

Species composition in California bighorn diets correlated poorly with forage quality (Wikeem and Pitt 1992). Crude protein (CP), often presumed to be a factor in grazing preference, was not included in any of 10 linear regression equations when the percent diet frequency of bluebunch wheatgrass, cheatgrass *(Bromus tectorum)*, prairie Junegrass, needle-and-thread, arrowleaf balsamroot *(Balsamorhiza sagittata)*, silky lupine, pasture sage, big sagebrush, snow buckwheat, and Wyeth buckwheat were regressed on forage CP levels for each plant species. Acid detergent fiber (ADF) negatively influenced diet composition ($P < 0.05$) only for arrowleaf balsamroot, and the calcium:phosphorus ratio negatively affected diet composition only for big sagebrush. Plant cover, however, was consistently correlated ($P < 0.05$) with bighorn sheep diet composition, explaining up to 62% of all variation in the diet (Wikeem and Pitt 1992).

Regression analyses suggest that bighorn sheep graze opportunistically, rather than seeking specific plant species or forage nutrients. This opportunistic foraging, based primarily on plant cover and availability, likely provides a nutritionally balanced diet. For herbaceous plant species, foliar cover normally peaks after flushes of plant growth and also yields large proportions of CP and phosphorus. As forage matures, nutritional quality declines and antiquality components such as ADF increase.

Bighorn sheep move constantly while grazing and therefore will likely encounter both widely distributed and common plant species. This grazing behavior can produce SIs either more or less than 1.0, depending on spatial distributions of the plants. Bighorn sheep, like most ruminants, evolved as grazing generalists, capable of digesting a wide variety of plant species. Grasses, forbs, and browse all contribute nutritionally to bighorn sheep diet during specific portions of the year. Such a generalized adaptation optimizes nutritive value of the entire diet, rather than maximizing acquisition of specific nutrients concentrated in a few plant species (Wikeem and Pitt 1992).

Growth

In this section, we have considered only body weight and horn dimensions, because linear measurements of body growth are much less available. Body weight data were compiled from many studies, but not every weight presented in the literature was used. We have not tried to develop growth equations because, except for two published studies (Jorgenson and Wishart 1984, Festa-Bianchet et al. 1996), data are lacking on the growth of individuals.

BODY WEIGHT

Most available lamb weights are for Rocky Mountain bighorn (fig. 3.2). The few data points for California lambs (Sugden 1961, Eccles and Shackleton 1979) suggest they fall well within the range for Rocky Mountain lambs, which weigh between 2.8 and 5.5 kg at birth. Young bighorn grow rapidly until about six months of age (fig. 3.2). Individual growth rates seem to follow a logistic growth curve during the summer period (Jorgenson and Wishart 1984), except for lambs and yearlings, whose growth appears to be linear (Festa-Bianchet et al. 1996). McEwan (1975) described this rapid growth phase in captive bighorn with the relationship: $W = 3.8e^{0.01804t}$, where W is weight (kg) at time t (days).

Following the rapid summer phase, growth during the subsequent late fall and winter appears to slow and stop. After this period, although lambs seem less prone to this, weight is usually lost over winter (Festa-Bianchet et al. 1996) until growth begins again the following spring, when the new vegetation growth provides a more favorable energetic and nutritional regime.

Male and female lambs differ slightly in weight, but males become increasingly heavier than females once they reach one year of age (Festa-Bianchet et al. 1996; fig. 3.2). However, differences in growth rates of lambs are found between years within populations (Jorgenson and Wishart 1984), and heavy parasite loads can also depress their body weight (Stelfox and McGillis 1970, Woodard et al. 1972, fig. 3.2). Weight differences between the sexes has been reported to range from 18% in two year olds to 65% in six year olds (Festa-Bianchet et al. 1996), with males being heavier.

The pattern of body growth in older bighorn in the data presented here (figs. 3.3, 3.4) must be treated with caution because older age classes had to be combined beyond a certain age. The heaviest weight recorded for a Rocky Mountain male was 137 kg from Jasper National Park, Alberta (Stelfox and McGillis 1970). Females reportedly reach a maximum of around 90 kg (D. A. Blood et al. 1970). Although Rocky Mountain bighorn are supposedly heavier (Cowan 1940), the few available weights for California bighorn indicate little difference between the subspecies (figs. 3.3, 3.4). The data also suggest that adult females cease growing at three to four years of age (fig. 3.4). Males continue to grow larger than females, at least up to six years of age (fig. 3.3), although the age at which growth ceases varies between populations and with latitude (Jorgenson and Wishart 1984).

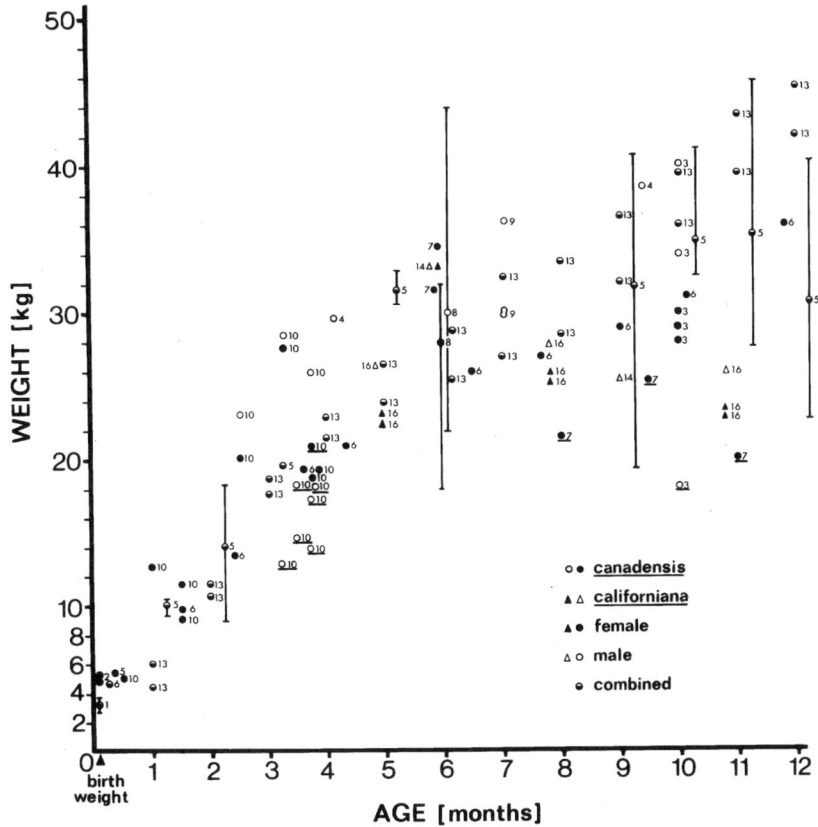

Figure 3.2. Weights (means and ranges) of Rocky Mountain and California bighorn lambs over the first year of life. All ages are estimates from an assumed birthdate in mid-May. Underlined points represent individuals with high parasite loads, except the 10-month-old male (3), which had a broken leg. Sources for figures 3.2–3.4: (1) Geist 1971; (2) Blunt et al. 1972; (3) D. M. Shackleton, unpubl. data; (4) Forrester and Hoffmann 1963; (5) D. A. Blood et al. 1970; (6) J. K. Morgan 1970; (7) Stelfox and McGillis 1970; (8) Schallenberger 1972; (9) Nash et al. 1972; (10) Woodard et al. 1972; (11) Berwick 1968; (12) K. G. Smith and Wishart 1978; (13) McEwan 1975; (14) Sugden 1961; (15) Good 1974; (16) Eccles and Shackleton 1979; (17) Eccles 1983.

Seasonal cycles in body weights of wild (Festa-Bianchet et al. 1996) and captive (McEwan 1975) bighorn sheep are not as evident as in captive deer (A. J. Wood et al. 1962). Captive Rocky Mountain sheep grew between April and mid-October, with weight decrease thereafter until the following April (McEwan 1975). Much of the weight change was attributed to deposition of fat reserves in summer and their subsequent depletion over winter (McEwan

1975). Unlike deer, bighorn do not fast during the rut but do reduce energy intake (McEwan 1975). Captive bighorn show reductions in body weight of between 7.1 and 8.3% for males 18 months of age and between 13.4 and 16.4% for older animals (McEwan 1975). In wild populations, weight losses of up to 23% body weight in males and 20% in adult females are reported. These losses are related to various factors including increased maintenance costs (e.g., homeothermy, locomotion), reproduction (e.g., rutting costs, gestation), lower food intake, low-quality food, and the costs of fetal growth and placental and amniotic fluids (Stelfox and McGillis 1970, Jorgenson and Wishart 1984, Festa-Bianchet et al. 1996). Also, a combination of low forage production and high endoparasite loads can depress body weight and increase overwinter

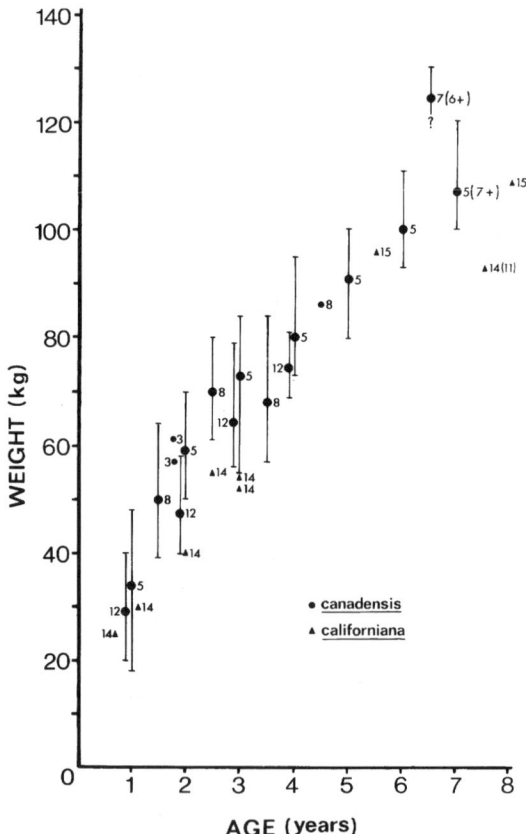

Figure 3.3. Weights (means and ranges) of male bighorn sheep. Numbers in parentheses refer to the maximum age class of each study. Refer to figure 3.2 for source numbers.

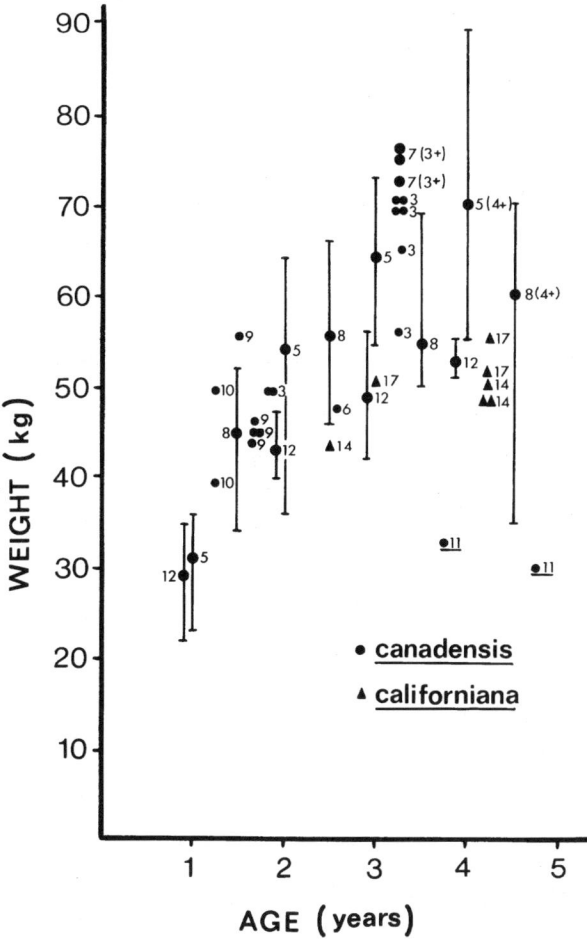

Figure 3.4. Weights (means and ranges) of female bighorn sheep. Numbers in parentheses refer to the maximum age class of each study. Underlined numbers refer to animals found dead. Refer to figure 3.2 for source numbers.

weight loss in adult females, as shown in the Canadian Rockies (Stelfox and McGillis 1970). The two adult female weights reported by Berwick (1968) indicate that they were in poor condition (fig. 3.4).

HORN GROWTH AND DEVELOPMENT

Individual horn growth is included here because it provides valuable information about individuals and about the populations sheep belong to.

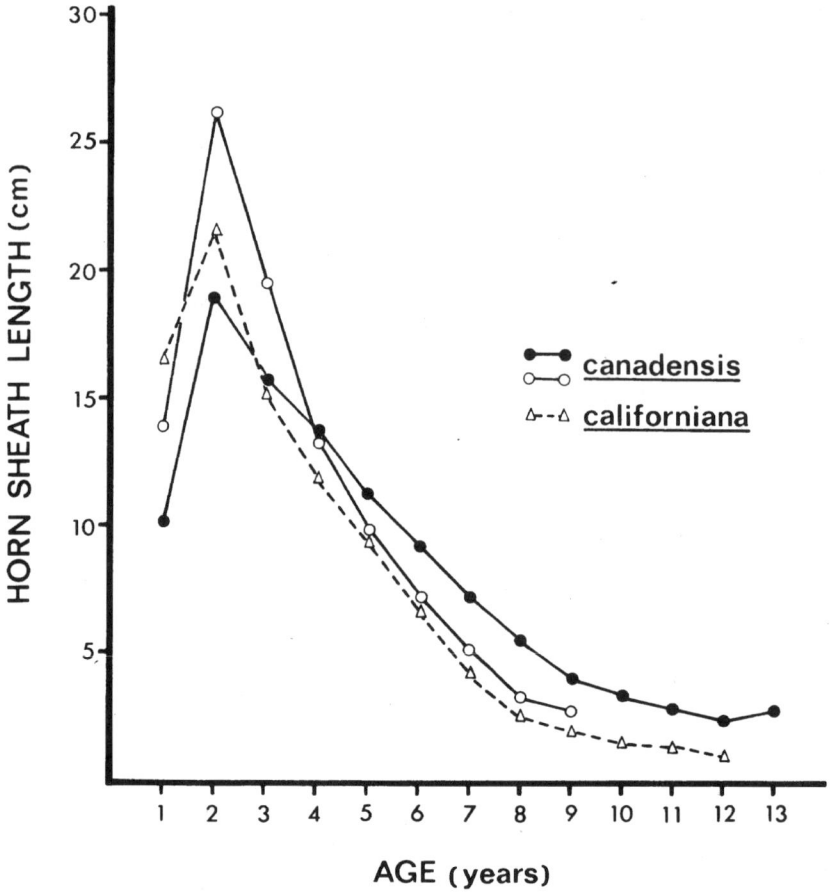

Figure 3.5. The typical pattern of annual horn sheath growth in bighorn males. Two populations of Rocky Mountain bighorn with different growth rates are compared to data from a California population (after D. M. Shackleton, unpubl. data).

Horns also have the advantage of being more readily available, in the form of hunter kills and heads, than are whole animals from which conventional morphometric measurements are taken.

How much horns and their growth rates are controlled by genetic (Castle 1940) and by environmental factors (Geist 1971; Shackleton 1973, 1976; Heimer and Smith 1974; Bunnell 1978; J. P. Elliott 1978; K. G. Smith and Wishart 1978) is unknown, but clearly both are important. We can only speculate on the relative significance of these two factors based on studies of antler growth in deer (French et al. 1955, Long et al. 1959) and of general body growth in wild and domestic ungulates (Brody 1945, A. J. Wood et al. 1962, McEwan 1975).

Like antlers, horns grow annually. Body growth is also seasonally cyclic, and these changes appear to be a response to seasonal availability and intake of food (Gray and Setna 1931, cited in Needham 1964; Bandy et al. 1970). However, we do not know exactly when or what factors control the annual onset and cessation of horn growth. Horns are composed of keratin deposited on bone horn-cores that are attached to the frontal bones of the skull. But horns, unlike antlers, are not shed each year. Instead, a new horn sheath grows each year over the horn-core and inside the preceding horn sheath. The result is like a series of "ice-cream cones" stacked one inside the other. The bone horn-core is essentially a template upon which the new horn sheath grows. As the animal grows so does its horn-core, and thus its horn sheath increases in size (R. A. Taylor 1962). What we see externally is only the exposed portion of the annual horn sheath. Only the first, or "lamb," sheath is completely visible, and even that portion may be lost by brooming (Shackleton and Hutton 1971). All subsequent sheaths are partially hidden beneath the previous year's growth, so measures of annual horn growth are only relative indicators of growth.

The general pattern of development (fig. 3.5) of the visible, annual horn sheaths is relatively consistent among mountain sheep populations and is independent of sex. However, individual horn-growth curves can depart from the general pattern, and such deviations may prove valuable indicators of environmental conditions, as has been found in other Caprinae. In Alpine ibex *(Capra ibex)*, a comparison of the pattern of an individual's annual horn sheath lengths with precipitation records showed years with above average precipitation were correlated with below average or atypical annual sheath growth (Nievergelt 1966). As a result, year of death could be relatively accurately determined. The pattern of annual horn growth in female Japanese serow *(Capricornis crispus)* provides a record of an individual's reproductive history (Miura et al. 1987). No comparable work has been done on bighorn, but it is certainly worth pursuing.

POPULATION QUALITY AND HORN GROWTH

Geist (1971) proposed a hypothesis to explain observed variation among mountain sheep populations. The hypothesis predicts that, along with greater annual horn growth (figs. 3.5, 3.6), high-quality or expanding populations are characterized by an integrated syndrome involving rapid population growth, fast individual growth rates, early maturation, short life spans, intense social interaction, and high milk production as compared to low-quality

Figure 3.6. Variation in annual horn growth between populations can be readily apparent in the field. The male on the left was approximately 33 months old (photographed in February), whereas the male on the right, from a different population, was estimated to be 18 months old (photographed in November). (Drawing by David Shackleton.)

populations that are stable or declining. This syndrome has been documented for some bighorn (Geist 1971; Shackleton 1973, 1976; K. G. Smith and Wishart 1978; Wishart and Brochu 1982; Gilchrist 1992) and thinhorn populations (Heimer and Smith 1974, Bunnell 1978, J. P. Elliott 1978), and for other wild Caprinae (Nievergelt 1966, Schaller 1977). Variation between years within a population has also been studied in caprins (Horejsi 1972, 1976; Grubb 1974; Bunnell 1978; K. G. Smith and Wishart 1978). The relationships predicted by Geist (1971) appear to hold, although the causal mechanism(s) still needs to be tested.

We can be relatively confident that the pattern of annual horn sheath growth and body growth can be used to make predictions about populations. Although Boone and Crockett measurements of mountain sheep are more of interest to hunters than to biologists and managers, there is a trend toward measuring annual horn sheath growth, particularly where compulsory hunter reporting is required by provincial and state game departments. However, care must be taken to correctly age animals and to ensure measures have sufficient background information about other population parameters, particularly measures of general productivity.

Mountain sheep can be aged by counting the annual rings that separate annual horn sheaths (Geist 1966a). The horn rings, or annuli, mark the end of one year's growth period from the start of the next. Aging by annuli counts assumes that these rings can be readily and accurately distinguished. Beyond approximately five years of age, it is generally impossible to determine accurately a female bighorn's age from her horns; however, in males, annual horn sheaths are absolutely larger and the annuli easier to separate from the many undulations in between. Even in males, problems can arise. Brooming (Shackleton and Hutton 1971) can remove up to three years' horn growth and thus hamper accurate aging. Also, the first year's growth, even if not broomed off in large males, can be difficult to determine because the first ring is usually indistinct, probably as a result of the way horns grow during the first year (D. M. Shackleton, unpubl. data). Even when the first ring can be distinguished, some observers find it difficult to accept that horns grown in the first summer and fall can be very short ($<$ 55 mm) in some individuals.

A second problem when aging by horn annuli is the "false" rings seen in some animals. They can be difficult to separate from true annuli. If the animal is relatively young, a check on the tooth succession can clarify the problem. In adult animals, another clue is if there are pairs of rings close together, separated by a longer distance from the next ring(s). The significance of these false rings and how they develop is not known, but the factor(s) causing them are probably important in the animal's life history. False rings are found more commonly in some populations and correspond to the same calendar year in animals of different ages (D. M. Shackleton, unpubl. data).

SEXUAL MATURITY

Based solely on physiological criteria (i.e., ovulation, spermatogenesis), bighorn reach puberty as early as 18 months of age (Woodgerd 1964, Geist 1971, Blunt et al. 1972), but full sexual maturity is not reached until later. Except in expanding populations (Woodgerd 1964, Shackleton 1973) or in isolated cases (Van Dyke 1978) where females have been observed to mate at 1.5 years and successfully produce lambs as 2 year olds, most females do not mate first until at least 2.5 years of age. By contrast, males in all populations do not breed until they are much older.

The sexes in bighorn have clearly different life history strategies. Females stop growing and begin breeding after they reach puberty. Geist (1968, 1971) proposed that female bighorn are pedogenic and are "frozen" physically and behaviorally upon reaching puberty, remaining in the "young male" stage for the rest of their lives (fig. 3.7). Female bighorn are iteroparous, usually producing a single young each year until old age or death. In contrast, males continue to grow well after puberty. They have essentially a semelparous reproductive strategy, typically beginning to mate only when seven or eight years of age (Geist 1971), though sometimes earlier in some populations (Shackleton 1973, 1991). Males successfully participate in the rut for only a

Figure 3.7. Adult Rocky Mountain female (left) and approximately 16-month-old yearling male (right) showing similar horn and body sizes. (Photograph by David Shackleton.)

few years, during which time they have the potential to sire many offspring. These differences between the sexes have potentially significant implications for the management of wild sheep, especially when trophy hunting is involved (see also F. J. Singer and Nichols 1992).

Activity Patterns and Activity Budgets

Diurnal activity patterns of Rocky Mountain and California bighorn are discussed by many authors, but data are limited (Mills 1937, David 1938, both cited in D. A. Blood 1963; Riggs 1977; D. R. Smith 1954; Woolf 1968, Kornet 1968, both cited in Van Dyke 1978). We found just two detailed studies of diurnal activity patterns over the year (Eccles 1978, 1983; Van Dyke 1978). Both were of California bighorn, with Eccles concentrating on captive adult females living in a 40-ha enclosure.

The general pattern of activity is one of alternating peaks of feeding and resting; other activities (e.g., social interactions, running) contribute little to the total daily activity budgets. Feeding is generally most frequent around dawn and dusk, although other major activity peaks can occur between these (Mills 1937, W. B. Davis 1938, both cited in D. A. Blood 1963; D. R. Smith 1954; D. A. Blood 1963; Eccles 1983).

In contrast to summer, in winter the proportion of daylight hours spent feeding increases while resting decreases (D. R. Smith 1954, Van Dyke 1978, Eccles 1983). In other words, resting during daylight hours is sacrificed to maintain feeding activity during periods of short day length in winter (Eccles 1983). This relative increase in the intensity of feeding activity during winter may be a response to at least one of several factors: (1) day length in winter approaches or is actually less than the average total daily feeding time; (2) forage plants in winter are of poorer nutritional quality and also are less available, especially with snow cover; and (3) thermal stresses in winter increase energy requirements (Van Dyke 1978, Eccles 1983). However, extremely cold ambient temperatures may result in a sudden reversal of this trend, and the proportion of time spent bedding is increased. Eccles (1983) observed that on very cold days, members of the maternal group started grazing later in the morning, rested in midday, and stopped feeding in the late afternoon; whereas on days with average winter temperatures, the group fed almost continuously throughout the daylight period. Such seasonal and daily variation in diurnal activity patterns is in marked contrast to the relative constancy reported for desert bighorn (Alderman et al. 1989).

In a captive herd, Eccles (1978, 1983) showed the average number of daily activity (feeding) peaks between dawn and dusk varied monthly. Between April and August monthly means ranged from 4.3 to 5.3 activity peaks/day, and between September and March means ranged from 1.3 to 2.3 activity peaks/day. Eccles proposed that these changes were related to forage conditions, energy requirements, physiological condition, and social organization of the animals. Contrary to these findings, D. A. Blood (1963) predicted that summer activity patterns would show a lengthening of activity peaks, rather than an increase in their number.

Van Dyke (1978) also found seasonal differences in daily activity budgets among age-sex classes. In summer, females, presumably due to lactation demands, spent proportionally more time feeding than did males, whereas lambs fed less and spent more time in other activities during spring and summer. This behavior of lambs probably reflects their dependence on milk early in life and their tendency to play. In the captive population that Eccles (1983) studied, California females mainly rested and fed very little on the day they gave birth, but returned to normal feeding levels (25 to 30% of daylight) by the time lambs were three to five days old (Shackleton and Haywood

1985). On their first day of life the newborn lambs played surprisingly often (5.6% total observation time), and perhaps their activity helped to develop their motor skills, particularly neuromuscular coordination. Activities, especially bedding, were generally very synchronized between a lamb and its mother in the first 14 days of life, but when she grazed the lamb either rested or was involved in some other activity independent of its mother (Shackleton and Haywood 1985).

Eccles and Shackleton (1986) found no significant differences in the daily activity patterns of females of different social status throughout the year. They did, however, find that adult females in extremely poor physical and physiological condition spent more time than did healthy females feeding during fall and more time resting during winter. Most of the females that were in poor condition died during late winter.

Available information on activity patterns and the ecological factors influencing them are still very limited for Rocky Mountain and California bighorn. There is great scope for more study, not only of the activity but of the environmental and physiological factors shown in studies of domestic sheep (Arnold and Dudzinki 1978) to influence an animal's behavior. Monitoring heart rates during various activities, as used in studies of bighorn harassment (R. A. MacArthur et al. 1979, 1982a,b; Stemp 1983; Geist et al. 1985), should also be considered in future studies of activity budgets.

Bioenergetics

Bighorn sheep occupy an environment of marked seasonal changes. Summers are usually mild with a surplus of high-quality, alpine vegetation, whereas winters are often harsh and characterized by high winds, low temperatures, snow, and forages low in quality, quantity, and availability. Consequently, adult bighorn sheep teeter on a precarious energy balance, gaining weight as stored fat each summer and losing it again through the winter as fat reserves and other body tissues are catabolized to provide energy to maintain life functions.

Bighorn sheep are well adapted to winter conditions. The lower critical temperature (the temperature at which metabolic rate begins to rise to compensate for heat loss) was found to be -20°C for well-fed bighorn sheep in winter pelage (Chappel and Hudson 1978). This temperature is low compared to the lower critical temperatures of many other ungulate species. However, decreasing the temperature from -20°C to -30°C caused a 37 to 39% increase in metabolic rate. Although winds up to 8 m/sec have little effect on a bighorn's metabolic rate when ambient temperatures are above the critical temperature, when temperatures drop below -20°c even a slight wind can substantially increase metabolic costs (Chappel and Hudson 1978).

Bighorn sheep exhibit a tendency to voluntarily restrict their level of activity during the winter (McEwan 1975). This functions to channel meager energy supplies toward more critical needs. On exceptionally cold mornings in Banff National Park, we (D. Shackleton and C. Shank, pers. obs.) often observed sheep to remain bedded until 1100, and Eccles (1983) observed similar responses in captive California bighorn. Petocz (1973) also found the level of social activity in males to be lower during winters with deep snow. Movements become limited and are largely restricted to the proximity of windswept areas where forage can be obtained with less energy expenditure.

We found only two studies detailing weight loss of wild bighorn sheep over winter (Stelfox and McGillis 1970, Jorgenson and Wishart 1984). These studies reported losses of 23 and 20% of body weight for males and females, respectively. Although some of the overwinter weight loss is due to rutting expenditures, increased energy demands (homeothermy, gestation), and decreased energy availability in the food, bighorn tend to voluntarily reduce their dry matter intake during the winter. Even in captivity when fed high-quality fodder *ad libitum,* dry matter intake may be reduced by 50% (Chappel and Hudson 1978). The reduction usually observed in early winter nutrient intake is most marked in male bighorn and coincides with the timing of the rut and peak testosterone levels (McEwan 1975). Some cervids show sharp increases in testosterone levels during the rut followed by a drop to undetectable levels. By contrast, bighorn males maintain measurable testosterone levels throughout the year and exhibit peak values of only about one-half those of reindeer and caribou (*Rangifer tarandus;* McEwan 1975). In captive bighorn males, winter weight losses may reach 16%, whereas male deer, with their more dramatic hormonal shifts and greater intake restrictions, may lose as much as 36% of their weight (McEwan 1975).

Chappel (1978) derived a multivariate analysis to determine which factors most affected winter metabolic rates in captive bighorn. Up to 87% of the variation in metabolic rate was explained by a combination of sex, date, nutritional status, trial temperatures, body weight, adaptation temperatures, and gross energy intakes. However, the interaction between these variables was complex, making it difficult to state which was the most significant.

Social Organization

Rocky Mountain and California bighorn typically inhabit grasslands (Cowan 1940). Like other ungulates occupying open habitat (Estes 1974, Geist

1974), they live in groups. Group living appears to provide two main advantages to individuals in many species, improving feeding efficiency and predation avoidance (R. D. Alexander 1974, Pulliam and Caraco 1984). For bighorn and other ungulates, predation is probably the main selective force for living in groups (Jarman 1974, Jarman and Jarman 1979). The risk of predation for individuals in groups may be decreased in several ways: cover seeking within the group (W. D. Hamilton 1971, Eshel 1978), more effective predator detection (Galton 1883, cited in Triesmann 1975), reduced detectability of a compact herd (Vine 1971, 1973, cited in Triesmann 1975), and cooperative defense (E. O. Wilson 1975, Shank 1977), with probably the most important being a reduced probability of attack for each individual (W. D. Hamilton 1971, Lima 1987, Dehn 1990). Although there is little doubt that antipredator strategies are the major factor responsible for group formation in bighorn (Berger 1978a, Shank 1979, Risenhoover and Bailey 1985), foraging efficiency can increase with group size (Risenhoover and Bailey 1985, Berger 1991). Such a relationship results from individuals in groups being able to spend more time foraging because each member has less need to be vigilant. Group living probably also allows bighorn to more efficiently exploit a renewable resource such as grass (Krebs and Davies 1987).

SEXUAL SEGREGATION

The degree and duration of sexual segregation in wild and feral sheep is generally related to the seasonality of breeding (Shackleton and Shank 1984). Mountain sheep inhabit regions with pronounced seasons. Males and females in most populations occupy separate seasonal ranges throughout most of the year, although there is some spatial and temporal overlap (Geist 1971, Geist and Petocz 1977, Morgantini and Hudson 1981, Ashcroft 1986a). Bighorn live in all-male groups (fig. 3.8) and in maternal or female groups comprised of females, lambs, and juvenile males.

All-male groups tend to wander further during their yearly migrations than do maternal groups. Even when they share or overlap a home range with a maternal group, they tend to remain separate. Why sexual segregation occurs in bighorn sheep and other ungulates is still unclear. Geist and Petocz (1977) suggested that segregation of the sexes serves to minimize competition and disturbance of females and lambs by males. However, neither Shank's (1979, 1982) nor Aschroft's (1986a) observations support the idea that sexual segregation minimizes competition or is due to different nutritional requirements. Although Geist and Petocz (1977) and Shank (1979, 1982) found that on common winter range males used open slopes more than did female groups, which used cliff terrain more often, this difference in habitat use most

Figure 3.8. Group of mature Rocky Mountain bighorn males. (Photograph by David Shackleton.)

likely reflects sex-related antipredator strategies rather than intrasexual competition (Shank 1979, Ashcroft 1986a, Berger 1991). Reviewing hypotheses for sexual segregation in ungulates, Main and Coblentz (1990) concluded that the differences in habitat use were due to different reproductive strategies, with females preferring secure areas for raising young and males selecting areas for maximal body condition. This explanation was accepted by Bleich et al. (1997) in their study of bighorn sheep. Males inhabited areas of higher predation risk but higher forage quality, whereas females preferred areas with more security cover even if this choice resulted in poorer foraging conditions.

We were unable to find any studies that provided data on dispersal in bighorn. However, some information is available on how individuals join groups. Young males leave their maternal group to join or form male groups when they are socially and physically dominant to adult females (Geist 1971). The actual impetus for this change in group preference is not known, but the age at which it occurs varies (Shackleton 1973, Festa-Bianchet 1991a) and appears related to individual growth and development rates of young males (Shackleton 1973). Males generally become socially independent of their mothers as yearlings, though they remain in females groups one to two years more (Festa-Bianchet 1991a). However, most females tend to remain in their mother's group, and although this could lead to kin selection, Festa-Bianchet (1991a) found no evidence that related females helped each other. In another population, where many females were observed nursing others' young as well as their own, relatedness of individuals was unknown (Hass 1984, 1990).

GROUP SIZE AND SEASONAL VARIATION

Bighorn spend most of their life in groups (D. R. Smith 1954, Sugden 1961, D. A. Blood 1963, McCullough and Schneegas 1966, Berwick 1968, Geist 1971, Shackleton 1973, Baumen and Stevens 1978, Ashcroft 1986a). These can vary in size from 2 to over 40 individuals (fig. 3.9), with the maximum reported being 110 individuals for California bighorn (Sudgen 1961). As we could find no apparent differences in group sizes between California and Rocky Mountain bighorn, environmental factors seem more important than subspecies.

Although solitary individuals are not uncommon (fig. 3.9), they represent only a very small proportion of all animals observed. Being alone is probably only a temporary condition, and the majority of bighorn observed alone are adult males (D. R. Smith 1954, D. A. Blood 1963, Geist 1971). Geist (1971) showed that the probability of a male being observed alone was directly proportional to its age or horn class. The frequency with which yearling sheep, particularly males, are seen alone varies between populations, whereas the proportion of lone adult females appears relatively constant (D. R. Smith 1954, Wishart 1958, D. A. Blood 1963, Geist 1971, Shackleton 1973).

There are several problems associated with studying and interpreting group size in bighorn, though they are not unique to these animals. The first lies with the criteria used to define groups. Authors either differ in their criteria or more often, simply fail to state them. For this review we assumed that the criteria to distinguish a group were comparable. A second problem arises if the sampling effort varies between studies, leading to underestimation of small, less-observable groups (Sugden 1961). The final limitation is that in most cases the only available statistical descriptor reported is mean (average) group size, and this is not the best measure because it usually underestimates the size of groups in which animals live (Jarman 1974, 1982; Ashcroft 1986a). Jarman's (1974, 1982) "typical" group size ($\bar{g}$) is a better measure of the social units bighorn live in because it is more representative of what the animals experience (Ashcroft 1986a). Even the frequency distributions of observed group sizes can be misleading if not interpreted carefully.

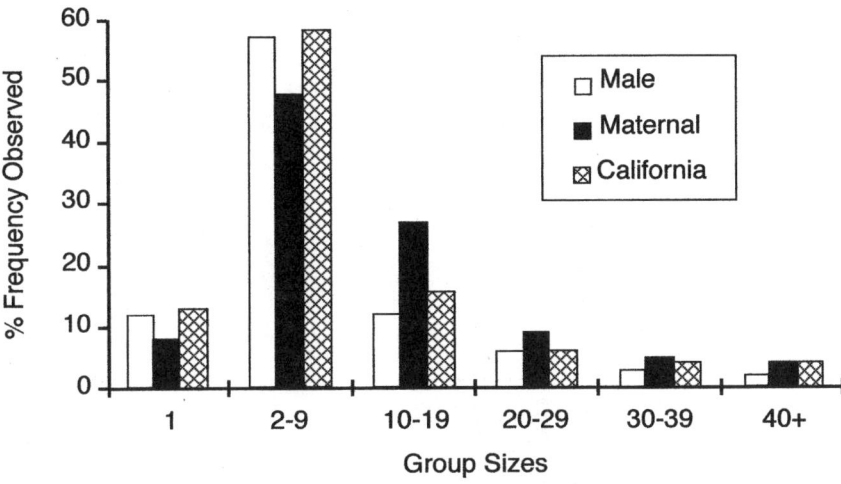

Figure 3.9. Percent frequency of sizes recorded for male and maternal groups of Rocky Mountain bighorn (after Shackleton 1973) and all groups (i.e., group type ignored) of California bighorn (after Ashcroft 1986a).

Ashcroft (1986a) compared the cumulative frequencies of groups and individuals and showed that although smaller groups are observed most frequently, most individuals live in much larger groups. Thus far, measures of typical group size are available only for two populations of Rocky Mountain (fig. 3.10) and one population of California (Ashcroft 1986a) bighorn.

Group size is affected by population size, by distribution and extent of available home ranges, season, and sex (fig. 3.10). Seasonal differences can generally be related to "social events" in the annual cycle. Maternal groups are small during the birth period, because most pregnant females leave their group and move to precipitous areas to have their young (Geist 1971, Festa-Bianchet 1988a). When they regroup after giving birth, group size increases and remains relatively large throughout the summer until the rut. Females with young lambs often group together for a period after lambing into what are referred to as "nursery groups." Male groups are largest in April and May (fig. 3.10) and smallest during the rut, when males wander between rutting groups (Geist 1971). The spring increase in male group size observed in some California (Ashcroft 1986a) and Rocky Mountain (fig. 3.10) populations could reflect the period when different males come together to interact (Geist 1971, Shackleton 1973). The size decrease in male and maternal groups during the rut is simply a consequence of a third group type being recognized (i.e.,

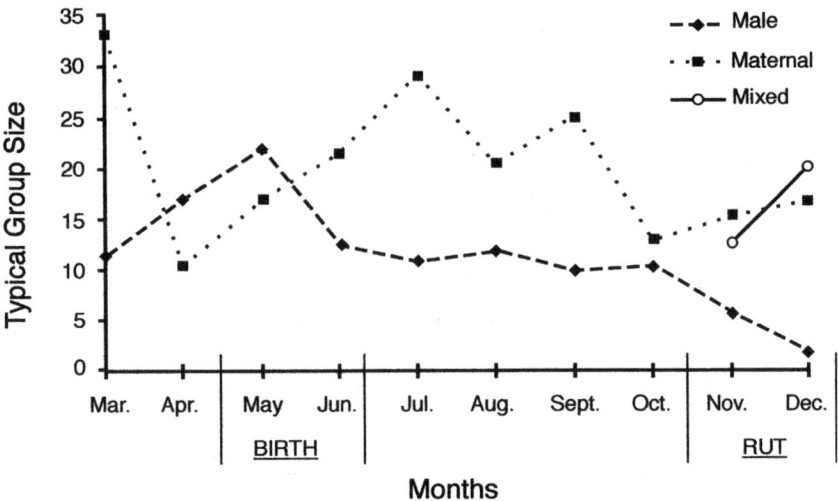

Figure 3.10. Typical group sizes of male, maternal, and mixed (rut) groups of Rocky Mountain bighorn. Data combined from Cascade Valley, Banff National Park, and Radium Hot Springs, Kootenay National Park, British Columbia, Canada (after Shackleton 1973).

mixed or rutting groups) when males and females come together for mating (fig. 3.10). If winter ranges are limited in extent, group size can increase because animals are forced to share a common area.

SOCIAL STATUS

Male social status is positively correlated with age (Hass and Jenni 1991) and/or horn size in bighorn (Geist 1966b, 1971). Status is also correlated with access to estrous females (Geist 1971; Shackleton 1973; Hogg 1984, 1987), thus supporting a positive relationship between social status and reproductive fitness of males. The picture is somewhat different in females. Hierarchies appear to be nonlinear (Eccles and Shackleton 1986, Hass 1991) and related to age (Hass 1991). Other attributes, such as horn size or body weight, have not been found to be consistently correlated with a female's social status (Eccles and Shackleton 1986). Nor is there evidence for a female's status being related to her reproductive fitness (Eccles and Shackleton 1986, Festa-Bianchet 1991a, Hass 1991), diet quality, activity costs (Eccles and Shackleton 1986), or to differential investment in male or female offspring (Festa-Bianchet 1988b). Only nursing rate has been shown to be directly related to social status (Hass

1991). It is worth noting, however, that the frequency of overt social interactions among females is very low (Eccles and Shackleton 1986), so the significance of females' less-obvious social interactions may not have been appreciated. This topic certainly deserves further study.

Mating

MATING SEASON

Mountain sheep are seasonal breeders, mating only between late fall and early winter. The timing, degree of synchrony, and duration of the mating season is dependent on the timing of the lambing season. The mating season usually begins in early November (McCann 1956, Buechner 1960, Geist 1971), but can start as early as late October in some Rocky Mountain (Buechner 1960) and California (D. A. Blood 1963, D. A. Demarchi and Mitchell 1973) herds. The mating season often extends into late December and sometimes early January (Wishart 1958), but the peak is usually between mid-November and mid-December (Honess and Frost 1942, D. R. Smith 1954, Wishart 1958, Buechner 1960, Sugden 1961, D. A. Blood 1963, J. K. Morgan 1970, Geist 1971, Shackleton 1973). Mating activities usually take place on female winter ranges, which may or may not be winter ranges for males (D. A. Blood 1963, J. K. Morgan 1970, Geist 1971, D. A. Demarchi and Mitchell 1973).

There are very few studies of the control mechanisms and physiological changes associated with ovarian activity in bighorn females (Ramsay and Sadleir 1979, Whitehead and McEwan 1980), so we must still refer to studies of domestic sheep to explain some processes. The assumption that domestic and wild sheep share a similar physiology is reasonable considering that both species are seasonal breeders (Hafez 1952), have a similar repertoire of sexual behavior (Pepelko and Clegg 1965, Geist 1971, Shackleton and Shank 1984), and have some common physiological processes (McEwan 1975, Ramsay and Sadleir 1979, Whitehead and McEwan 1980).

The mating season in domestic females is normally stimulated by decreasing photoperiod (Hafez 1952, Thibault et al. 1966), but also is affected by low ambient temperatures (Dutt and Bush 1955, Godley et al. 1966, Lees 1966) and nutrition (Hafez 1952). Captive Rocky Mountain females begin their ovarian cycles with one or more initial ovulations before their first behavioral or overt estrus (Whitehead and McEwan 1980). A similar pattern of silent heat preceding behavioral estrus is found in domestic (Walton et al. 1977) and feral Soay (Grubb and Jewell 1973) sheep. During silent heat, estrous females are not usually detected by males and additional stimuli may be required for expression of behavioral estrus. The presence of males, or merely the sight, sound, or odor of them, can serve to stimulate behavioral estrus in domestic females (Schinkel 1954, R. H. Watson and Redford 1960) and to promote synchronization of estrus within a flock (Thibault et al. 1966). These influences, especially the estrous-synchronizing effect of the "sudden" addition (Thibault et al. 1966) of males into the female herd, are also probably relevant to bighorn, which live for most of the year in sexually segregated groups. We suggest that when males and females form the rutting groups, this sudden influx of males into maternal groups acts as a similar stimulus for behavioral estrus and possibly for estrous synchronization among females. It would be interesting to find if estrous synchrony is greater within than between rutting groups, as it is in feral Soay sheep (Jewell and Grubb 1974).

The single study of hormone levels in bighorn males found that testosterone levels in captive Rocky Mountain males fluctuated throughout the year, with a peak in October followed by a decrease to low levels in December (McEwan 1975). Testicular volume in desert bighorn *(O. c. nelsoni, O. c. cremnobates)* can increase by a factor of two to three times before the mating season, and spermatogenesis is strongly seasonal, being quiescent in December and January (Turner 1976). However, males of another desert subspecies *(O. c. mexicana)* have been seen to court year-round (Lenarz 1979), and motile sperm were found in a captive, 20-year-old California male in February (D. M. Shackleton, unpubl. data).

MATING BEHAVIOR

One or two months before the onset of the rut, bighorn males in some populations congregate on fall or winter ranges. Members of different summer range groups may join to interact with each other and with members of their group. These interactions probably establish and reinforce dominance relationships between individuals. The more physically similar the rivals, the longer the interaction can be before relative social status is established (Geist 1971). Not only do pairs of males interact, but also groups or "huddles" (Geist 1971) of three or more individual males frequently interact. Fights between males may be drawn out, with a high frequency of display behaviors interspersing the dramatic exchanges of horn clashes (Geist 1971, Shackleton 1973). It is during this prerut gathering of males, rather than in the rut itself, that the spectacular clashes and associated display behavior of males are most often observed (fig. 3.11). These dominance interactions are in marked contrast to the aggressive encounters between males during the rut, when they compete and attempt to court estrous females (Shackleton 1973).

During the rut, males are rarely involved in ritualized battles over estrous

Figure 3.11. Spectacular horn clashes between bighorn males are more frequent prior to the rut. During the rut, ritualized displays and interactions between males are greatly reduced as males concentrate on finding and courting females. (Photograph by David Shackleton.)

females, though this impression has been given in many popular articles and films. Most dominance relationships are determined well before mating begins. The most important tasks for the males during the mating season are the courtship and successful insemination of females. To ensure mating success, males should not waste time in prolonged disputes over social status.

Aggression during the rut in the presence of estrous females is noticeably less ritualized and is typically limited to short interactions involving overt aggressive patterns (Shackleton 1973). At this time, the males drop the more usual rules of conduct associated with male dominance interactions (Geist 1971).

The rut usually begins in early November when the sexes form mixed groups. Although it is often stated that the males seek and join maternal groups, we can find no clear support for this. Estrous domestic females will seek out males (Lindsay 1965, Tompkins and Bryant 1972), and captive California bighorn females have been observed to do the same a few days before they come into full behavioral estrus (A. Bottrell, Univ. British Columbia, pers. comm.).

The earliest signs of the rut are the males moving through the female groups in "low stretches" (Geist 1971), approaching the rear of a female, and sniffing and occasionally kicking her. The female invariably responds to this approach by squatting and urinating. The male then sniffs the urine as it is expelled, noses and sniffs the urine on the ground, and then raises his head and lip curls (Geist 1971) for several seconds, often moving the head from side to side. Lip curling, common in most ungulates, allows a male to detect estrus with his vomeronasal organs (Estes 1972, Ladewig and Hart 1980) using a chemical sensing pathway parallel to, but separate from, the olfactory network. Anestrous females typically use the time while the male is lip curling to move away and avoid further interaction (Geist 1971, Shackleton 1973), whereas estrous females tolerate the proximity of a courting male.

Males do not seem to compete during these preliminary investigations of females (Geist 1971), but once females come into estrus, the picture changes dramatically. Dominant males gain preferential access to estrous females and defend them against all other males. This tactic has been referred to as "courtship" (Geist 1971; Shackleton 1973, 1991) or "tending" (Hogg 1984, 1987). Dominant males, however, cannot rely solely on their superior body and horn size, or on social status, to subdue competitors. Where there is a large number of adult males in a population, an estrous female guarded and courted by a dominant, tending male will invariably be surrounded by a group of subordinate males eager to gain access to her (Spencer 1943, D. R. Smith 1954, McCann 1956, Wishart 1958, Buechner 1960, Sugden 1961, D. A. Blood 1963, Geist 1971, Shackleton 1973).

Although subordinate bighorn males tend to be inhibited by the dominant, courting male, invariably one or more of them rushes toward the female and attempts to copulate with her. No courtship patterns are observed during these attempts. The subordinates simply rush at the female and attempt to mount; a mating strategy Hogg (1984, 1987) calls "coursing" and others call "rape chases" (Geist 1971, Shackleton 1973). In response to this sexually aggressive approach, the female simply runs away, only to be followed by her new suitors. The dominant male will attempt, often successfully, to rebuff the subordinates aggressively, smashing into them with his horns on any accessible part of their body. Chases can last several minutes, with the females running back and forth across the mountainside and with males in pursuit and attempting to mount her at every opportunity or clashing briefly and viciously with each other. Eventually the group tires and exhausted males stand behind the females, while the old, dominant male comes slowly up behind and usually takes control again (Geist 1971, Shackleton 1973). This performance may be repeated several times during courtship. However, despite the numerous attempts of the subordinates to mount the females, she rarely, if ever, accepts them. It is unknown how often they succeed, but Hogg's (1984, 1987) behavioral data suggest they sometimes do, because the tending male increases his copulation rate significantly after each coursing episode. In addition, bighorn males have some of the heaviest testes relative to their body weight of bovids, also indicative of sperm competition (Hogg 1988). In populations with low male:female ratios, this coursing/rape-chasing behavior is rare and courtship may only occasionally be interrupted by these interactions (Shackleton 1973). Pulling (1945) suggested chasing by males may reduce female fertility, but other studies do not support this hypothesis.

A third tactic, "blocking," has been described by Hogg (1984, 1987) in a small, introduced population of Rocky Mountain bighorn. Here a male attempts to keep one or more females from reaching the main rutting area and away from the dominant males, even before the female is in estrus. If a male succeeds, he later courts and copulates with the female(s) just like a tending male. Hogg (1984, 1987) found that blocking was a more successful strategy for copulating with females than was coursing, but was about five times less successful than normal tending. Blocking males threaten and actually attack females. In this way the behavior is similar to a fourth tactic called "herding." So far this has been described only for a free-ranging population of Rocky Mountain bighorn with low male:female ratio (Shackleton 1973). Unlike blocking, herding always occurred when more than one male was in a group. Also, though only one male at a time performed herding, it was not always the largest or dominant male that herded females. A final difference was that when a female came into estrus, herding ceased and was replaced by courting and tending by the dominant male.

Courtship or tending is used primarily by dominant males and is the typical courtship tactic of bighorn sheep (Geist 1971; Shackleton 1973, 1991; Hogg 1984, 1987). For an estrous female to accept copulations from any male, he must perform specific courtship behavior patterns (Geist 1971; Shackleton 1973, 1991). These consists of a series of patterns that gradually decrease the distance while increasing the degree of physical contact between the pair, until ultimately the female accepts copulation. In the approximate order performed, these precopulation patterns include: (1) nosing her flanks and rump, often while twisting his head, flicking his tongue in and out, and vocalizing; (2) kicking; (3) standing with his chin on her rump; (4) pushing his chest against her rear; and (5) rising up in a premount.

The female is not totally passive in the presence of a courting (tending) male. When no other males are nearby, estrous females may also actively court the male (Geist 1971) and even mount him in an attempt to stimulate his interest (Shackleton 1973). The frequency of female courtship varies between populations and is more often seen in high-quality herds (Geist 1971; Shackleton 1973, 1991). At other times, the estrous female may try and evade the attentions of a courting, dominant male, using various tactics from simply running away to lying down or seeking a rock cleft in which to stand and face the male (Geist 1971).

Toward the end of the rut, subordinate males may have greater opportunity to court estrous females because dominant males are exhausted.

Dominants expend much energy guarding and courting females, but often spend much less time feeding than normal (Geist 1971), with the result that Rocky Mountain males may lose between 13 and 23% of their body weight during the rut (Stelfox and McGillis 1970, McEwan 1975). The consequences of depleting their energy reserves early in winter carry over throughout winter and may decrease their chances of surviving.

Where male:female ratios are low, either due to heavy hunting pressure or from other factors, young, subordinate males may have the chance to copulate with females throughout the rut (Shackleton 1973, 1976, 1991). In domestic sheep, male:female ratios have a significant effect on productivity. Though low male:female ratios result in a greater number of females serviced per male, there are fewer copulations per female and a greater number of females are bred during a second estrous cycle. In bighorn, even though there appears at least some short-term accommodation to small numbers of mature males (Shackleton 1991), the potential impacts of sex ratio should be considered when setting harvest limits, especially where hunting significantly alters sex ratios or where it reduces or eliminates the large males. We do not know the long-term effects on highly skewed population sex ratios and they need to be determined (see also F. J. Singer and Nichols 1992).

Rearing Young

BIRTH SEASON

Variation in the timing and duration of birth season in bighorn is correlated with latitude, being delayed and of shorter duration with increasing latitude (Bunnell 1982, Thompson and Turner 1982). As we have noted, most California and Rocky Mountain bighorn live at latitudes characterized by marked seasonal fluctuations in climate and forage production. Consequently, timing of the birth season is a function of the requirements of newborn lambs and of lactating females for favorable local environmental conditions at the time of birth. To ensure maximal lamb survival, birth must occur when the neonate can survive thermal stresses and when there is adequate food for lactation (Sadleir 1969; Geist 1971; Festa-Bianchet 1988*a,b*). The lamb also must have sufficient time to grow before it is weaned and has to face the rigors of its first winter, because late-born lambs suffer higher mortality than those born at the beginning of the birth season (Festa-Bianchet 1988*c*). These problems are shared by most northern temperate ungulates and have led to seasonal synchrony of parturition and, hence, the timing of mating

(Sadleir 1969). The optimal birth period in the northern latitudes is early spring, when vegetation is beginning its annual growth and climatic conditions are improving.

The bighorn's birth period begins in early spring, when sheep can capitalize on available high-quality, growing vegetation for lactation and climatic conditions that are more favorable for neonate survival (Geist 1971, Bunnell 1982, Thompson and Turner 1982, Festa-Bianchet 1988a,c). However, as Festa-Bianchet (1988a) found, predation constraints may override choices for optimum forage conditions around the birth period. Near-term females may choose secure habitats with poorer forage conditions over higher risk areas with better forage (Festa-Bianchet 1988a).

Most bighorn lambs are born between May and June (W. B. Davis 1938; Honess and Frost 1942; Spencer 1943; Packard 1946; Couey 1950; D. R. Smith 1954; Wishart 1958; Sugden 1961; Moser 1962; D. A. Blood 1963; Berwick 1968; J. K. Morgan 1970; Geist 1971; Horejsi 1972, 1976; D. A. Demarchi and Mitchell 1973; Shackleton 1973; Festa-Bianchet 1988c). However, in some California populations lambing is earlier and April births occur (Sugden 1961, D. A. Blood 1963, D. A. Demarchi and Mitchell 1973, Van Dyke 1978, Eccles and Shackleton 1979), whereas a late birth season has been reported in a high-elevation Rocky Mountain bighorn population (Stewart 1982).

GESTATION PERIOD

Gestation periods estimated from dates of peak rutting and appearance of new lambs are between 180 and 189 days (Honess and Frost 1942, Spencer 1943, D. R. Smith 1954, Sugden 1961). Geist (1971) speculated that these were generally overestimates because they did not account for the seclusion period; he revised the estimate to about 175 days. The gestation period of four captive Rocky Mountain females ranged between 173 and 176 days (Blunt et al. 1972, Whitehead and McEwan 1980) and in a captive group of California sheep the mean was 174.2 ($\pm$ 1.2 SD, n = 20) days (Shackleton et al. 1984). All these estimates are longer than the mean of 152 days gestation of domestic sheep (Asdell 1964).

BIRTH AREAS

Female bighorn usually leave their group just prior to parturition (Geist 1971, Shackleton and Haywood 1985). Anywhere from a few hours to several days before birth, preparturient bighorn females move to steep, rugged areas to give birth to their lambs in seclusion (Spencer 1943, Kennedy 1948, D. R.

Smith 1954, McCann 1956, D. A. Blood 1963, Geist 1971, Horejsi 1976, Becker et al. 1978, Van Dyke 1978), although occasionally, lambing is seen in quite different habitats (Geist 1971). Though data are limited, the actual seclusion period seems to vary between five to seven days (Geist 1971); Shackleton and Haywood (1985) recorded isolation periods in captive California females lasting from zero to 72 hours, with most falling between 12 and 60 hours.

Lambing grounds are located either on the winter range (McCann 1956, D. A. Blood 1963, Berwick 1968, Geist 1971, D. A. Demarchi and Mitchell 1973, Akeson and Akeson 1992) or on a specific lambing range (D. R. Smith 1954, Wishart 1958, Berwick 1968, J. K. Morgan 1970, Geist 1971, Becker et al. 1978, Akeson and Akeson 1992, Hengel et al. 1992). The same lambing grounds are often used each year by the local maternal group (Geist 1971, Becker et al. 1978).

The isolated nature and ruggedness of lambing sites probably provide for at least three functions: (1) a relatively predator-proof habitat; (2) shelter during inclement weather; and (3) isolation necessary for the development of the mother-young bond. However, the latter two are mainly necessary for only the first one to two days postpartum, and longer isolation periods suggest that predation is a key factor in this behavior in bighorn (Shackleton and Haywood 1985), as in other ungulates (Lent 1974, Arnold and Dudzinski 1978). The female is extremely vulnerable during parturition and the lamb continues to be so several days after birth. Isolated instances of mothers defending or attempting to defend their lambs from predators are known (Kennedy 1948, McCann 1956, Hornocker 1969, Geist 1971, Berger 1978c, Ashcroft 1986b), but the only consistently effective antipredator strategy at this stage seems to be the female's selection of an isolated and precipitous area to give birth. Later, when the lamb has developed sufficient locomotory skills, speed, agility, and proximity to escape terrain can be used to avoid predators. Females with two- to three-week-old lambs have been seen to go directly to timbered areas when golden eagles *(Aquila chraesytos)* were overhead, suggesting such habitat may provide security against aerial predators (D. M. Shackleton, unpubl. data)

Cliff birthsites, particularly if located on south-facing aspects, also provide some shelter and solar radiation. Even when such rugged terrain is not available, shelter seeking is often observed before birth (Shackleton and Haywood 1985) and has been reported in free-ranging domestic sheep (Winfield et al. 1969, Lynch and Alexander 1976). It is not known whether such microhabitat selection is advantageous for the neonate, whose relatively large heat dissipating surface may make it susceptible to hypothermia (G. Alexander 1968, Geist 1971).

PARTURITION

Little is known of the birth process in free-ranging bighorn, but some data are available for captive California bighorn (Shackleton and Haywood 1985). No diel pattern of births could be detected in 20 births of captive lambs in the 40-ha enclosure (Shackleton and Haywood 1985). Also, depending upon breed, this seems to be the case in domestic sheep births as well (George 1969, J. K. Morgan 1970). After moving to the birthsite, the first signs that parturition is imminent is that the female becomes restless; she paces back and forth, frequently stopping to lie down for brief periods or to paw the ground. This distress behavior was observed in 12 of 13 females as much as 48 hours before birth. Once labor commences, the female alternately rises and lies down until the moment the lamb is about to be expelled, when she usually stands and the lamb falls to the ground (Shackleton and Haywood 1985).

Duration of labor appears short in wild sheep (Pitzman 1970), as it is in domestic breeds, in which the majority of labors are less than 20 minutes (Arnold and Morgan 1975). Primiparous domestic females or females with dystocia (prolonged or difficult labor) may have labor extending for more than two hours (Arnold and Morgan 1975). Dystocia in domestic sheep usually increases the probability of lamb mortality (Gunn 1968), and Geist (1971) suggested the same is probably true for wild sheep. This is supported by observations in a captive California group (Shackleton and Haywood 1985).

Immediately following birth, the female licks and nuzzles her lamb and usually eats the placenta. In three captive California bighorn births, lambs first stood anywhere from 1.5 to 56 minutes, suckled from 28 to 166 minutes, and walked from 19 to 96 minutes after birth (Shackleton and Haywood 1985). The longest durations were for a lamb that received little attention from its mother.

Data on birth weights of bighorn are scarce (fig. 3.2); most are from captive animals. Birth weights ranged from 2.8 to 5.5 kg, and this larger value is the weight that Geist (1971) predicted should be the maximum for bighorn sheep. He suggested that, as with domestic sheep, an above-average birth weight would result in difficult births for female bighorn.

ISOLATION AND THE MOTHER-YOUNG BOND

Bighorn are a "follower" species (Lent 1974, Walther 1984). The young follow their mothers shortly after birth, usually staying in close visual and/or auditory contact at all times for the first weeks of life. Establishment of a mother-young bond is critical for subsequent survival and development of the young. Formation of the bond involves the mother and neonate learning to recognize each other without interference from other sheep. Otherwise, problems can arise such as the mother rejecting her lamb, the lamb failing to attach itself to its mother, or stealing of the lamb by another female. In domestic sheep, the critical period for the female to recognize her lamb starts at parturition and may last 5 hours (Arnold and Dudzinski 1978), although after only 5–30 minutes contact, females may be able to discriminate their young (F. V. Smith et al. 1966). The newborn domestic lamb, however, may require up to five days experience before reliably recognizing its mother (Arnold et al. 1975, Shillito 1975).

In domestic sheep, mothers and young use auditory, visual, and olfactory cues to recognize each other, but differ in their relative reliance on these stimuli. Immediately following birth when the female is cleaning the placental membranes and drying the lamb, mother and young constantly vocalize and the female learns the odor of her lamb, thereafter relying heavily on this for identification (F. V. Smith et al. 1966, B. A. Baldwin and Shillito 1974, P. D. Morgan et al. 1975). After about one week, visual and auditory stimuli play an increasing role in recognition; when female and lamb are separated, visual cues are particularly important (G. Alexander and Shillito 1977). The lamb relies more on auditory cues, with vision and olfaction being of secondary importance (Arnold et al. 1975, Shillito 1975).

Our field observations (D. Shackleton and C. Shank, pers. obs.) of wild bighorn suggest similar recognition patterns as those in domestic sheep. A bighorn female can be observed moving through a group bleating and sniffing each lamb she comes across. In contrast to this searching by the mother, her lamb will usually run directly to her. As the lamb reaches her and starts to suckle, she will invariably turn to sniff it, as if in final confirmation of its identity. At times of alarm, the female vocalizes and her offspring runs to her and may nurse briefly, the milk perhaps serving as reinforcement for the young to respond to its mother (Shackleton and Haywood 1985).

SUCKLING BEHAVIOR

The suckling behavior and related activities of California and Rocky Mountain bighorn lambs has been studied by several authors investigating differences between populations (Geist 1971; Shackleton 1973; Berger 1979 a,b; Hass 1990) and within (Horejsi 1972, 1976; K. G. Smith and Wishart 1978; Shackleton and Haywood 1985; Festa-Bianchet 1988b; Hass 1990). Although perhaps not all suckles are for nutrition (Shackleton and Haywood 1985), the critical assumption in these studies is that quantitative measures of suckling and associated behaviors can be used as indices of milk production, which, in turn, reflect a female's level of nutrition (Geist 1971, Shackleton 1973, Horejsi 1976). Indices measured in the field include mean durations of successful suckle bouts, rates of suckling, total suckling times (bout duration × rate), proportion of refused or unsuccessful suckles, and grazing times.

Most variables associated with suckling change with the age of lambs, and trends are similar in all studies of wild and domestic sheep. The duration of suckling bouts, rates, and estimates of total suckling times all decrease, whereas grazing times increase, with lamb age (Ewbank 1964, 1967; Fletcher 1971; Geist 1971; Horejsi 1972, 1976; Shackleton 1973; Berger 1979b; Shackleton and Haywood 1985; Festa-Bianchet 1988b). At least during the first two weeks of life, suckles following bedding periods are significantly longer than those during active bouts, when many of the short suckles terminated by lambs probably have a social rather than nutritional function (Shackleton and Haywood 1985). For any given age, the inverse relationship between milk intake and solid food (i.e., forage or concentrates) intake shown in domestic lambs (Hodge 1966, Joyce and Rattray 1970) and white-tailed deer *(Odocoileus virginianus)* fawns (Robbins and Moen 1975) is found in bighorn lambs (Horejsi 1972, 1976). Maternal condition is also known to affect nursing during midlactation. Young females and females with late-born lambs or with high lungworm (*Protostrongylus* spp.) counts permit shorter suckles and nuzzle their lambs less while they nurse (Festa-Bianchet 1988b). Suckling data suggest female bighorn do not appear to invest more in male than in female offspring (Festa-Bianchet 1988b; D. M. Shackleton, unpubl. data).

The most comprehensive study of suckling behavior and other related behaviors of bighorn lambs is Horejsi's (1972, 1976). He investigated annual variation between years of high and low lamb survival in a population of Rocky Mountain bighorn in western Alberta, Canada. Compared to those in high-survival years, lambs in the low-survival year had shorter mean suckle durations, fewer successful suckles, shorter total suckling times, were more easily refused or stopped by their dams, and had increased grazing times and intensities of grazing. The lambs from the low-survival years were also smaller in size.

Shackleton (1973) found that lambs in an expanding population had longer suckle durations, suckled less frequently but for longer total periods per day, were refused less often, and began grazing later than did lambs from a stable population. These differences were predicted from Geist's (1971) hypothesis of population quality. In Berger's (1979b) study of three populations, data from the desert bighorn population appeared contradictory to Geist's (1971) predictions. However, the extreme ecological conditions and nutritional regime under which these animals live may impose different restrictions on milk production and weaning strategies. For example, water intake restriction significantly influences milk production (R. L. Baldwin 1969) and is probably important in the desert sheep's economy. A unique situation has been described by Hass (1990) in which females frequently, but selectively, nursed lambs other than their own and in two cases fostered others' lambs.

It is assumed that alloparenting is rare in ungulates because lactation costs are so high (Oftedal 1985) that females cannot afford to nurse young other than their own without reducing their reproductive fitness. Hass (1990) believed that the behavior of the females she observed was a response to the high level of predation in her study population. She argued that by allowing strange lambs to suckle, group cohesiveness would be promoted and larger nursery groups could be maintained, which, in turn, would be less susceptible to predation. However, Hass (1984) offered several alternative explanations for the allomothering, including nasal botfly *(Oestrus ovis)* infestations causing anosmia and hence potentially eliminating the mothers' ability to learn their lambs' odor.

Many studies have shown variation in the factors affecting suckling both among (Geist 1971; Shackleton 1973; Berger 1979a,b; Hass 1990) and within (Horejsi 1976; Thorne et al. 1979, cited in Hass 1990; Hass 1990) bighorn populations. However, the value of suckling measures as indices of other production parameters has yet to be demonstrated, even though there are correlations between studies. What is clearly needed is a study of both suckling and grazing behavior, together with measures of milk production and growth rates in lambs. Such a study has been made with captive red deer *(Cervus elaphus)* and supports the assumption made by sheep researchers, showing identical trends in the suckling variables to those in bighorn (Louden et al. 1983).

Population Biology

In the following section, we treat the individual birth and death processes and then briefly discuss their interacting effects in limiting the size of bighorn populations.

POPULATION PARAMETERS

Of the four primary population parameters, fecundity, mortality, immigration, and emigration, the latter two appear relatively insignificant in bighorn populations because of the species' range fidelity (Geist 1971, Festa-Bianchet 1991a, Jorgenson et al. 1997). This simplifies population analysis because populations can be effectively considered as naturally bounded.

FECUNDITY/NATALITY

Age-specific fecundity is the mean number of live births in each female age class. It is determined by the proportion of females mating and conceiving, the proportion of fetuses dying *in utero,* twinning rates, and sex ratio at birth. Each of these can be influenced by the age of the female. Fecundity is a specific formulation of natality, namely the number of individuals born into a population each year.

Bighorn females typically first mate at 2.5 years of age. However, in favorable environments, they can produce lambs near their second birthday (Woodgerd 1964, Jorgenson and Wishart 1984, Festa-Bianchet 1988*d*). A female's first estrus also can be affected by social conditions (K. G. Smith and Wishart 1978, Heimer and Watson 1982). Fecundity in a bighorn population in Alberta increased up to at least five years of age (K. G. Smith and Wishart 1978; fig. 3.12), whereas in domestic sheep females reach their reproductive peak at about eight years of age, after which there is a decline in performance (Turner and Dolling 1965, cited in Geist 1971). Festa-Bianchet (1988*d*) observed a similar decline in fecundity of bighorn females after eight years of age, but it was not statistically significant. Caughley (1977) notes that any

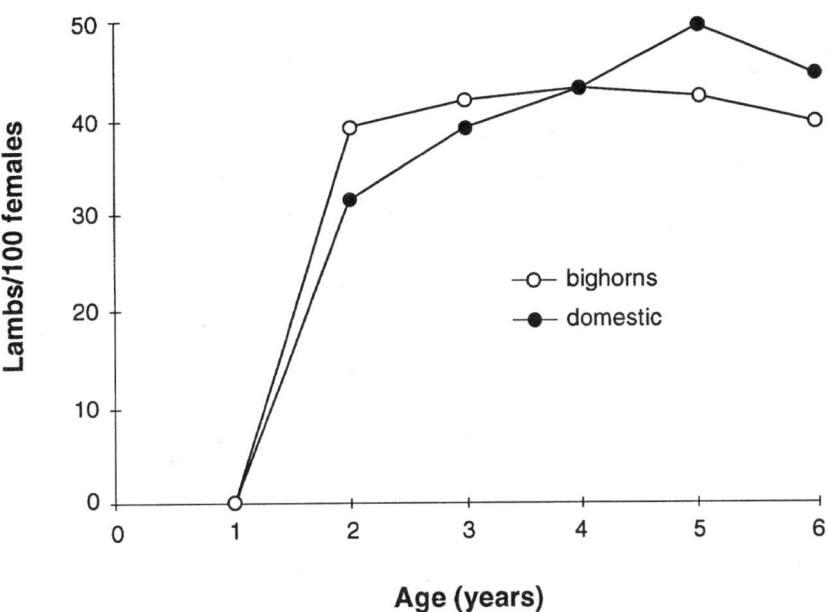

Figure 3.12. Fecundity of female Rocky Mountain bighorn sheep (K. G. Smith and Wishart 1978) and domestic sheep (after F. Hickey 1960).

decline in fecundity at older ages is so slight that a very large sample size is required to detect it statistically.

Reported pregnancy rates in bighorn are high, with 100% pregnant in a small sample of California females (Harper and Cohen 1985) and over 90% pregnant in Rocky Mountain bighorn (Hass 1989, Jorgenson 1992). However, techniques used to determine pregnancy, such as rectal palpitation, laparotomy, ultrasound, and hormonal assay do not always produce accurate results (Brunidge et al. 1988).

Conception may also be affected by population sex and age ratios, particularly ratios of adult males and females. However, the available evidence does not support this for bighorn sheep. Buechner (1960) suggested that a single bighorn male can service 10 to 20 females. However, observations that only one female at a time usually comes into estrus (Wishart 1958) suggest that a few males have ample opportunities to breed all available females. Finally, Shackleton (1973, 1991) showed that mature courtship behavior by males is a function of physical development and relative social status, not merely chronological age, and that most females were bred in a hunted population with no class IV males and a male:female ratio of 1:8. Except in extreme cases, it seems unlikely that bighorn pregnancy rates are greatly affected by age structure of the male population, despite some evidence from Dall's sheep (F. J. Singer and Nichols 1992).

Bighorn sheep typically produce only one young per year. Although females have been seen suckling two lambs and maternal groups can be observed with more lambs than females, these observations are not explicit evidence for twinning because females will suckle others' lambs (Eccles and Shackleton 1979, Hass 1990). Also, classified counts are subject to biases in relative observability of age-sex classes. However, substantiated reports exist of observations of twin births (Gammill 1941, cited in Buechner 1960), of twin fetuses *in utero* (Spalding 1966), and in cases where researchers had close familiarity with the population (Van Dyke 1978, Eccles and Shackleton 1979). Most reports of twins are for California bighorn (Spalding 1966, Van Dyke 1978, Eccles and Shackleton 1979), indicating the subspecies may be particularly disposed to twinning. Specific populations might have a genetic propensity for twinning, but twin mortality is likely significantly higher than for single lambs (Eccles and Shackleton 1979).

There are few references to fetal mortality in bighorn sheep. In other species, maternal weight influences fetal weight (e.g., caribou, Adamczewski et al. 1987) and neonatal survival (e.g., reindeer, Skögland 1984), but the effect on fetal mortality is unknown. Overwinter weight losses of up to 23% in Rocky Mountain females have been found (Stelfox and McGillis 1970, Jorgenson and Wishart 1984), but even with such losses, lamb production can remain high (Jorgenson and Wishart 1984). Sadleir (1987) concluded that postimplantation mortality is also rare in cervids. In bighorn, maternal condition is more likely to have an effect on early postnatal survival than on fetal death.

The intrinsic rate of natural increase (r_m) represents the exponential rate of population growth when no resource is limiting, that is, when the population is increasing at its genetic potential (Caughley 1977). Following the assumptions of one lamb per year, birth of the first lamb when females are three years old, a unitary sex ratio, a stable age distribution, and no mortality, Buechner (1960) calculated an r_m for bighorn of 0.258, but there is empirical evidence for r_ms considerably higher than this. Between 1922 and 1929, the National Bison Range herd in Montana increased at a rate of 0.288 (Hass 1989), and the Fort Peck herd had an r_m of 0.265 over 11 years and an r_m of 0.305 over four years (Buechner 1960). The most conservative of Buechner's assumptions is first breeding at three years of age; however, incorporating the assumption that two-year-old females can produce lambs, the r_m value becomes about 0.308 (doubling time = 2.25 years). Observed rates of increase approaching this value might be suspected of having an unstable sex-age distribution skewed toward adult females.

SEX RATIOS

There are few data on sex ratios of bighorn at birth. Woodgerd (1964) found a 1:1 sex ratio of young lambs and Geist (1971) assumed a unitary sex ratio at birth. Sex differences in juvenile survivorship may be related to environmental conditions. Geist (1971) stated that male lambs show better survival in "low quality" populations. The principles developed by Werren and Taylor (1984) are in accord with this generalization; survivorship of females should be greater than that of males in good recruitment years. This is supported by the observed shift in the yearling sex ratio on Wildhorse Island in favor of males as population growth rate declined (Woodgerd 1964). Both Geist (1971) and Shackleton (1973) also found more male than female yearlings in a "poor quality" population and the opposite in a "high quality" population. This is contrary to expectations, because we might expect that in poor years females would be selected to invest in the least "costly" sex. However, Festa-Bianchet (1989a) showed that male lambs are more costly to rear because males grow faster and require greater investment during lactation. This suggests that the cost of reproduction hypothesis for sex-ratio adjustment is not appropriate.

In unmanipulated populations, adult sex ratios are usually near unity (Buechner 1960, Geist 1971). The wide variation in sex ratios reported by Buechner (1960) are partially census artifacts caused by spatial segregation of the sexes. Distorted sex ratios in an unhunted population normally should be viewed with suspicion.

LAMB MORTALITY

Bighorn sheep are particularly susceptible to death during their first year of life. The lamb:female ratio is of questionable value as an indicator of population condition in bighorn (Festa-Bianchet 1992, Jorgenson 1992), although these ratios calculated in June (but not January) can predict yearling recruitment (Festa-Bianchet 1992). Yearling:adult female ratios may provide an estimate of total lamb mortality if it is assumed that almost all adult females bear lambs. The range of yearling:female ratios varies from as low as 8:100 (J. K. Morgan 1970, Akeson and Akeson 1992) to 61:100 (D. A. Blood 1961); therefore, first year mortality is at best nearly 40% and at worst over 90%. It is possible that neither lamb:female nor yearling:female ratios may have much value for predicting population changes (Festa-Bianchet et al. 1996).

Clutton-Brock et al. (1987) warn against treating first year mortality as a homogeneous rate because early postnatal mortality and first winter mortality are distinct. In bighorn sheep, lamb mortality is normally concentrated in these two periods. Once into summer, lambs normally suffer little further mortality until winter (Geist 1971), although midsummer lamb mortality can be severe during lungworm pneumonia epidemics (Woodard et al. 1972, Spraker 1974). Overwinter lamb mortality varies from almost none (Wishart 1958) to about 70% (Woodgerd 1964, Berwick 1968), and the causes are numerous and interrelated.

Within bighorn sheep populations, lamb:female ratios vary between years and decline with season, being lowest in late winter (Jorgenson 1992). Mortality is often highest during the first few weeks of life. Spraker (1974) and Van Dyke (1978) reviewed the earlier literature; more recent studies all report very high early mortality of lambs (Stewart 1980, Harper 1984, Hass 1989, Akeson and Akeson 1992). By contrast, Festa-Bianchet (1988c) recorded 96–97% survivorship of lambs over the first two weeks after birth, and Woodgerd (1964) recorded no early mortality in a rapidly expanding population. Festa-Bianchet (1988c) suggested that high early mortality of lambs occurs when predation is a major factor, with death from undernutrition occurring later in the summer. Indeed, Harper (1984) and Hass (1989) both cite predation as the primary cause of the early mortality they observed and point out that this could have been due to the lack of adequate security cover (escape terrain) in their study areas.

In a central Idaho population, high summer mortality of lambs resulted from pneumonia caused by *Pasteurella haemolytica* (Akeson and Akeson 1992). Besides disease, Hass (1989) cites inclement weather, inbreeding depression, poor maternal nutrition, poor mothering, human disturbance, and predation as proximal causes. Underlying factors influencing these causes include birth date, range condition, population density, and suitability of security cover. Based on studies of domestic animals, Geist (1971) suggested that inadequate maternal condition results in a low birth weight, which leads to lambs being susceptible to hypothermia. However, there is little published evidence that inclement weather is a direct cause of lamb death, and we suspect that it is a relatively minor proximal mortality factor. Only Akeson and Akeson (1992) have found evidence to suggest that young bighorn lambs may die of exposure during inclement weather. Further work is required to determine if hypothermia is a common proximal cause of early postnatal death. However, as shown for Dall's sheep (Heimer 1978), desert bighorn (C. L. Douglas and Leslie 1986, Wehausen et al. 1987), and red deer (Albon et al. 1983), weather almost certainly is indirectly important through its effect on the production, quality, and availability of forage, which, in turn, affect maternal condition.

Whatever the direct causes of death, low birth weight is almost certainly an important contributory factor in lamb mortality. In many ungulates, maternal condition and birth weight are linked to first-year mortality (Geist 1971; Clutton-Brock et al. 1987, 1992). Although the only study we located that relates bighorn birth weight to mortality found no such relationship (Hass 1989), it should be remembered that there are very few data for birth weights of bighorn (fig. 3.2).

Birth date has an important influence on first-year survival in ungulates (Bunnell 1982, Thompson and Turner 1982, Clutton-Brock et al. 1987). Bighorn births are concentrated in the first few weeks of the lambing period. Although lambing extended over 66 days, Festa-Bianchet (1988c) found 71% of bighorn lambs were born in the first 15 days and their birth date affected their subsequent survival. Lambs born during the first two weeks were more likely to live to five months of age than those born later. Only 5% of lambs born later than 10 June (about 25 days after the first birth) survived to reproduce (Festa-Bianchet 1988c). Late-born lambs may miss the peak of forage nutrition between late June and early July and eventually die from factors related to inadequate nutrition (Festa-Bianchet 1988c). In contrast, in a less-inclement climate, Hass (1989) found no relationship between birth date and survival. Late birth may be related to population density on the winter range in bighorn (Festa-Bianchet 1988c) and other ungulates (Clutton-Brock et al. 1987).

Coyotes appear to be the most important predator of bighorn lambs, though reported predation levels vary among populations. In the Junction California bighorn population of south-central British Columbia, up to 80% of the lambs are reported lost each year to coyotes (Hebert and Harrison 1988), and high levels of coyote predation have also been recorded by others (Harper 1984, Hass 1989). However, the significance of predation in lamb mortality may depend upon the availability and steepness of cliff terrain used for security cover. Some populations in which predation is believed to be a major cause of lamb mortality tend to have limited security cover (Harper 1984, Hass 1989) and are not historical sheep ranges (Hass 1989).

Sausman (1984, cited in Hass 1989) found higher rates of mortality and preferential male survival among inbred captive lambs, as compared to outbred populations, which support Werren and Taylor's (1984) conclusion that sex ratios should be shifted to males during periods of poor recruitment. Inbreeding has the capacity to be a significant factor in small bighorn sheep populations because of their islandlike nature and low rates of dispersal. A promising line of inquiry might be a comparison of heterozygosity between bighorn populations of differing size and demographic vitality. Studies of genetic and mitochondrial variation between bighorn populations of differing size and demographic vitality are also providing valuable insights into the genetic histories of bighorn populations (Luikart and Allendorf 1996, Luikart and Cornuet 1997).

JUVENILE MORTALITY

Using data from skulls found in the field, Geist (1971) found no mortality among male yearlings and only 3% mortality among two year olds. Although this corresponds closely to Murie's (1944) classic findings for Dall's sheep, these data contrast with Festa-Bianchet's (1989b) results. Using declines in cohort numbers from several populations, he showed that mortality rates for yearling and two-year-old males are consistently higher than those determined in studies using pickup skulls to estimate mortality. Festa-Bianchet (1989b) found mortality rates for bighorn yearling males of 33% and 18% for two year olds. By following live cohorts, Stewart (1980) found similar mortality rates of 33% and 41% for yearling males and females, respectively, and 41% and 16% for two-year-old males and females, respectively.

Cohort-derived yearling mortality data not cited by Festa-Bianchet (1989 b) include the 11% for males and 8% for females given by Wishart (1958) and the 8% for yearling males estimated by Woodgerd (1964). Although more comparative data are required, Geist's (1971) generalization that juvenile mortality is low is probably not valid. The discrepancy between mortality estimated from skulls and that determined from comparing cohorts is seen not only in sheep but also in other species (Festa-Bianchet 1989*b*). This probably results from the skulls of young ungulates being less robust and hence deteriorating more rapidly than those of older animals.

ADULT MORTALITY

Data on adult mortality are largely restricted to males because of difficulties of aging females five years and older from their horns. Estimates of adult male mortality of Rocky Mountain bighorn have been presented by Festa-Bianchet (1989*b*), Geist (1971), Jorgenson and Wishart (1986, cited in Festa-Bianchet 1989*b*), Stewart (1980), and Woodgerd (1964). These estimates were obtained either from pickup skulls (Caughley 1977, method 5), by survivorship of marked individuals (Caughley 1977, method 3), or by comparing the age distribution of a population in successive years (Caughley 1977, method 4).

Calculating mortality rates from pickup skulls has difficulties (Murphy and Whitten 1976, Caughley 1977) and, as discussed earlier, can lead to an underestimate of juvenile mortality. The technique also requires a known, constant rate of increase and a stable age distribution. These requirements are probably never met in large, long-lived animals (R. Miller 1976) like bighorn. Nevertheless, we think that some general conclusions can be drawn from mortality schedules derived from skull collections.

Caughley (1966) has suggested around 150 as a minimum sample size for accurately estimating mortality rates from skulls. If we assume the skulls collected in Banff National Park by Geist (1971) from the Palliser Range and by Shackleton (1973; D. M. Shackleton, unpubl. data) from the Palliser Range and the Panther River represent a single functional population, we can use the combined sample of 122 to estimate the male mortality rate (table 3.4). The population remained approximately stable during the time period between the two collections, and there is a correlation coefficient of 0.98 for the mortality rates of the two data sets. However, the mean mortality rate between ages 3 and 14 years was 6.2 ± 7.1% (SD) higher in Shackleton's sample, indicating slightly higher turnover rates of males in later years (table 3.4).

A comparison of age-specific male mortalities for several Rocky Mountain bighorn populations shows that populations exhibit similar death rates of about 3 to 14% from ages 3 to 5, after which mortality rates diverge sharply (fig. 3.13). Stable, nutritionally limited, unhunted populations exhibit slowly increasing mortality rates and long life expectancy. Increasing, hunted populations on excellent range tend to show rapidly increasing natural male mortality rates and short life expectancy. Beyond this, few generalizations can be drawn.

Table 3.4. Calculation of age-specific mortality in Rocky Mountain bighorn males from collection of skulls aged by counting horn annuli.

AGE (YEARS)	GEIST SAMPLE	SHACKLETON SAMPLE	TOTAL NO. SKULLS	NO. ENTERING AGE CLASS	% MORTALITY
3	1	3	4	122	3
4	3	0	3	118	3
5	4	3	7	115	6
6	0	2	2	108	2
7	6	5	11	106	10
8	6	5	11	95	12
9	8	7	15	84	18
10	9	5	14	69	20
11	8	6	14	55	25
12	13	7	20	41	49
13	9	2	11	21	52
14	3	1	4	10	40
15	3	1	4	6	67
16+	2	0	2	2	100

Sources: Geist (1971) and D. M. Shackleton (unpubl. data) for the Palliser and nearby Panther River areas, Banff National Park, Alberta, Canada.

Note: Data for ages 0–2 years are biased due to skull fragility and are not presented.

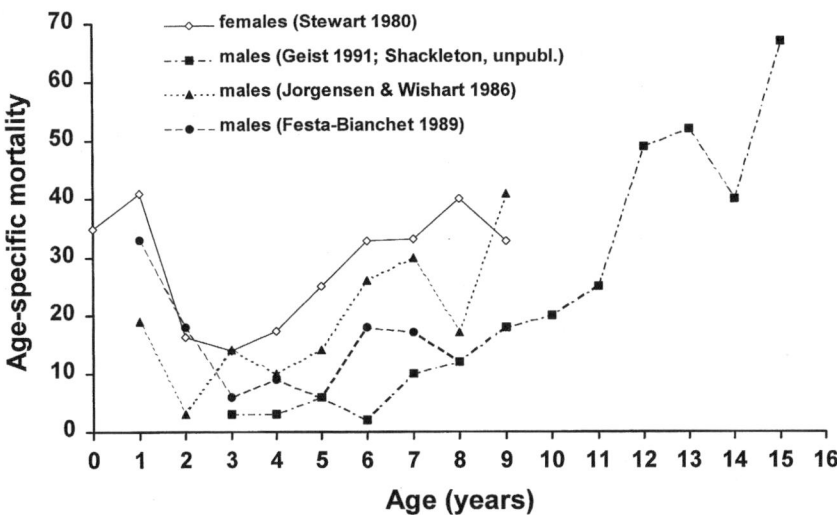

Figure 3.13. Percentage age-specific mortality rates for males in three populations and females in one population of Rocky Mountain bighorn. Data derived from Geist (1971) and D. M. Shackleton (unpubl. data) are from skull collections and are presented in table 3.4. Yearling and two-year-old mortalities are not presented in this data set because of the biases mentioned in the text. The mortality for the ninth year of the Jorgenson and Wishart (1984) data set is an average for all animals nine years old and older.

Several authors have suggested that mortality rates of female mountain sheep are higher than those of males (Sugden 1961, Woodgerd 1964, Bradley and Baker 1967, C. G. Hansen 1980c). The best data on Rocky Mountain bighorn female mortality are Stewart's (1980), which, using Caughley's (1977) method 4, show that the age-specific death rate of females is higher than that of males. Using classification counts of 15 radiocollared adult females and their associates, Hengel et al. (1992) estimated the mortality rate of adult females to be 10.8% in a bighorn population in Wyoming. As a *post hoc* argument, it seems reasonable that females incur earlier mortality because typically they begin breeding at 2.5 years of age versus at 7–8 years in males and hence are subject to all the stresses of reproduction some 5–6 years before most males.

Population Limitation and Regulation

As a broad generalization, populations of bighorn sheep are numerically stable but occasionally subjected to catastrophic die-offs. In this section, we consider what is known about what keeps population numbers in check. To do so, we must differentiate between the terms "regulation" and "limitation." Regulation is defined by Skögland (1991) as any density-dependent process tending to stabilize population numbers over time. By contrast, a limiting factor is one that tends to reduce the population. A population limiting factor can also be population regulating if the factor acts in a density-dependent manner.

LIMITING FACTORS

It seems likely that shortage of food is a common factor limiting bighorn populations. However, there is no direct evidence to support this contention. Skögland (1991) could find only six studies that documented ungulate food resources as limiting or that measured their regulatory effects. During the summer months, bighorn food tends to be superabundant and of excellent quality, suggesting that only winter forage has the capacity to be limiting.

The role of interspecific competition in population limitation of bighorn is not clear. There is a considerable diet overlap between sheep and elk, cattle, and, to a lesser extent, mule deer, mountain goats, and horses (Streeter 1969). Picton (1984) has shown that lamb:female ratios decline in the spring after elk heavily use sheep winter range. In contrast, Berger (1986) found no effects of interspecific competition on desert bighorn populations.

Predation is likely to be a major limiting factor of bighorn populations inhabiting ranges without adequate escape terrain (Harper 1984, Hass 1989). In habitats where sheep can escape into cliffs, predation is expected to be much less important. In one population, cougars *(Puma concolor)* preyed heavily on mature California bighorn males immediately following the rut; the level of predation was possibly increased by coyotes who chased cougars from their kills, thus forcing the cougars to kill again (Harrison and Hebert 1988, Harrison 1990). Humans can be a major predator of bighorn and are responsible for the near total destruction of the oldest male cohorts in most areas where sheep are not protected. There is no evidence demonstrating that predation regulates large mammal populations (Sinclair 1989).

Accidents are a minor mortality factor. Bighorn inhabit dangerous terrain and occasionally fall to their deaths. In an extensive study, Festa-Bianchet (1987) found only one lamb and one adult male to have died from falls. Falling rock and spring avalanches are probably more important hazards. Mountain sheep have been observed to be alert and nervous when rock-fall or avalanche danger is great (Pitzman 1970, Geist 1971), and on the Palliser Range in Banff National Park, Alberta, bighorn have often been observed moving from the path of avalanches.

Disease is an infrequent but major limiting factor. Large ($> 50\%$) and sudden (< 12-month) declines, termed "die-offs," are common and have been reported since the 1800s (Buechner 1960, Stelfox 1971, Ryder et al. 1992). The exact etiology of die-offs is still poorly understood, but seems to originate from stress, which is not always nutritional (Ryder et al. 1992), interacting in some uncertain manner with endemic lungworm *(Protostrongylus stilesi* or *P. rushii)* infestation (Spraker et al. 1984, Dunbar 1996). The resistance of bighorn to disease organisms such as *Pasteurella* spp. bacteria is reduced as a result of the stress. An ensuing infection can result and cause acute bronchopneumonia and die-offs involving over 50% of the population. Reflecting the uncertainty and the many pathological agents identified, these die-offs have been labeled as "pneumonia-lungworm" or "respiratory disease" complexes (Onderka and Wishart 1984, Schwantje 1988) and can occur in less than a year. Schwantje (1988) suggests that these respiratory die-offs have arisen approximately every 20 years in the East Kootenay region of British Columbia. It is still unclear to what extent die-offs can be attributed to density-dependent factors such as range condition (Dunbar 1992, 1996) and to density-independent factors such as the proximity of domestic sheep (Onderka and Wishart 1988, Coggins and Matthews 1992). Festa-Bianchet (1991*b*) has argued that lungworm infection is a normal state for bighorn and that higher levels of infection do not predict pneumonia epizootics or indicate poor population health. L'Heureux et al. (1996) found that contagious ecthyma, a common viral disease of bighorn, had no effect on the dynamics of their study population.

POPULATION REGULATION

Jorgenson et al. (1997) have examined the effects of density on population parameters in two populations having differing numerical histories. They found that in a population allowed to increase to equilibrium numbers, juvenile female survival was negatively affected by population density, whereas survival of all other classes was unaffected. In the other population, which went through a die-off and never fully recovered numerically, there were no density affects on survival detected because the population never reached levels at which the density limitations would become operative. Festa-Bianchet et al. (1995) showed that age of first female reproduction switched from two to three years as female numbers increased dramatically. In another population, where ewe numbers varied much less, no density effect was seen on age of primiparity.

The above results suggest that bighorn populations are regulated through density-dependent feedback on fecundity and on lamb survival and that density dependence only begins at "intermediate" population levels. Adult survival appears to be inconsequential as a population regulating mechanism.

Because changes in fecundity and female lamb survival became evident at intermediate population levels (Jorgenson et al. 1997), bighorn populations tend to be stable (Clutton-Brock et al. 1997), at least in the absence of catastrophic die-offs. The mechanism by which population density feeds back on fecundity and lamb survival remains uncertain.

Geist (1971, 1985a, 1987c) speculates that bighorn populations are regulated through nutrition. He characterizes populations as either stable, food-limited populations with low recruitment rates and low adult mortality (i.e., "low quality" or "maintenance" phenotype) or increasing populations with high recruitment and high adult mortality (i.e., "high quality" or "dispersal" phenotype; see Population Quality and Horn Growth). The suggested mechanism relates maternal food limitation to lamb survival and development and to subsequent behavioral and mortality patterns. Nutritionally stressed females give birth to small lambs that tend to die young. If they survive, the lambs receive little milk, are consequently slow in development, and remain behaviorally lethargic throughout a prolonged life span. Despite the attractiveness of Geist's proposal, there are no data directly demonstrating that food acts to limit or regulate bighorn sheep populations.

POPULATION PERSISTENCE

Berger (1990) suggested that extinction of bighorn populations with fewer than 50 animals is highly probable and that the causes are unlikely to be food shortages, weather changes, predation, or interspecific competition. To avoid short-term loss of genetic variability, Fitzsimmons and Buskirk (1992) recommended a minimum effective population size for bighorn of over 150 individuals (but see also Hartl et al. 1990; Luikart and Allendorf 1996; Fitzsimmons et al. 1997; Hartl, in press). Catastrophes such as epizootics appear to be a proximal cause of some population extinctions. Die-offs represent a major natural perturbation in the otherwise relatively stable population numbers of bighorn and a significant threat to populations already shrunk by habitat loss due to human encroachment.

four

NATURAL HISTORY OF DESERT BIGHORN SHEEP

Paul R. Krausman, Andrew V. Sandoval,
and Richard C. Etchberger

Introduction

We review the relevant literature related to the ecology of the desert races of bighorn sheep *(Ovis canadensis nelsoni, O. c. mexicana, O. c. weemsi,* and *O. c. cremnobates).* The overview of these data is designed as a reference for source material and as a guide for future research. For those interested in the conservation and management of desert bighorn sheep, we hope these data provide a comprehensive framework of available information that can be refined and built upon. The future of the species in the desert ranges depends on it.

Habitat Components

DISTRIBUTION

Historically, desert bighorn sheep ranged from Nevada south to Coahuila, Mexico, and from western Texas, southern New Mexico and Arizona, western Colorado, and Utah west to California (Monson 1980). Although this region encompasses several ecologically distinct life zones, desert bighorn sheep have adapted to the rugged, arid, and sparsely vegetated desert environment. They have the ability to maintain their water requirements through

Figure 4.1. Desert bighorn (Kofa National Wildlife Refuge, Arizona) are found sparsely scattered in isolated mountain ranges characterized by deep canyons, rock outcrops, and numerous cliffs providing a high degree of visibility. (Photograph by Bob Furlow.)

metabolic pathways (Turner 1973) or use of succulent vegetation (Warrick and Krausman 1989, Krausman and Etchberger 1996). Also, desert bighorn exhibit extended mating and lambing seasons to compensate for unpredictable and often unreliable forage availability (Thompson and Turner 1982).

The current distribution of desert bighorn encompasses seven deserts: the Great Basin, Painted, Mojave, Sonoran, Colorado, Vizcaino-Magdalena, and Chihuahuan Deserts (C. G. Hansen 1980b). Throughout this vast region, desert bighorn are found sparsely scattered in isolated areas characterized by deep canyons, rock outcrops, and numerous cliffs that provide a high degree of visibility (Risenhoover and Bailey 1985; fig. 4.1). This is typified by the cliffs and benches within the Grand Canyon, Arizona (Bendt 1957); the mesas and canyons of southeastern Utah (L. O. Wilson 1968); the highly dissected, north-south trending mountain ranges of Arizona (Russo 1956), Nevada (McQuivey 1978), and California (Weaver 1972); and the massive fault blocks in southern New Mexico (Sandoval 1979b), western Texas, and Baja California and Sonora, Mexico (Alvarez 1976).

Desert bighorn are most numerous in southwestern Arizona, southern

Nevada, southeastern Utah, southern California, and Baja California and Sonora, Mexico. Isolated populations with questionable viability are found in southeastern California, southern New Mexico, and western Texas.

COMPONENTS

At least four habitat components are essential to bighorn sheep: food, water, escape terrain, and open space. The proper placement of each component is important. Areas that provide unrestricted visibility and high-quality, palatable forage but lack escape terrain generally are not used. Bighorn are specialized to negotiate precipitous terrain. The heavy musculature of the front shoulders is suited for climbing, and their hooves are adapted to clinging to steep surfaces that provide a minimum amount of footing.

Although desert bighorn have behavioral patterns and adaptations that minimize their dependence on water, Turner (1973, 1979a) indicated that metabolic water formed by oxidation is not sufficient to meet their physiological requirements. Bighorn have an efficient digestive system that allows them to sustain their nutritional needs on dry, abrasive forage common to the desert. Bighorn have distinguished themselves as a wilderness mammal inhabiting rugged, isolated terrain. With few exceptions, bighorn tend to be intolerant of humans and their activities.

Early habitat studies, although qualitative in nature, are valuable. However, the demand for quantitative ecological data are critical. Sandoval (1979a, 1982) and Watts (1979) used a mathematical approach to describe desert bighorn habitat in the San Andres, Peloncillo, Alamo Hueco, and Big Hatchet Mountains, New Mexico. Holl (1982) described habitat evaluation models to determine the quantity and quality of bighorn habitat in the San Gabriel Mountains, California. In Arizona, desert bighorn sheep habitat has been measured and modeled in a number of regions (Gionfriddo and Krausman 1986; Krausman and Leopold 1986; Cunningham 1989; Etchberger et al. 1989, 1990; Krausman et al. 1989; Wakeling and Miller 1989, 1990).

The modeling of habitat use by desert bighorn sheep has been facilitated using geographic information system (GIS) technology (W. Dunn 1991, T. S. Smith and Flinders 1992, L. K. Harris et al. 1995). Also, enclosures have been used to evaluate responses of desert bighorn to various habitats. In an enclosure in Nevada, vegetation and topography were evaluated, C. G. Hansen's (1980b) model was evaluated (Berner and Krausman 1992), habitat use by desert bighorn was determined (Berner et al. 1992), and resource use was examined (Zine et al. 1992). The sheep in the enclosure used habitat similar to free-ranging sheep studied in other areas, and more than 98% of the locations of sheep were in habitat rated as important to desert bighorn (Berner and

Krausman 1992, Berner et al. 1992). Although enclosure studies have limitations, they provide opportunities to make observations that would not otherwise be possible. As the semi-free-ranging animals studied by Berner and Krausman (1992), Berner et al. (1992), and Zine et al. (1992) responded to habitat as did free-ranging animals, there are implications for further studies on low-density populations.

TOPOGRAPHY AND ELEVATION

Desert bighorn preference for rock outcroppings, precipitous cliffs, and rough topography is well documented (L. O. Wilson 1968, Welch 1969, Merritt 1974, McQuivey 1978, Leslie and Douglas 1979, Sandoval 1979a, DeForge 1980, Holl and Bleich 1983, Etchberger et al. 1989, Krausman et al. 1989, Wakeling and Miller 1990). In Utah, L. O. Wilson (1968) found a direct correlation between the amount of escape terrain and bighorn sheep use. As the rock cover increased from 0 to 100%, the probability of sighting sheep increased proportionally, with approximately 61% of the sightings occurring where rock cover varied between 71 and 100%. Sandoval (1979a) compared the distribution of different desert bighorn group types (i.e., ewes, rams, female-lamb, mixed) in relation to habitat components to determine if habitat use was independent of different group types. He found the probability (P) of randomly locating a female-lamb group on slope gradients between 21 and 40% to be 0.43, but only 0.14 for slopes between 5 and 20%. Barren female groups preferred relatively gentle topography, (P = 0.38 for slopes between 21 and 40%). Male groups also preferred more gentle topography (P = 0.37 for slopes between 5 and 20% and P = 0.33 for slopes between 21 and 40%).

Steel and Workman (1990) examined microhabitat use relative to forage availability in Capitol Reef National Park, Utah. No differences were found between used and unused plots in terms of percent cover of shrubs, grasses, forbs, or forage species. Forage quantity did not appear to influence microhabitat use (Steel and Workman 1990).

Holl (1982) employed two models to study the effects of topography and elevation. The first, a linear regression equation, predicted the number of females in an area based on the amount of escape terrain available. When the estimated female population size was regressed against hectares of escape terrain, the relationship was significant (P = 0.01). This indicates that without escape terrain there should not be any females and the size of the female population is directly proportional to the amount of escape terrain available. The second model employed probability theory to determine habitat quality and required knowledge of habitat components needed by bighorn sheep and a standard with which these components could be compared. The second model was based on the probability that an area will support a high- or low-density population. Using both models, the approximate size of the female segment of a population and whether it will be a high- or low-density population can be predicted (Holl 1982).

In Arizona, Krausman and Leopold (1986) quantified habitat of desert bighorn sheep and compared areas used and not used by sheep. Bighorn sheep habitat was farther from bajadas and had fewer large boulders and less sheetrock than areas not used by sheep (Krausman and Leopold 1986). Gionfriddo and Krausman (1986) characterized sheep habitat as steep and rugged in the Pusch Ridge Wilderness. In this same area, Etchberger et al. (1989) quantified the differences between range desert bighorn have abandoned and those areas still being used. Both areas were steep and rugged, with average slopes being greater than 55%. However, abandoned areas showed the effects of fire suppression, which allowed vegetation to grow tall and obstruct the vision of sheep (Etchberger et al. 1989, 1990).

Application of these objective analyses of bighorn habitat provides a quantitative description and shows habitat preferences relative to availability. For a detailed description of these computerized techniques for habitat evaluation, refer to Sandoval (1979a, 1982), Watts (1979), Holl (1982), and Cunningham (1989).

In the San Andres Mountains, New Mexico, cliffs were used by sheep in greater proportion than their availability. Cliffs accounted for 70% of the total bighorn use, yet comprised only 23% of the available habitat (Sandoval 1979a).

Welch (1969) studied the distribution and density of sheep beds in various habitats in the San Andres Mountains and found bed densities to be five times greater in cliff habitats. He found densities of 10.07 and 2.00 in cliff and noncliff types, respectively, expressed as the number of beds per 929 m^2. Merritt (1974) compared bighorn bedsite densities relative to canyons, grazing plateaus, rocky hills, and escape terrain in the Santa Rosa Mountains, California. Areas overlooking or close to plateaus received 45.6 and 55.2% use during winter and summer, respectively. Merritt (1974) used bedsite densities as an index of habitat use by desert bighorn in the Santa Rosa Mountains, California, and determined 76% use on slope gradients greater than 33%. In the Superstition Mountains, Arizona, Wakeling and Miller (1989) compared bedsites used by sheep to random sites and concluded that sites on canyon terraces, on escape terrain, and close to water were preferred. Vegetation structure around bedsites was 70% bare ground, 14% herbaceous cover, 15% shrubs, and 1% trees (Wakeling and Miller 1989).

Desert bighorn range from 78 m below sea level in Death Valley, California (Welles and Welles 1961), to over 4,267 m above sea level in the White Mountains, California (Kovach 1979). Consequently, it is difficult to generalize about elevational preferences. A wide variety of factors, both exogenous and endogenous, influence elevational use of a particular range. Elevational preference is correlated with numerous variables that reflect the animals' environmental, physiological, and behavioral preferences.

In the Cabeza Prieta National Wildlife Refuge, Arizona, the location of water sources was the major factor influencing elevational distribution of bighorn, particularly during summer. When temperatures dropped, rendering sheep less dependent on surface water, sheep would disperse to the summit of the range, where temperatures were coolest (Simmons 1969). Eustis (1962) reported different elevational distribution between female-juvenile and all-male groups in the Kofa National Wildlife Refuge, Arizona. Whereas 84% of the females and 93% of the lambs were observed in the upper third of the mountain range, 62% of the males were seen in the middle third and 9% were observed in the lower third of the mountains.

Sandoval (1979a) reported that different group types of desert bighorn in the San Andres Mountains, New Mexico, preferred different elevational spans. His data revealed that between 1,525 and 1,676 m, the probability of randomly locating a female-lamb group was 0.42 but only 0.01 for locating an adult male group. Conversely, between 1,830 and 1,981 m, the probability of locating an adult male group was 0.63 and only 0.18 for a female-lamb group.

CLIMATE

The climate in desert bighorn habitat ranges from alpine conditions on mesic mountain ranges to arid desert conditions below sea level. Precipitation may range from less than 2.54 cm in the Mojave Desert to more than 50.8 cm in the White and San Gabriel Mountains, California (C. G. Hansen 1980b). Precipitation is relatively low and unpredictable. Consequently, the erratic rainfall patterns reduce the significance of any precipitation measurements taken at one specific locality. Distribution and annual production of forage are determined by the quantity of seasonal precipitation. Rainfall, therefore, is an important physical requisite to life for desert bighorn sheep in southwestern North American (Russo 1956). Snow is usually not a limiting factor, provided that it does not affect forage availability for extended periods. However, deep snow may precipitate movements of animals to lower elevations.

Seasonal and daily ambient temperatures in desert bighorn habitat fluctuate considerably. Seasonal temperatures may range from -29°C (Sandoval 1980) to above 49°C (C. G. Hansen 1980b). Desert bighorn activities are affected by temperature. Bighorn in the Cabeza Prieta National Wildlife Refuge, Arizona, bedded in the shade an average of seven hours each day when wet bulb temperatures exceeded 19°c (Simmons 1969). During the hot, dry summer months, one of the most serious obstacles to desert bighorn survival is heat stress. Desert bighorn avoid heat gain by minimizing energy-expending activities, by bedding in the shade during the hottest part of each day, and by feeding and watering when slopes are shaded.

GROUND-COVER HEIGHT

Visibility is an important habitat feature of bighorn sheep because their predator-evasion strategy involves foraging diurnally in large, dispersed groups on open habitat close to escape terrain. Predators are detected visually, and a large, dispersed group of sheep would be more alert to potential predators over a large area. Risenhoover and Bailey (1985) compared bighorn foraging efficiency in Colorado, relative to group size, behavior, and habitat visibility. They concluded that average foraging efficiency was greater when sheep were in large groups and when sheep were in habitats providing greater visibility. In contrast to foraging efficiency, the number of alert postures exhibited by females each minute was negatively correlated with group size.

In the San Gabriel Mountains, California, bighorn use heavy cover only during times of stress or for thermal cover (DeForge 1980). As chaparral matures and reduces visibility, suitable habitat for bighorn is reduced, thus lowering carrying capacity potential and eventually resulting in loss of range. In the San Andres Mountains, New Mexico, the piñon (*Pinus* spp.)–juniper (*Juniperus* spp.) community received little use by bighorn (2% of 225 observations). The few occasions that sheep were seen in areas with an overstory of large shrubs or trees were when they were traveling along well-established trails close to escape terrain (Sandoval 1979a).

Krausman and Leopold (1986) concluded that a reason sheep did not use some areas of the Harquahala Mountains, Arizona, was that large boulders obstructed their vision. Also in Arizona, Etchberger et al. (1989, 1990) and Krausman et al. (1996) found that desert bighorn abandoned areas after fire suppression allowed vegetation to grow and obstruct the vision of sheep.

Caution must be exercised when attempting to measure visual obstruction (McCarty and Bailey 1992). Methods used to measure visual obstruction in habitat of desert bighorn are not clearly described in the literature, permit interobserver differences and observer expectancy bias, and may produce data with poor statistical properties (McCarty and Bailey 1992). McCarty and Bailey (1992) present a computer program and protocol for measuring visual obstruction.

HABITATS

Biotic communities are determined by latitude, precipitation, elevation, exposure, and land use practices. The relationship of biotic communities and desert bighorn habitat has been discussed by Bradley and Deacon (1965), L. O. Wilson (1968), McQuivey (1978), C. L. Douglas and White (1979), Kelly (1979), Leslie and Douglas (1979), Sandoval (1979a,b), Watts (1979), Krausman et al. (1989), and R. C. Etchberger and P. R. Krausman (1999) and was summarized by C. G. Hansen (1980b).

Throughout much of the habitat occupied by desert bighorn the vegetation associations are predominantly adapted to dry, rocky, or sandy soils and plants are characterized by thickening of the epidermis and reduction of the leaf surface (E. C. Jaeger 1957). In general, the vegetation is uniformly sparse, and the plants are widely spaced due to demands of their root systems in shallow, rocky soils; a rounded canopy results from equal exposure to solar radiation from all sides. Finally, the plants must withstand severe drought, which may last from several months to several years.

Mountains taller than 1,500 m support stands of pine, fir (*Abies* spp.), Douglas fir *(Pseudotsuga menziesii),* and juniper. Bighorn use is confined to open woodlands or forests where trees are not so dense as to adversely affect visibility. Alpine zones characterized by sparse vegetation associations consisting of low, perennial, herbaceous plants and grasses are used mainly during the summer (Weaver and Mensch 1970, M. C. Jorgensen and Turner 1975, McQuivey 1978, Kovach 1979).

Throughout much of the bighorn habitat of the Great Basin Desert, sagebrush (*Artemisia* spp.), shadscale *(Atriplex confertifolia),* black brush *(Coleogyne maleosissima),* and cliffrose *(Cowania stansburiana)* comprise the major browse species. The major grasses include wild rye (*Elymus* spp.), Indian rice *(Oryzopsis hymenoides),* galleta *(Hilaria jamesii),* bluegrass (*Poa* spp.), and fescue (*Festuca* spp.; Bradley 1964, C. G. Hansen 1980b). In the lower elevations of the White, San Gabriel, San Jacinto, and Santa Rosa Mountains, California, bighorn habitat is characterized by lowland browse types dominated by sagebrush, redberry *(Rhamnus crocea),* and chaparral white-thorn *(Ceanothus leucodermis).* Bighorn habitat extends through piñon-juniper and ponderosa pine *(Pinus ponderosa)* associations. The summer range includes the subalpine and alpine biotic communities (Weaver and Mensch 1970, McQuivey 1978, Kovach 1979, DeForge 1980). In the Painted Desert, sagebrush, black brush, shadscale, galleta, and piñon-juniper comprise the major vegetation associations used by desert bighorn (L. O. Wilson 1968). Characteristic plants of desert bighorn habitat in the Sonoran Desert include paloverde (*Cercidium* spp.), ironwood *(Olneya tesota),* saguaro *(Carnegiea gigantea),* and organpipe cactus *(Lemaireocereus thurberi)*. The major grass species include grama *(Bouteloua* spp.), galleta, and sacaton *(Sporobolus* spp.; Mendoza 1976, Seegmiller and Ohmart 1981). Throughout the Chihuahuan Desert, bighorn habitat is characterized by few trees, agave *(Agave* spp.), yucca *(Yucca* spp.), small cacti, and numerous spiny shrubs (Moore 1958, Sandoval 1979a, Watts 1979). In general, volcanic soils support relatively homogeneous grasslands, and sedimentary parent material produces creosote *(Larrea tridentata),* mesquite *(Prosopis* spp.), tarbrush *(Flourencia cernua),* and cactus savannas and agave thickets. Typical indicator species of this region include sotol *(Dasylirion wheeleri),* agave, and ocotillo *(Fouquieria splendens;* Sandoval 1979b). Guzman (1961) and Alvarez (1976) reported that agave, ocotillo, ironwood, cholla *(Opuntia* spp.), acacia *(Acacia* spp.), and numerous cacti characterized much of the bighorn habitat in Baja California and Sonora, Mexico.

WATER

The importance of surface water to desert bighorn has been investigated by many (Turner 1973, 1979b; Campbell and Remington 1979; C. L. Douglas and White 1979; Leslie and Douglas 1979; Sandoval 1979a; Witham and Smith 1979; Turner and Weaver 1980). Desert bighorn can survive on preformed water found in their food and metabolic water formed as a result of oxidative metabolism. However, increased day length, high ambient temperatures, reduced forage moisture content, and mating activities during summer require additional intake of water, resulting in a dependence on surface sources of water (Turner 1979b, Turner and Weaver 1980). During summer, bighorn may go without water for 5 to 15 days, resulting in a loss of more than 20% of the hydrated body weight (30% total body water; Turner 1979b).

Desert bighorn require a minimum of 4–5% of their body weight in water per day during summer and 1–2% of their body weight in water during winter, although the amount of water consumed by individuals varies. Turner (1970) measured amounts of water consumed by bighorn at a water source during midsummer in the Santa Rosa Mountains, California. He found that during one visit to the water source, a male consumed the largest amount (18.7 L), representing 23% of his estimated body weight. However, consumption of large amounts of water was not confined to older males. A female, a male lamb, and a four-year-old male were found to consume more than 20% of their estimated body weights.

A close relationship between bighorn distribution and water sources has been quantified in southeastern Utah (Irvine 1969), the River Mountains, Nevada (Leslie and Douglas 1979), and the San Andres Mountains, New Mexico (Sandoval 1979a). Irvine (1969) observed 82% of the sheep within a

1.6-km radius of water during summer. Leslie and Douglas (1979) found 84% of the sheep within a 3.2-km radius of water during summer. Sandoval (1979a) compared bighorn distribution and habitat availability in relation to water sources by calculating the mean distance from each 200-m grid within the study area to the nearest available water source used by bighorn. During summer, 70% of the sheep were observed between 400 and 1,500 m from water, yet only 24% of the habitat was situated this same distance from water. During winter, 93% of the sheep were observed within 2,000 m of water, yet only 37% of the habitat was situated this same distance from water. In the San Andres Mountains, New Mexico, bighorn sheep were not significantly closer to water in summer versus winter (Sandoval 1979a). The probability of randomly locating sheep more than 1,000 m from water was 0.33 during winter and 0.41 during summer. These data suggest that the water requirements of this population were more closely related to vegetative succulence than to ambient temperature (Sandoval 1979a).

This contradiction to the available data on water requirements of desert bighorn may be explained by different climatic regimes in the vast region occupied by desert sheep. The San Andres Mountains receive greater precipitation, contain more sources of surface water and experience lower annual mean temperatures relative to many ranges occupied by desert bighorn. In contrast, the Sonoran, Mojave, and Colorado Deserts are generally characterized by wet winters and dry, hot summers. This subjects bighorn to a water imbalance during the period of maximum heat and water stress. Because vegetative succulence is dependent on available soil moisture, under dry climatic conditions desert sheep cannot obtain preformed water normally found in the diet. This results in a greater affinity to water sources during the hot and dry summer months. However, Alderman et al. (1989) reported the moisture content of key forage species was greater than 32% for each hour of the day for each season. They did not detect relationships between activity patterns of desert bighorn sheep and moisture content of plants.

Turner (1973) reported that desert bighorn must have free water above that received from vegetative succulence. The loss of water from normal body functions was not offset by water gained through forage intake and oxidation. He concluded that bighorn may subsist on preformed water when their diet contained 1.0 to 1.5 mL (0.03 to 0.05 fluid ounces) of water per gram dry weight of forage.

The hypothesis that desert bighorn can exist without permanent sources of water is not widely accepted. Nevertheless, in the Big Hatchet Mountains, New Mexico (Watts 1979), and northwestern Sonora, Mexico (Mendoza 1976), there is no record of bighorn utilizing free water. Watts (1979) found a high occurrence of cactus in the bighorn diet, which he attributed to an adaptation aimed at increasing the amount of preformed water in the diet during periods of greatest heat stress. In the Coxcomb, Pinto, Hexie, Cottonwood, and Granite Mountains, California, bighorn also have survived on temporary pothole water (Weaver and Hall 1971, Weaver 1973). Krausman et al. (1985) also documented sheep existing on ranges in southwestern Arizona without permanent sources of free-standing water. Sheep did not use water catchments to any extent, even when they were added in the Little Harquahala Mountains, Arizona (Krausman and Etchberger 1993, 1995). Further, Broyles (1995) questioned the value of artificial water sources for bighorn sheep in the Cabeza Prieta National Wildlife Refuge, Arizona.

Home Range and Movements

Movements of desert bighorn sheep cannot be attributed to a particular factor, since movements are dictated by the individual's response to a variety of stimuli (Leslie 1977). Robinette (1966) reported that movements and home range in mule deer (*Odocoileus hemionus*) were affected by age, sex, heredity, population density, topography, season, food and water availability, cover, and reproductive activities. These same generalities can be applied to desert bighorn. Males travel more widely than females, and the home range of lactating females is considerably reduced.

Through radiotelemetry monitoring, movements and distribution of select desert bighorn populations have been studied in Utah (Bates et al. 1976, Dalton et al. 1978, Jense et al. 1979, King and Workman 1982, Bates and Workman 1983), Nevada (Leslie 1977, Leslie and Douglas 1979), New Mexico (Sandoval 1979a, Watts 1979, Munoz 1981, Elenowitz 1983), Arizona (Witham and Smith 1979, Chilelli and Krausman 1981, Krausman et al. 1989), and California (C. L. Douglas and White 1978, 1979; DeForge 1980). These studies indicate that the home ranges of desert bighorn usually consist of a summer range close to permanent water, a fall-winter range that can overlap with the summer range, and a spring range characterized by precipitous topography close to water. These studies also indicate that behavioral, physiological, and environmental factors influence movements of desert bighorn. Home range knowledge, water and forage availability, lambing and mating activities, season, topography, and age and sex differences were found to influence movements.

In Utah, Jense et al. (1979) found that the mean home range was 61 km^2 for males and 24 km^2 for females. Furthermore, movement between relocations of radiocollared sheep was larger during each season of the year for males than for females. Relocation distances indicated that fall movements of males exceeded movements during winter and summer, but these movements did not exceed spring movements. Fall movements of females exceeded those of all other seasons and spring movements exceeded movements during the winter.

Witham and Smith (1979) analyzed movement patterns of radiocollared bighorn sheep in western Arizona to determine areas of range overlap using home ranges as an indicator of sheep concentration. They found that areas of greatest home range overlap corresponded with perennial water sources, areas of spring female-lamb concentration, or both. Male movements, particularly during the mating season, encompassed a series of several mountain ranges, with a maximum straight-line distance traveled of 56 km. In contrast, the apparent lack of long-range movements and home range loyalty of female-juvenile groups in contiguous sheep habitat was attributed to limited and widely dispersed perennial watering sites. Krausman (1985), Krausman et al. (1989), and Krausman and Etchberger (1993) also concluded that the large home ranges used by desert bighorn sheep (e.g., > 40 km^2 in summer) were in response to widely scattered resources.

Home range and movement patterns have been altered by providing artificial water sources, mineral licks, and by supplementing remnant populations with additional sheep. The reestablishment of an introduced female-juvenile population in an unoccupied portion of the Big Hatchet Mountains, New Mexico, resulted in the native males expanding their summer-fall range by approximately 9.6 km^2 (Bavin 1980). The construction of water sources in the Buckskin Mountains, Arizona, allowed bighorn to use previously dry portions of their range on a perennial basis, and was effective in manipulating water-related movements from areas of adverse impact (Campbell and Remington 1979).

Migration patterns vary considerably; most populations remain in isolated areas whereas others move considerable distances between mountain ranges. Migratory patterns of desert bighorn may be classified as seasonal elevational movements within the same range, long-distance annual migration between mountain ranges that include elevational movements and dispersal from water sources, and dispersal from and return to water sources depending on the time of year (McQuivey 1978).

Throughout most of the year, female-juvenile and adult male groups remain spatially and sexually segregated. Females have a relatively small home range. In contrast, adult male groups usually have larger ranges (Krausman et al. 1989, J. E. Scott et al. 1990). Most desert bighorn populations are constricted during the hot summer months due to less than optimum water distribution. Therefore, local rainfall patterns affect home range and movements of desert bighorn. Summer showers that fill natural water catchments allow sheep to use areas not normally available during dry periods.

Diet

Studies of food preferences and diet of desert bighorn provide critical information necessary for better management. Data on nutritional requirements, the impact on the vegetation portion of the habitat, and the degree of competition based on dietary overlap with native and exotic wildlife are now available.

Diets of desert bighorn have been determined by direct observation (L. O. Wilson 1968), inferred observation (Irvine 1969), stomach analysis (Barrett 1964; Bradley 1964; Yoakum 1964; Irvine 1969; K. W. Brown et al. 1975, 1977; Sanchez 1976), and analysis of fecal material (Walters and Hansen 1978, Sandoval 1979a, Watts 1979, Seegmiller and Ohmart 1981, Krausman et al. 1989, Bleich et al. 1997). L. O. Wilson (1976) found it difficult to correlate food preferences to forage availability due to insufficient sample size. Combining diet data of different sex and age classes also could result in significant biases.

The correlation between digestibility and energy intake of dry matter has been investigated with captive bighorn (Krausman et al. 1988). The relationship corresponded with results obtained with domestic sheep and patterns observed for other ungulates (Krausman et al. 1988). Mazaika et al. (1992) found similar results with captive bighorn. The nutritional composition of forage of desert bighorn in western Arizona has been tabulated for 32 plant species (Seegmiller et al. 1990). Woody and succulent plants contained the highest levels of protein throughout the year, followed by forbs and grasses, respectively (Seegmiller et al. 1990). In the Santa Catalina Mountains, Arizona, forbs were consistently higher in protein than browse or grasses in all seasons (Mazaika et al. 1992). Morgart et al. (1986) and Bleich et al. (1992a, 1997) describe the quality of forage consumed by desert bighorn in the Mojave Desert.

INVESTIGATIONAL BIASES

Direct observation entails observing bighorn while they feed and recording the plant species eaten. The major problem with direct observation is that the investigator is usually biased toward recording browse. When a sheep is browsing, the forage is readily visible and identifiable; however, it is more difficult to recognize herbaceous species because vegetation, topography, or other sheep can obscure the observer's vision. Inferred observation refers to assumed use of plant species by bighorn that are close to feeding areas (B. M. Browning and Monson 1980). This method also is subject to numerous biases as noted by L. O. Wilson (1976).

Stomach analysis involves the collection of rumen contents that are usually obtained by sportsmen participating in monitored hunts. Rumen analysis of harvested sheep reveals only what a sheep has eaten recently and usually includes samples from males only; consequently, adequate data identifying seasonal food preferences and trends are not obtained. L. O. Wilson (1976) attributed differences in diets among females, lambs, immature males, and mature males to differences in plant availability and abundance, which he correlated with plant distribution and habitats used by the different age and sex groups.

Microscopic analysis of fecal material provides a valid index to diets of numerous ruminants (Free et al. 1970, R. M. Hansen 1971, Todd and Hansen 1973). This method has been used as the major technique to study diets of desert bighorn and other herbivores. It involves microscopic examination of discernible plant fragments (i.e., trichomes, pollen grains, silica, and epidermal cells) in fecal material collected in the field. The frequency of fragments of each plant species is then converted to relative frequency. Microscopic analysis of fecal material, however, does not provide an exact percentage of dietary composition because soft, fragile succulents and green forbs may be thoroughly digested and not appear in discernible amounts in the feces. Also, errors in plant identification may arise (Sandoval 1979a).

Researchers should be aware of other disadvantages of fecal analysis, including the high cost involved in laboratory analysis, the time involved in collecting only fresh bighorn droppings where bighorn and the ranges of other ungulates overlap, and the time involved in collecting and identifying the indigenous plant species to serve as a reference aid for laboratory microanalysis.

SEASONAL DIET CHANGES

Bighorn sheep are primarily grazers. However, considerable amounts of woody plants and forbs are consumed by desert bighorn. In some areas, they actively seek browse and forbs even when grasses are plentiful. Geographical and elevational differences in their habitat and availability of food plants, in addition to biases inherent in diet analysis methods, make comparisons difficult.

Desert bighorn are able to subsist on a variety of plants, and under adverse conditions they can subsist on desiccated and decadent forage. Available data suggest that desert bighorn are to a certain extent opportunists and widely adaptable foragers (B. M. Browning and Monson 1980). They can survive under varying environmental conditions as long as high-quality and palatable forage is available. Food selection can be attributed in part to habitat and climatic characteristics (i.e., parent material, plant phenology, and water availability). Areas of volcanic or granitic origin, especially those exhibiting high relief, are usually dominated by grasses, whereas areas of sedimentary origin are usually dominated by shrubs.

Water, including seasonal precipitation, is a major influence in food selection. Welles and Welles (1961) reported that seasonal diets of desert bighorn are determined to a great extent by water availability and rainfall and their effect on forage availability, quality, and palatability. Russo (1956) found that in Arizona seasonal growth was important in determining preferences, palatability, and forage requirements of desert bighorn, but added that ephemeral plant species used by desert sheep represent a small part of their diet. Because ephemeral plants grow only under specific climatic conditions, the benefit realized from the temporary use of these plants is probably negligible.

Grasses comprised the bulk of the diet of desert bighorn examined by Yoakum (1964), Barrett (1964), Bradley (1964), and K. W. Brown et al. (1977). However, Sandoval (1979*a*), Watts (1979), Seegmiller and Ohmart (1981), Bavin (1982), Ginnett and Douglas (1982), King and Workman (1982), Elenowitz (1983), and Holt et al. (1992) found that shrubs comprised the majority of preferred foods. In contrast, Sanchez (1976) reported that forbs constituted the bulk of the fall diet of desert bighorn in Baja California, Mexico. Cunningham and Ohmart (1986) reported grasses as only 2% of the diet in Carrizo Canyon, California, with the rest comprised of 57% shrubs, 32% forbs, and 4% cacti. This is similar to the findings of Krausman et al. (1989) in the Harquahala and Little Harquahala Mountains, Arizona, where browse dominated the diet and less than 5% was comprised of grasses. Holt et al. (1992) found that shrubs comprised 92% of the summer diet in the Superstition Mountains, Arizona. In the Dome Rock, North Plomosa, and Kofa Mountains, Arizona, annual diets of sheep averaged 64, 23, and 13% browse, forbs, and grasses, respectively (G. D. Miller and Gaud 1989). However, in the Pusch Ridge Wilderness, Arizona, grasses comprised greater than 30% of the diet in all seasons of the year, followed by forbs and browse (Mazaika et al. 1992). In the Virgin Mountains, Arizona, the diets of sheep consisted of 53.5% forbs, 27.3% browse, 16.6% grasses, and 2.4% succulents (D. R. Smith and Krausman 1987).

In an enclosure in Nevada, desert bighorn selected forbs and grasses and avoided browse in all seasons (Zine et al. 1992). In winter, females used more forbs than males; however, sites used by females had more forbs available than sites used by males (Zine et al. 1992). Bleich et al. (1997) also reported different use of forage by males and females. During 1987–1988, males ate more annuals, shrubs, and succulents than females, and females ate more grasses and forbs than males when they were segregated. During aggregation, males ate more annuals and grasses than females, and females ate more forbs, shrubs, and succulents than males.

K. W. Brown et al. (1977) analyzed rumen contents (primarily from legally harvested males) during fall and winter from 17 different areas in Nevada. He identified 120 plant species consisting of 17 grasses, 61 shrubs, and 42 forbs, representing 37 families. Rumen composition was 64.3% grasses, 28.2% shrubs, and 6.5% forbs. B. M. Browning and Monson (1980) reported numerous species used by desert bighorn and that grasses were generally preferred, with more than 70 species represented.

In the San Andres Mountains, New Mexico, Sandoval (1979a) found that preferred habitats had 41% ground cover, with shrubs comprising 24%, grasses 14%, and forbs 3%. Diets determined by microscopic analysis of fecal material revealed that 25% of the plants collected ($n = 181$) were in their diet. Forty-two percent of the plants consumed were shrubs or half-shrubs, whereas forbs and grasses each constituted 27% of the diet. Mountain mahogany *(Cercocarpus montanus)* was the most frequently consumed plant, comprising 48% relative density of the annual diet.

Seasonal use of different forage classes was attributed to precipitation patterns. The decline in forb consumption during early summer was correlated with below normal precipitation for this period. In contrast, the increase in forb use during late summer corresponded with high levels of precipitation during this period. The decrease in browse consumption during summer and early fall was found to be inversely related to use of forbs and grasses. However, the decline in use of browse also could be due to a decline in palatability. Decreased palatability of browse might be due to new growth becoming coarse, woody, and possibly exceeding an acceptable diameter for bighorn sheep (Sandoval 1979a).

Saguaro, prickly pear *(Opuntia* spp.), pincushion cactus *(Mammillaria* spp.), and barrel cactus *(Ferocactus* spp.) have been reported in the diet of desert bighorn (Simmons 1963, Sanchez 1976, Sandoval 1979a, Watts 1979, B. M. Browning and Monson 1980, Seegmiller and Ohmart 1981, Krausman et al. 1989). Watts (1979) reported that prickly pear was the third most common food item used in the Big Hatchet Mountains, New Mexico. Cacti were found in 10 of 11 monthly fecal samples, comprising 18% of the annual diet.

The overlap in diets between desert bighorn sheep and sympatric ungulates has been examined in a number of locations. Dodd and Brady (1988) compared the diets of sheep and cattle in Aravaipa Canyon, Arizona, and found a similarity of 39%; however, annual use of all forage classes, except cacti, was different. Overall, sheep preferred shrubs, cattle preferred grasses, and the potential for competition appeared to be low (Dodd and Brady 1988). In the Harquahala Mountains, Arizona, desert bighorn, desert mule deer *(O. h. crooki)*, and burros had similar forage species in their diets; however, dietary overlap between the animals was insignificant and competition was not observed (Krausman et al. 1989).

MINERAL REQUIREMENTS

It is not unusual for bighorn sheep to eat periodically small quantities of soil during certain times of the year. However, little is known regarding their mineral requirements. The occurrence of large quantities of clay in Rocky Mountain bighorn *(O. c. canadensis)* droppings in Wyoming was attributed to sodium and phosphorus deficiencies (Honess and Frost 1942). Packard (1946) reported bighorn sheep ingesting mud containing high concentrations of calcium in Rocky Mountain National Park, Colorado. Similar observations were reported by D. R. Smith (1954) in Idaho. During a six-week period, captive bighorn in Texas used two 15-kg blocks of Moorman nitrate salt (Hailey 1964). L. O. Wilson (1968) reported bighorn consuming large amounts of phosphorus-enriched clay in southeastern Utah. Sandoval (1979a) analyzed mineral lick soil from the San Andres Mountains, New Mexico, and showed the soil contained high sodium and calcium concentrations. However, bighorn there would not use sulfurized and plain salt blocks (Sandoval 1980). Watts (1979) found that in the Big Hatchet Mountains, New Mexico, desert bighorn frequently crossed 4 km of flat, open land to use mineral licks, mainly for their sodium content. He attributed these movements to an increased sodium appetite in response to increased fecal sodium loss and correlated mineral-lick use with maximum occurrence of emergent vegetation. In the San Gabriel Mountains, California, Holl et al. (1980) observed that approximately 90% of the bighorn were less than 2,000 m from known mineral licks. During spring and summer, 73% of the sheep (n = 292) were actually observed using the licks. The elements measured in the mineral licks were ranked in order of decreasing availability as calcium, magnesium, sodium, potassium, and chloride. At nonlick sites, the mineral content of the soil was ranked as calcium, magnesium, potassium, sodium, and chloride. They concluded that sodium was the main attractant and that the location of mineral licks may affect distribution and habitat use by bighorn sheep, particularly during spring and summer.

Sodium is a required element for all mammals, comprising approximately 0.1% of the live weight (Botkin et al. 1973). Sodium is the main constituent of blood plasma and is important in osmotic regulation of body fluids. This great sodium requirement of mammals, coupled with low sodium levels in plants

(Weeks and Kirkpatrick 1976), forces many herbivorous mammals to supplement their sodium intake by soil ingestion (Wiener 1975).

Data on sodium requirements of bighorn sheep are not available. The National Academy of Sciences (National Research Council 1964) recommended a daily intake of 9 g of salt per day for a 45 kg domestic female sheep. However, domestic stock have been found to actually have lower sodium requirements than that recommended by the National Research Council (McClymont et al. 1957).

Growth

BODY WEIGHT

Desert bighorn are relatively short-legged, blocky animals, but not as heavy bodied as the Rocky Mountain bighorn (C. G. Hansen 1980*a*). Males are normally larger than females, although some females may be taller and have a larger body than the smallest males.

Weight varies seasonally. Rams weighing 90.7 kg in early summer may weigh only 64.0 kg by winter. The poorer condition of males compared to females during the fall season can be attributed to stress from mating activities and to deteriorated range conditions during late summer and early fall, particularly in the Sonoran and Mojave Deserts. Males continue to gain weight through eight years of age, with a slight decrease in weight after age 10. This increase in weight may be attributed to increased horn size, whereas weight loss after age 10 is attributed to poorer body condition resulting from old age.

Live weight of males (n = 20) from the Desert National Wildlife Range, Nevada, averaged 70.8 kg (range = 57.6–82.0 kg). Dressed weight of harvested males (n = 25) from the same area averaged 47.4 kg (range = 38.5–56.2 kg). Weight loss of field-dressed males averages 33% (range = 28–38%; C. G. Hansen 1980*a*). The average weight of females (n = 15) from the Desert National Wildlife Range was 43.7 kg (range = 33.5–51.7 kg; Aldous et al. 1958). In Utah, mature males estimated to weigh 77.1 kg weighed 58.9 kg field dressed (John 1968). Caution must be exercised when analyzing data collected from harvested animals because the size is probably skewed toward large males (i.e., trophy animals) and is not representative of the population.

In the Kofa National Wildlife Refuge, Arizona, the average live weight for seven males was 78.2 kg (range = 65.0–90.7 kg). For five adult females, the average live weight was 52.2 kg (range = 45.8–57.1 kg; Russo 1956). In the San Andres Mountains, New Mexico, the average live weight of five adult males in early fall was 67.1 kg (range = 52.1–102.0 kg). Females (n = 37)

Figure 4.2. Ram horns are curling, massive, and diverging, consisting of a bony core with a permanent corneous sheath formed by the deposition of keratin. (Photograph courtesy of Nevada Division of Wildlife.)

during the same period weighed 50.9 kg (range = 36.2–61.2 kg). This population was experiencing a psoroptic scabies *(Psoroptes ovis)* epizootic, so the weight of these animals may not be indicative of otherwise healthy individuals (Sandoval 1980).

C. G. Hansen and Deming (1980) summarized the weights of desert bighorn from a number of locations and found that the average adult female weighed 48 kg and the average adult male weighed 82 kg. Remington (1982) concluded that the live weights of male and female desert bighorn from northern and southern Arizona were not different; these weights were consistent with the summary of C. G. Hansen and Deming (1980).

HORN GROWTH AND DEVELOPMENT

Both sexes possess horns, and considerable variation exists in horn size, shape, and growth rates. Bighorn sheep horns are nondeciduous and grow throughout life. Male horns are curling, massive, and diverging (fig. 4.2). In contrast, female horns are relatively thin, curved, and much smaller. The

Figure 4.3. One-year-old ram (second from right) appears equivalent in size to adult ewes (third and fourth from right). The most obvious characteristics of males are the presence of external genitalia; the shorter rostrum, which gives the head a blocky appearance; and the thicker horn base. (Photograph by James R. DeForge.)

horns project upward and outward from the head, forming a sickle (fig. 4.3). The horns consist of a bony core, or osicone, with a permanent corneous sheath formed by the deposition of keratin. The horn sheaths are epidermal in origin and grow from the base and from the inside around the long osicones of the frontal bones (Geist 1971).

Horn growth is variable. Maximum growth for females occurs at three to four years and for males at five to six years. In females, the horns grow rapidly for the first two years, reaching a length of 30 cm. After the second year, the rate of increase in horn diameter declines rapidly, becoming insignificant by the fourth year (L. R. Baker 1967, Geist 1971). In males, horns may reach a length of 30 cm in the first year. Three-year-old males possess horns greater than 51 cm along the outside curl, whereas mature male horns may be greater than 89 cm. Horn diameter in males increases rapidly for the first three years. After the fifth year, there is a pronounced decline in the rate of increase; growth appears to become negligible by the sixth year.

Growth rings (annuli) are formed during fall and are associated with hormone and diet changes. Because lambs are born during spring, age estimates during fall surveys or from hunter returns should be shown in 0.5-year increments. Annuli are usually absent or not well defined at 1.5 years of age. The first visible annulus is formed at 2.5 years and consists of a series of several growth rings close together. The fourth-year annulus is normally the first heavy, single, dark ring formed. The distance between the annuli becomes less each year until 7 or 8 years of age. Thereafter, the annuli become relatively compressed. The new horn growth in young males is blue-gray in color and relatively smooth. A protective knob of erectile tissue develops posteriorly at the base of male horns and becomes distinct as males age (C. G. Hansen and Deming 1980). The knob expands during prolonged butting bouts (Welles and Welles 1961), suggesting that it serves a shock-absorbing function.

Horns tips often become broken, splintered, or broomed. Broomed horns in males are usually the result of intrasexual clashes and horning of shrubs and rocks. Converging, tightly curled horns are usually broomed more extensively than diverging horns. The former configuration probably restricts visibility, and brooming might alleviate this (Sandoval 1979a). L. R. Baker (1967) found that males 10 to 12 years of age usually have broomed 95% of their first year's horn growth and some individuals may broom into the second year's growth. Brooming in older individuals commonly exceeds horn growth, resulting in a reduction of horn length with age. Frequently the widest tip-to-tip spread is found in animals younger than six years of age.

LAMB GROWTH AND DEVELOPMENT

Pregnancy has been diagnosed in desert bighorn females using real-time ultrasound (Bunch et al. 1986). Ultrasound techniques were 100% accurate in diagnosing pregnancy after 35 days gestation in adult females (Bunch et al. 1986). Progesterone levels also are a positive test for pregnancy, providing that samples are collected at a time when the breeding season is over and most females are either pregnant or anestrus (Bunch et al. 1986).

At birth lambs weigh approximately 2.9 kg. Lambs grow rapidly and by six months of age weigh approximately 29 kg. Except for facial characteristics, one-year-old males appear equivalent in size to adult females; the most obvious differences are a thicker horn base and a shorter rostrum for the male, which gives the head a block appearance. Yearling males stand about 91 cm at the shoulder and weigh between 52 and 59 kg. Yearling females are noticeably smaller than adult females, standing about 69 cm at the shoulders and weighing between 41 to 50 kg.

A complete set of teeth is acquired by age 4 (Cowan 1940, Dalquest and Hoffmeister 1948). However, considerable variation from the normal dental formula, frequency of damaged and lost teeth, and other aberrations of the tooth row have been reported by Deming (1952), Allred and Bradley (1965), and Bradley and Allred (1966). Deming (1952) reported maxillary canines in 4 of 11 lamb skulls from the Desert National Wildlife Range, Nevada. Bradley and Allred (1966) found vestigial upper canines in 9.1% of adult males from Arizona and concluded that vestigial canines were more prevalent in adult males than in females. The incidence of absent second premolars was 29% from Nevada and 19% from Arizona.

ISOLATION AND THE MOTHER-YOUNG BOND

The initial female-lamb bond is established soon after birth when the mother ingests the placenta and licks the newborn lamb (Etkin 1964). There is pronounced tendency for lambs to remain together during the first part of the lambing season (Welles and Welles 1961). As lambs increase in size and numbers, females will often leave their lambs alone or with a foster mother for extended periods during the day (Welles and Welles 1961, L. O. Wilson 1968, C. G. Hansen and Deming 1980). These nurseries consist of one female remaining with the lambs while the other females continue their daily activities. This system allows females to seek high-quality, palatable forage. However, L. O. Wilson (1968) suggested that this behavior trait might also enhance lamb survival by minimizing exposure of lambs to potential predators.

PELAGE

Considerable variations in color of pelage exist within and among populations on the Desert National Wildlife Range, Nevada. Bighorn are generally medium gray-brown with white on the rump, back of the legs, and the muzzle. Females do not exhibit wide variation in color. In some areas females are grayer than brown, whereas in other areas the opposite is true. Males may be blackish or even rusty red. The amount of white on the rump, legs, and nose also varies (C. G. Hansen 1980a). In Utah, desert bighorn are generally dark slate gray on the neck, back, and sides, whereas the brisket tends to be slightly darker (L. O. Wilson 1968). Male horns vary in color from a yellowish brown to a dark brown.

Albino desert bighorn have been reported from the Spring, Eldorado, and Muddy Ranges of Nevada (McQuivey 1978), Grand Canyon National Park, Arizona (Walters 1979), and Baja California, Mexico (Cossio 1976). C. G. Hansen (1980a) reported white-spotted bighorn from the Pintwater Range, Nevada. Examinations of photographs of white desert sheep show pigmentation on the nose pad, lips, and eyes, indicating they are not true albinos but rather a genetic strain endemic to that area (McQuivey 1978).

Young lambs are grayish with light brown hair on their neck. The "mane" intergrades into a light brown stripe along the top of the back extending to the tip of the tail. The rump is light cream colored and is not strikingly dissimilar from the remainder of the back (C. G. Hansen and Deming 1980).

The pelage is composed of a dense outer layer of coarse, brittle hair underlaid with a loose layer of fine fleece. The hair grows longer on the sides and top of the body; when combined with the undercoat of fleece it is quite dense. Hair length varies from sparse, short hairs a few centimeters long on the face and the junction of the legs and body to several centimeters long on the nape. The body fleece appears to be a fine quality wool about 1.3 cm long. It is kinky-curly and when stretched can be 2.54 cm long. The fleece is shed with the hair during the annual summer molt and is not replaced until the fall season (C. G. Hansen 1980*a*).

During the summer, fat deposits and hair function as insulation from heat, and the molt pattern may reflect this (C. G. Hansen 1980*a*). During winter, subsequent to the mating season, most males are relatively thin; therefore, pelage probably is the primary source of insulation.

MOLTING

Molting usually takes place during June and July. However, in warm years, some sheep may complete molting by May. During other years, molting may take place during late summer. This variation has been attributed to environmental factors (L. O. Wilson 1968), age, and physical condition of animals (C. G. Hansen 1980*a*). Sheep from lower and more xeric areas shed before those living in higher, more mesic habitats. In southeastern Utah, L. O. Wilson (1968) found that during years of average precipitation, acquisition of the summer coat was completed in July. During drought years, molting occurred much later.

Molting progresses from the posterior dorsal region, with the chest region being the last to shed. Molting pattern may be an indication of the age and condition of sheep. Degree of molting has been used to distinguish individual animals during waterhole counts in the Desert National Wildlife Range, Nevada. C. G. Hansen (1980*a*) found that healthy males and barren females shed first and old or diseased animals shed last.

Activity Patterns

The activities of desert bighorn are determined to a large extent by the availability of food and water, which is governed more by pluvial cycles than by calendar years. Although it is difficult to generalize about daily activities, desert bighorn are basically diurnal, with activity beginning at dawn. Feeding periods separated by periods of resting and loafing characterize the daily pattern (L. O. Wilson 1968, Augsburger 1970, Olech 1979, Sandoval 1979*a*, Chilelli and Krausman 1981).

Augsburger (1970) described and quantified the behavior of desert bighorn in the San Andres Mountains, New Mexico, and found that desert bighorn behavior patterns were similar to those described by Geist (1971) for northern mountain sheep. Three social behavior patterns described by Geist (1971), the neck fight, head shake, and ejaculation, were not observed by Augsburger (1970). Augsburger (1970) surmised that these behavior patterns may have been exhibited and were not observed or that they were not performed by desert bighorn. However, the head shake and ejaculatory display patterns have been observed in desert bighorn by Welles and Welles (1961) and Sandoval (1979*a*). Head shaking was displayed by subordinate to dominate males and between subordinate males. The ejaculatory posture was displayed by males in both bachelor and mixed groups during the early phases of the rut. No other significant variations of individual behavior have been reported for desert bighorn.

L. O. Wilson (1968), Lopez and Orihuela (1976), Olech (1979), Sandoval (1979*a*), and Chilleli and Krausman (1981) found that desert bighorn in Utah, Baja California, California, New Mexico, and Arizona, respectively, exhibited similar daily activity patterns. However, Olech (1979) noted many variations in summer activity for different group types in Anza-Borrego Desert State Park, California. Male groups became active around 0600, and this period of activity continued until approximately 0845. The second and highest activity periods were between 0900 and 1130, during which time males usually watered. A maximum of 23% of the individuals observed at a given time were active in male groups. Mixed groups became active at approximately 0915. This activity period lasted until around 1330 and included a high level of interactions (i.e., courtship) and maintenance activities (e.g., moving, feeding, drinking, and standing). A peak of 80% activity was observed at 1145. Ewe-juvenile groups usually became active around 1545. The primary activities included foraging and moving to and from water. A maximum of 48% activity was recorded at 1400.

Periods of little activity were most apparent during the summer and least apparent during winter. Increased activity during winter could be attributed to greater effort needed to obtain a given quantity of forage, higher dietary requirements due to lower nutritional quality of forage, and fewer daylight hours in which to obtain food.

In the Harquahala Mountains, Arizona, Chilleli and Krausman (1981) noted that daily activity patterns changed seasonally, primarily in bedding and feeding peaks. During winter, feeding was the primary activity from 0545 to 0755 and from 1400 to 2005. Chilelli and Krausman (1981) found two periods, 0900 to 1055 and 1200 to 1355, in which bedding comprised more than 50% of the winter activity. During summer, the majority of feeding was done prior to 0800 and after 1600. Sheep bedded more than 50% of the time between 0800 and 1155 and between 1400 and 1555. Aggression, contact greeting, and dominance displays were seen 1% of the time. During late summer, sheep fed from 0545 to 0755 and from 1500 to 2005 and feeding was interspersed with inactive periods of bedding and standing. From 0800 to 1455, when sheep were not bedded, their main activity was standing. Chilleli and Krausman (1981) interpreted these inactive periods as an adaptive strategy for conserving energy during the hot summer months. During autumn, they found that feeding was the primary activity prior to 0900 and after 1400. Only between 0900 and 1055 did bedding constitute more than 50% of the activity.

Social interactions observed by Olech (1979) and Chilelli and Krausman (1981) between desert bighorn were less intense than those described by Geist (1971) for northern mountain sheep. Chilleli and Krausman (1981) found that sexual, aggressive, contact greeting, dominance display, and resource competition (the act of pushing another animal away from food, water, shade, or other resources) behaviors generally constituted less than 1% of the activity within each hour. Berger (1978a) attributed this lower behavioral activity level in desert bighorn to smaller group sizes, less social facilitation, and reduced play among lambs.

Nocturnal activity is another method by which desert bighorn can avoid solar radiation, minimize water needs, and still obtain sufficient forage. In the Cabeza Prieta National Wildlife Refuge, Arizona, Simmons (1969) found that during hot, dry summer months desert bighorn fed at night to combat heat gain and to minimize energy-expending activities. Krausman et al. (1985) reported on the diel activity of ewes in the Little Harquahala Mountains, Arizona, during the hottest part of the year and found sheep active 34% of any hour. In the Little Harquahala Mountains, Arizona, Alderman et al. (1989) examined diel activity of females relative to the moisture content of preferred forages, but could find no correlation. However, the moisture content of forage species was high during each hour of the day in all seasons and may not have been a factor in selection (Alderman et al. 1989).

Monson (1964) and Simmons (1980) reported nocturnal movement by desert bighorn in Arizona. L. O. Wilson (1968) reported that desert bighorn in Utah seldom moved on dark nights, but occasionally moved on moonlit nights. However, in Death Valley, California, Welles and Welles (1961) reported little nocturnal activity. Sandoval (1979a) obtained an index of nocturnal activity by monitoring radiocollared sheep and concluded that bighorn are primarily diurnal but that some activity is nocturnal. During some sessions, activity was recorded more than two hours after dark with activity periods throughout the night; however, renewed activity would not necessarily begin at dawn. Other sessions revealed that sheep bedded before dark, which was followed by little or no activity throughout the night, and then activity commenced before dawn. Sheep were usually relocated less than 300 m from where last seen the previous nightfall; however, movements greater than 1,000 m were documented on moonlit nights.

In the Peloncillo Mountains, New Mexico, Elenowitz (1983) documented nocturnal movements by direct observation on moonlit nights and by radio-signal fluctuations on dark nights. Sheep fed and engaged in cross-country travel during both moonlit and dark nights. The greatest nocturnal movement recorded was 5 km during a dark night.

McCutchen (1987) compared activity patterns of females in a captive herd at Zion National Park, Utah, prior to and after release. In both cases, feeding dominated the diurnal period with short, intermittent periods of bedding. Feeding activity primarily was crepuscular, and there was no evidence of nocturnal activity (McCutchen 1987).

Social Organization

SEXUAL GROUPING

Bighorn, particularly males, are gregarious and maintain a rigid social hierarchy; however, group integrity remains flexible throughout the year (Leslie and Douglas 1979). In general, group composition consists of spatially and sexually segregated groups of males, females, lambs, and offspring from the previous one or two years. During the mating season, adult males join female-juvenile groups for the duration of the rut, which usually takes place within the females' home range. Segregation of the sexes reduces intraspecific competition for forage and the nutritional stress of lactation resulting from male-female interaction. Sexual segregation also likely results from the different reproductive strategies of males and females. By using habitats with superior forage, males may enhance their fitness while incurring greater predation risks than do females. Ewes appear to enhance their fitness by selecting protected sites and thus minimizing risks to their offspring (Bleich et al. 1997).

Bighorn sheep may be classified into four group types according to their composition: (1) mating or mixed groups consisting of mature females (older than two years) and mature males (older than four years); (2) female-juvenile groups consisting of adult females accompanied by lambs and yearlings, however, they also may contain socially immature males younger than three years of age; (3) male groups, usually more than four years of age; and (4) groups of barren adult females, yearlings, and socially mature males (Sandoval 1979*a*).

GROUP INTEGRITY

Persistent family association in desert bighorn has been reported by Welles and Welles (1961), L. O. Wilson (1968) and Elenowitz (1983). However, persistent family groups, related individuals, and home range groups (individuals that may or may not be related sharing a common home range) were not evident in the River Mountains, Nevada (Leslie and Douglas 1979), or Harquahala Mountains, Arizona (Chilelli and Krausman 1981). Using marked females and their marked offspring, Leslie and Douglas (1979) found that males begin to disassociate from their mothers earlier than female lambs. Termination of persistent association was related to mating activities. Leadership also was found to be highly variable in contrast to the matriarchal leadership reported by Welles and Welles (1961) and L. O. Wilson (1968). Leslie and Douglas (1979) attributed the lack of singular matriarchal leadership to the lack of group constancy and related this social strategy to the high density of sheep increasing the probability of group association and interchange. They concluded that persistent family groups may be a density-dependent phenomenon in that group integrity may be enhanced at lower population densities, where intergroup association is minimized. Group integrity at lower densities also would be an adaptive strategy because predator detection is enhanced when several individuals can forage while dispersed over habitat permitting good visibility (J. A. Bailey 1980). In the Little Harquahala Mountains, Arizona, group size of bighorn increased as terrain ruggedness decreased, and in the Harquahala Mountains, foraging increased with group size, which suggests antipredator strategies (Warrick and Krausman 1987).

GROUP SIZE AND SEASONAL VARIATION

Mean size of bighorn sheep groups could be indicative of habitat quality and population density. In Arizona, Simmons (1969) found that the average sizes of female and mixed groups were larger in the higher-quality Kofa-Ajo Mountains than in the Cabeza Prieta Range.

Leslie and Douglas (1979) documented a mean group size of 4.5 ($n = 305$) in the River Mountains, Nevada, after excluding observations at water sources, which attract concentrations of sheep during summer. Mean monthly group size ranged from 3.5 in December to 6.3 in April. The decline in mean group size was concurrent with a general decrease in forage availability and an increase in ambient temperatures. During periods of limited forage availability it would be advantageous for desert bighorn to disperse to minimize their impact on the vegetation. Mean group size did not vary significantly during the mating and lambing seasons. Female groups during the lambing season were larger than male groups during both periods due to the presence of lambs and socially immature males. During the lambing season, males were alone more often than were females. These fluctuating group sizes may be a behavioral response to changing habitat conditions in an unpredictable environment (Leslie and Douglas 1979).

Group sizes in the San Andres Mountains, New Mexico, ranged from 1 to 49 individuals with a mean of 5.9 ($n = 346$). Mean group size varied from 6.5 during the mating season to 5.6 during the lambing season. Male groups ranged in size from 1 to 11, and mean sizes varied from 1.9 during the mating season to 3.4 during the lambing season. Sandoval (1979a) attributed larger mean group size of males during the lambing season to an increased gregariousness of males as they migrated to their respective winter ranges. Mean female group size ranged from 5.0 during the mating season to 8.9 during the lambing season. The largest female groups were found during February, March, and April with means of 15.0, 9.1, and 8.8, respectively. Mixed groups formed during late summer and had a mean of 12.5 individuals. As mating activities subsided by December, the disassociation of males from female groups was pronounced, as revealed by abrupt changes in the mean group size between December (11.6) and January (7.8).

Reproductive Behavior

SEXUAL MATURITY

McCutchen (1976), Sandoval (1981), and Morgart and Krausman (1983) reported 1.5-year-old captive females giving birth and successfully rearing lambs in Utah and Arizona. In contrast, females from temperate populations of wild mountain sheep breed first at 2.5 years of age and have their first lamb at 3.0 years of age (Buechner 1960, Geist 1971).

Yearling males in captive populations in Utah (McCutchen 1976), California (Blaisdell 1976), and New Mexico inseminated all females present. This early sexual maturation in desert bighorn males was further substantiated by Turner (1976). He found that male lambs began spermatogenesis at 26 to 28

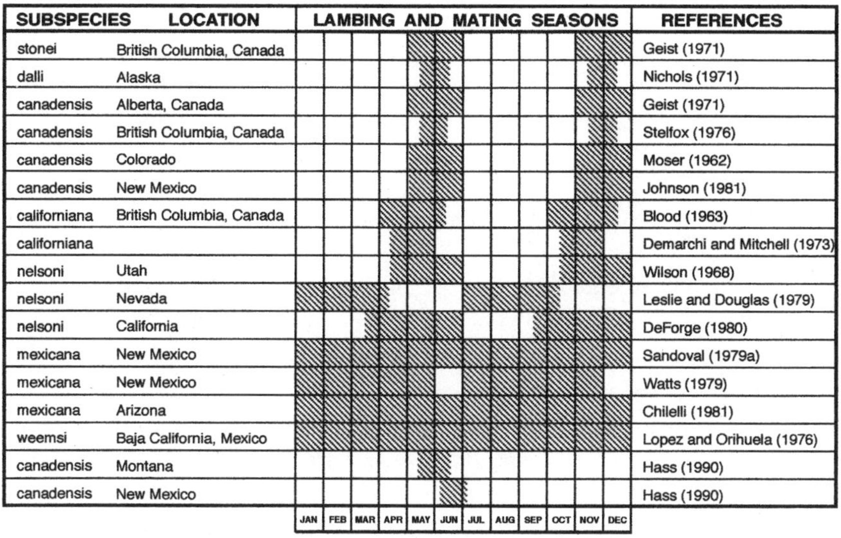

SUBSPECIES	LOCATION	LAMBING AND MATING SEASONS	REFERENCES
stonei	British Columbia, Canada		Geist (1971)
dalli	Alaska		Nichols (1971)
canadensis	Alberta, Canada		Geist (1971)
canadensis	British Columbia, Canada		Stelfox (1976)
canadensis	Colorado		Moser (1962)
canadensis	New Mexico		Johnson (1981)
californiana	British Columbia, Canada		Blood (1963)
californiana			Demarchi and Mitchell (1973)
nelsoni	Utah		Wilson (1968)
nelsoni	Nevada		Leslie and Douglas (1979)
nelsoni	California		DeForge (1980)
mexicana	New Mexico		Sandoval (1979a)
mexicana	New Mexico		Watts (1979)
mexicana	Arizona		Chilelli (1981)
weemsi	Baja California, Mexico		Lopez and Orihuela (1976)
canadensis	Montana		Hass (1990)
canadensis	New Mexico		Hass (1990)
		JAN FEB MAR APR MAY JUN JUL AUG SEP OCT NOV DEC	

Figure 4.4. Variation in mating and lambing seasons of wild sheep in western North America.

weeks and exhibited a seasonal spermatogenic cycle after reaching 21 months. Irvine (1969) evaluated the degree of testicular activity and reproductive activity of desert bighorn males ($n = 20$) from Utah, Nevada, and Arizona relative to age. He found no apparent decrease in spermatogenesis with increase in age and concluded that older males were capable of breeding. In wild populations, successful mating is not entirely a function of sexual maturation; social hierarchy and behavior play a major role (Geist 1971).

The estrous cycle in desert bighorn females lasts 28 days, with a receptive period of about 48 hours. Observations of captive females indicate that the gestation period is 179 ± 6 days (Turner and Hansen 1980).

MATING SEASON

Turner and Hansen (1980) defined the rutting season as that season in which mating activities result in 70% of the lamb production for the following season. The duration of the mating season is longer at lower elevations and southern latitudes and shorter at higher elevations and more northern latitudes (Thompson and Turner 1982; fig. 4.4).

Desert bighorn exhibit an extended mating period that encompasses several months (L. O. Wilson 1968, Lenarz 1979, Leslie and Douglas 1979, Sandoval 1979a). Witham (1983) studied desert bighorn reproduction in the

Sonoran Desert in southwestern Arizona and documented the birth of lambs in all months except October. However, DeForge (1980) found that in the San Gabriel Mountains, California, the duration of the mating season was short and lambing occurred later in spring than reported in most desert bighorn populations.

PARTURITION

It is generally accepted that parturition in bighorn, particularly in temperate populations, occurs during that time of year that offers the greatest opportunity for lamb survival (D. R. Smith 1954, Russo 1956, Welles and Welles 1961, Moser 1962, Geist 1971, Shackleton 1973, Thompson and Turner 1982). This hypothesis is based on the relationship of periodicity and predictability of forage production to the reproductive strategy of bighorn sheep. However, Lenarz (1979) and Sandoval (1979a) found that reproductive behavior was not seasonal in the Big Hatchet and San Andres Mountains, New Mexico, respectively. Apparently, periodicity and predictability of forage production was not the primary ecological factor determining the timing and duration of lambing. Lenarz (1979) attributed the nonseasonal reproductive behavior in the Big Hatchet population to a reduction in the postpartum interval and the relationship between cessation of lactation and estrus based on the interval between consecutive births. Lenarz (1979) concluded that the reproductive strategy exhibited by this population could represent a response to low population density.

Thompson and Turner (1982) found poor correlation between the inception and duration of the growing and lambing season for desert bighorn. They concluded that the extended lambing season is a manifestation of unpredictable precipitation patterns, which are essential for maternal and neonatal survival. The extended lambing season is a function of young females entering estrus for the first time having receptive periods displaced from the peak in estrus; late-estrous, older females; and reproductively mature, noncyclic lamb and yearling males servicing anestrous females.

Throughout much of the range of desert bighorn, plant productivity is related directly to temporal and spatial precipitation patterns that vary considerably. Nonseasonal reproductive behavior exhibited by some populations of desert bighorn is an adaptive strategy that ensures lamb survival during periods of varying and unpredictable forage production. An extended lambing season would increase the probability that late gestation and early lactation would coincide with a period of adequate precipitation and forage availability (Leslie and Douglas 1979, Sandoval 1979a, Thompson and Turner 1982).

However, J. A. Bailey (1980) questioned the adaptive significance of extended reproductive seasons in desert bighorn. He stated that perhaps their gestation period fits no seasonal precipitation pattern, so that mating and lambing cannot both coincide with any reasonably predictable seasons of adequate precipitation and forage availability. He further stated that prolonged mating and lambing seasons appear to be unsuited to the social system of desert bighorn because a prolonged rut brings estrous and lactating females and rutting males together. This association was considered by Sandoval (1979a) to be a major cause of lamb mortality in the San Andres Mountains, New Mexico. J. A. Bailey (1980) concluded that "the opposing selective forces of unpredictable precipitation which tends to lengthen the reproductive seasons, and of social disruption due to prolonged rutting, seem to create a situation in which the biology of bighorn does not fit the environment of the desert."

Population Dynamics

Populations of wild animals are dynamic with respect to total numbers and age structure. A thorough understanding of population changes under different environmental conditions and at different levels of population density, termed population dynamics, is the principal objective in population analysis. Some of the factors that influence population dynamics are intrinsic or fixed, such as the maximum reproductive potential and longevity of the species. Intrinsic factors can be quantified because they are usually constant and characteristic of a species. Extrinsic factors, such as weather, changes in habitat, and predation, are more difficult to evaluate. Reproduction and lamb survival in desert bighorn are determined to a large extent by precipitation and resultant quantity and quality of forage, the availability of free water, and disease.

FECUNDITY

Bighorn sheep show low fecundity; single lambs are the rule and twinning is infrequent (Buechner 1960, Geist 1971, Turner and Hansen 1980). Twinning in wild populations has yet to be observed and has not been documented in 137 lambs born in captivity in New Mexico (A. Sandoval, unpubl. data). However, McQuivey (1978) reported that of 60 lambs born in captivity in Nevada, one set of twins was produced. Moore (1961) reported one set of twins born in captivity in Texas. Twins are so infrequent in bighorn sheep that the rate of population increase is not influenced by the incidence of twinning.

The age that females first give birth has been documented for a number of desert bighorn populations. McQuivey (1978) and Berger (1982) observed yearling females with lambs in Nevada and California, respectively. Leslie and Douglas (1979) and Berger (1982) concluded that females were sexually mature at 13 months of age. Morgart and Krausman (1983) observed that semi-free-ranging females could copulate successfully as early as 9.7 months of age and produce young at 15.4 months. Berger (1982) speculated that age of sexual maturation was related to body size and that genetic differences accounted for subspecific variation in breeding ages among desert bighorn populations.

The breeding and parturition periods of desert bighorn sheep are more lengthy than those of more northern populations. Thompson and Turner (1982) assessed the temporal variation in parturition seasons for 22 populations of bighorn sheep. In populations from northern latitudes, parturition seasons were shorter, later, and cued to brief, relatively predictable periods of vegetation growth. Bunnell (1982) came to similar conclusions and added that departures from a strong central tendency with regard to peak of parturition resulted from predictability of vegetation rather than thermal stress or predation pressure. Lenarz and Conley (1982) hypothesized that the wide temporal variation in parturition dates of desert bighorn was a form of reproductive gambling resulting from the unpredictable nature of vegetation growth.

Although temporal variation in parturition seasons of desert bighorn has been studied, the characteristics of sites used for parturition have received little attention. Evidence from the Sonoran Desert, the Great Basin, and Badlands National Park suggests that pregnant bighorn shifted broad-scale habitat use to more rugged areas containing less-nutritious forage because potential predation pressure overrode increased nutritional needs during the late stages of pregnancy (Berger 1991, Bleich et al. 1997). Etchberger and Krausman (1999) compared microsites used for parturition to microsites used by females during other times of the year. Parturition microsites were comprised of steep, rugged terrain and had significantly fewer barrel cacti for up to eight weeks following parturition than microsites used by females without lambs. Parturition site fidelity was strong between years. The mean distance between parturition sites for individual females was 450 m. Etchberger and Krausman (1999) concluded that the traditional use of parturition sites in rugged terrain was an important but complex relationship between females, lambs, and habitat.

SEX RATIOS

Sex ratios in desert bighorn vary considerably. Equal sex ratios at birth have been reported by Welles and Welles (1961), L. O. Wilson (1968), Welch (1969), and Sandoval (1979a). However, Russo (1956) found a male:female ratio of 123:100 in Arizona, whereas Alvarez (1976), McQuivey (1978), De-Forge (1980, 1984), Leslie and Douglas (1979), Remington (1983), Guymon and Bates (1984), Vandenberge et al. (1984), Delaney (1984), and Olding (1984) reported sex ratios skewed toward females.

Some of the variability of sex ratios may be attributed to biases in the census technique, season of survey, and behavioral differences between males and females. Unusually high male ratios obtained during waterhole counts on the Sheep Range, Nevada, were the result of counts being conducted in September, during the peak of the mating season (McQuivey 1978). Waterhole censuses conducted later in the summer had a greater number of males in the sample, particularly if it had not rained. In contrast, waterhole counts conducted in spring resulted in a preponderance of females in the sample because females were more likely to begin using water earlier in the year due to the demands of lactation (Turner and Weaver 1980). Surveys conducted during fall provide the most accurate sex ratios (Sandoval 1979a). Fall surveys correspond to the mating season, when sheep are concentrated in large, mixed groups; are easier to locate; and provide a representative sample of the population structure.

The mean male:female ratio for bighorn populations in Nevada was 60:100 and has varied from 53:100 in 1974 to a high of 70:100 in 1971 and 1972 (McQuivey 1978). In an unhunted population in Utah, L. O. Wilson (1968) reported a fall male:female ratio of 98:100, including yearlings of both sexes. In the San Andres Mountains, New Mexico, Sandoval (1979a) obtained a male:female ratio of 91:100 during the fall mating season and 58:100 during the spring lambing season. These differences were attributed to the influx of mature males into the female-juvenile home ranges during the mating season.

Foot surveys usually produce higher male:female and lamb:female ratios than aerial surveys. A generalization from foot and aerial counts is that more males are observed during the mating season. This is the result of males being consistently less observable during the lambing season due to their greater mobility and solitary tendencies.

Aerial counts facilitate the location of the majority of a given population in a relatively short time span, thus reducing duplicate sightings. Ground counts usually produce more accurate sex and age data; however, the possibility exists that the proportion of males to females during the mating season can be inflated by duplicate counts of males that frequently travel between female groups in search of estrous females. The proportion of lambs to females is significantly lower when based on aerial counts. This could be attributed to classifying yearling sheep as females from the air. Even to an experienced observer, yearlings cannot be distinguished from females with complete certainty, except at close range and if observation time is sufficient. If this misidentification did occur, the female segment of the population would be inflated, thus erroneously suggesting proportionally lower lamb production and survival.

There are discrepancies in the literature regarding sex composition in hunted and unhunted populations. C. G. Hansen (1967) and Turner and Hansen (1980) reported that during a 10-year period of hunting on the Desert National Wildlife Range, Nevada, the sex composition dropped from 100:100 to 23:100. Using C. G. Hansen's (1967) data, Bradley and Baker (1967a) concluded that the factor(s) responsible for sex-ratio changes was unknown. McQuivey (1978) attributed these discrepancies to inaccurate population estimates before and after hunting. C. G. Hansen's (1967) sex-ratio data were obtained from June waterhole counts; sampling error arose because a greater amount of time was spent surveying lambing areas where males were not present. McQuivey (1978) found no significant difference in sex ratios between hunted and unhunted populations of desert bighorn in Nevada. The average male:female ratio for unhunted populations was 57:100, as compared to a mean statewide ratio of 60:100.

AGE STRUCTURE

The average mortality rate for males is relatively constant after one year of age (McQuivey 1978). McQuivey found that 46.4% of the male population were between one and three years, 32.3% were between four and six years, and 21.3% were older than seven years of age. In the River Mountains, Nevada, Leslie and Douglas (1979) reported an average of 47% of the males to be one to three years, 34% to be four to six years, and 19% to be older than seven years of age, which compares closely with McQuivey's (1978) statewide data. McQuivey (1978) attributed the discrepancy between his data and C. G. Hansen's (1967, 1980c) to the fact that Hansen used estimated female ages in computing his mortality table. Only minimum ages can be determined for females (Geist 1971), which may result in erroneous conclusions. In addition, the mortality rates for males used by C. G. Hansen (1967, 1980c) were based on 120 skulls collected over 17 years, representing an insignificant percentage of the total male mortality during this period.

LAMB SURVIVAL

Few births have actually been observed in free-ranging desert bighorn sheep; therefore, measures of lamb production are difficult to obtain. This problem is compounded because lambing usually occurs over a period of six months. Lambs observed in the field represent only the survival rate at that point in time, rather than the documentation of overall production. Therefore, females observed during spring months may be females with lambs, gravid females, females that were never gravid or that had resorbed or aborted their lambs, or females that gave birth but lost their offspring prior to the time they were observed.

The relationship between lamb production and survival, termed recruitment, has been used as an indicator of population quality (fig. 4.5). Geist (1971) found that the number of lambs per females was related to the percentage of lambs surviving to yearling age. In warm climates where mild weather during parturition results in a negligible loss of lambs to hypothermia, high lamb production would be expected to be followed by heavy mortality before yearling status is reached. In contrast, lambs born in temperate climates usually experience high mortality soon after birth (D. R. Smith 1954, Wishart 1958, Woodgerd 1964, Berwick 1968, Nichols 1978a).

Lamb production and survival in desert regions is closely related to habitat quality; however, McQuivey (1978) questioned habitat quality as a measure of population quality. Even in a healthy, viable population, unpredictable precipitation and resultant phenology may result in poor lamb survival.

The relationship of nursing by lambs to population quality and survival has been examined in only a few instances for desert bighorn. Berger (1978a) believed that the high rate of suckling duration and short weaning time in a herd in the Santa Rosa Mountains, California, were adaptations to desert environments. In later analyses, Berger (1979a) noted that only 5% of the lambs survived and suggested that the short nursing period and weaning were the result of disease or poor nutrition. McCutchen (1988) presented data on suckling duration and lamb survival for a captive herd in Zion National Park and also pointed out that Berger's (1978a, 1979a) interpretations may be incorrect due to insufficient data collection.

Data presented by L. O. Wilson (1968), Sandoval (1979a), and DeForge (1980) describe stable populations of low quality as characterized by high lamb production, poor survival, and long life expectancies. However, life expectancies were short in the population studied by DeForge (1980). His data do not support Geist's (1971) conclusion that long life expectancies are typically characteristic of stable, low-quality populations. DeForge (1980) reasoned that the population in the San Gabriel Mountains, California, was compensating for high juvenile mortality by exhibiting a high reproductive rate as an adaptive response to long-term human disturbance in the area. Krausman et al. (1989) determined productivity as 70% and recruitment as 21% in the Harquahala and Little Harquahala Mountains, Arizona. However, the cost of raising lambs must have been high because females that

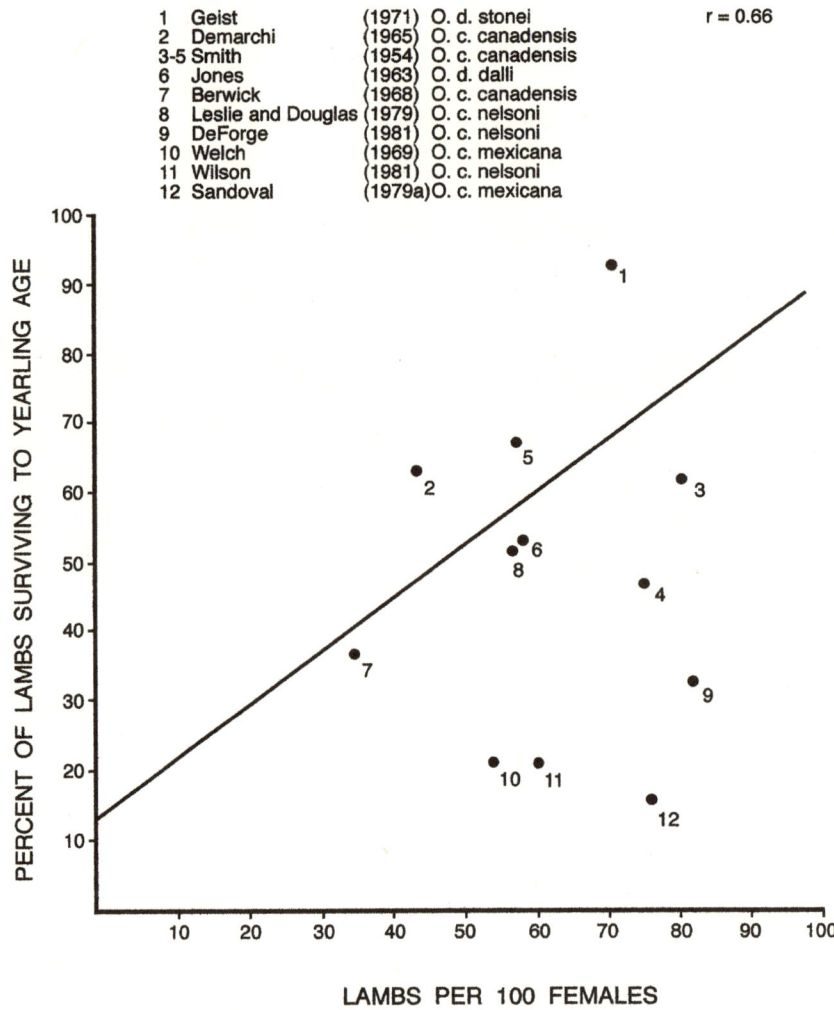

Figure 4.5. Relationship between lamb crop (expressed as lambs per 100 ewes) and survival of lambs in selected populations of North American wild sheep (adapted from Geist 1971).

successfully raised a lamb to older than six months of age were rarely successful in raising a lamb the following year. In a follow-up study in the Little Harquahala Mountains, Krausman and Etchberger (1993) documented recruitment of lambs at 32%. In contrast, Leslie and Douglas (1979) found an expanding, high-quality population in the River Mountains, Nevada, and correlated the health of this population with precipitation patterns.

Correlations between precipitation and lamb survival have been investigated by a number of authors. C. L. Douglas and Leslie (1986) concluded that precipitation during gestation in autumn of the previous year had a positive effect on lamb survival and that herd density was inversely related to lamb survival. Most of the variability (87%) was accounted for by autumn precipitation and herd density (C. L. Douglas and Leslie 1986). In the Santa Rosa Mountains, California, Wehausen et al. (1987) found that increasing herd density probably caused a slow decline in the recruitment rate. Precipitation during November, January, and February showed a positive influence on recruitment of lambs in the Santa Rosa Mountains (Wehausen et al. 1987). Berger (1982) found similar trends.

The effects on populations from the genetic composition of desert bighorn lambs have not received much attention in the literature. Sausmann (1982) analyzed the survival of captive lambs and concluded that strongly inbred lambs did not survive. Sausmann (1982) recommended that strict attention be paid to parentage and that efforts to use small populations of desert bighorn for reintroductions should be controlled to maximize genetic diversity.

LIFE TABLES

A life table is a theoretical model for the dynamics of a population based on estimates of the various parameters used to measure the performance of a given population. Parameters indicative of negative or positive responses to environmental factors include the age structure, sex composition, numerical abundance, natality, and mortality rates.

Excluding immigration and emigration, population increase is equivalent to the natality rate minus the mortality rate. In a high-quality population, the maximum rate of increase may be closely achieved. After reaching relative stability, the average rate of increase equals the average rate of loss, fluctuating around a mean population level. Population increase for bighorn sheep can range from near maximum to zero or to less than zero if the population is decreasing (Buechner 1960).

Life tables for desert bighorn have been constructed from data on age at death of a random or representative sample of a population (Kdx series) and age structure of a living population or observation of a single age cohort throughout its life (Klx series), as detailed by Bradley and Baker (1967b), McQuivey (1978), Leslie and Douglas (1979), and C. G. Hansen (1980c). The age of death (Kdx) and age structure of a living population (Klx) series assume a relatively stable population and age distribution. Both assumptions are open to criticism because precise data usually are not available indicating relatively stable populations over a given period of time.

Survivorship data presented by C. G. Hansen (1980c) and Bradley and Baker (1967b) were determined from skulls found in the field (*Kdx* series). These skulls were aged by counting horn annuli, which provided accurate ages for males but only minimum age for females (Geist 1971). McQuivey (1978) and Leslie and Douglas (1979) obtained survival rates from observations of living individuals (*Klx* series) and mortality rates were inferred from the shrinkage between successive ages.

Both methods for obtaining survivorship data are subject to different biases. Data for a known age-at-death series is open to criticism because the probability of locating skulls of all ages and sex categories is not the same due to more rapid deterioration and difficulty in locating skulls of subadults and females (Bradley and Baker 1967b). To compensate for younger age intervals, age at death and data on age structure of living individuals are combined using ratios of lambs and yearlings per 100 females from field observation. Life tables constructed from skull collections also could be biased if the sample size is insufficient relative to the actual mortality. Life tables calculated from age structure of living individuals may be biased if each age group is not sampled equally. This is particularly true if population data are not obtained during the mating season, when the sexes are together. Secondly, the validity of age structure data obtained from aerial surveys may be questioned. Even to an experienced observer, aging males from a helicopter cannot be done with complete certainty, except at close range and if observation time permits. Mortality rates calculated from *Kdx* and *Klx* series are subject to different biases and may not directly compare.

MORTALITY

Life tables have been based on combined age at death (*Kdx* series) and age structure of living individual (*Klx* series) from the Desert National Wildlife Range, Nevada. Due to younger age classes being misrepresented when using age-at-death series, C. G. Hansen (1980c) corrected the calculations by computing average ratios based on herd composition counts as follows: (1) a lamb:female ratio of 70:100 for lambs from birth to one month old; (2) a lamb:female ratio of 35:100 for lambs six months old; and (3) a yearling:female ratio of 15:100.

High lamb mortality and low postyearling mortality followed by increasingly higher mortality in the old age structure is evident. Once a bighorn sheep reaches age 2, mortality is negligible until age 7. Mortality rates increase for both sexes after 7 years. Life expectancy after 10 years of age is less than 2 years. However, a few individuals may live to be 15 years of age. The life table for females shows the same general mortality pattern. However, mortality rates are higher for most age classes of females than for males.

C. G. Hansen (1980c) found 47% adult mortality in age classes 10, 11, and 12. In contrast, McQuivey (1978) reported that the average age structure of males ($n = 1,939$) in Nevada declined gradually through successive years of life. He found no single mortality factor affecting only one age group for males older than one year of age. These data closely agree with Leslie and Douglas (1979), who found relatively constant mortality rates for males from the River Mountains, Nevada. Leslie and Douglas (1979) hypothesized that increased mortality of young males could be due to involvement in rutting activities, especially during the prerut period. This prerut hyperactivity of socially immature males also was reported by Sandoval (1979a). Krausman et al. (1989) found that survivorship of females did not differ between the Little Harquahala and Harquahala Mountains, Arizona; however, the survivorship curve for males was consistently higher for males in the Harquahala Mountains. Mortality factors were similar between the mountain ranges, except that mountain lion *(Puma concolor)* predation was greater on female sheep in the Harquahala Mountains (Krausman et al. 1989).

If one assumes equal sex ratios exist at birth (Murie 1944, Geist 1971) and that females have a shorter life expectancy than males, the sex ratio among adult sheep should favor males. Nevertheless, the conclusion that females experience higher mortality rates than males is open to criticism. The higher mortality rates for most age classes of females is based on age determination by counting horn annuli. This method of aging sheep is reliable for males, but tends to underestimate the age of females, thus providing only a minimum age (Geist 1971).

Mean length of life or population turnover is another population characteristic that can be calculated from life tables. In a stable population, turnover rate would be equivalent to the annual replacement rate and mean annual adult mortality rate. Therefore, mortality rates are a measure of the flow of individuals through a population (Buechner 1960).

ACCIDENTAL MORTALITY FACTORS

Considering the nature of bighorn habitat, accidents and injuries must occur. Numerous researchers have documented bighorn mortalities resulting from accidental falls, entanglement in fences, entrapment in deep potholes or forks of trees, fighting, and collisions with motor vehicles. Other examples of mortality are unclear. For example, Witham et al. (1982) documented the deaths of 14 desert bighorn, including 5 adults, found in a cave in the Plomosa

Mountains, Arizona. However, there was no indication that a general die-off had occurred.

Welsh (1971) summarized mortality factors over a 20-year period ($n = 141$) afflicting desert bighorn in Arizona and southeastern Nevada. He found that poaching accounted for 41% of the mortality, road kills 20%, natural accidents 15%, drownings 12%, and fence mortalities 12%.

McQuivey (1978) documented unusual mortalities of desert bighorn in Nevada. These include a male that was found suspended by the horns on a cable that crosses the Colorado River below Hoover Dam. Another male was found wedged by the head and horns in the fork of a tree apparently after becoming entangled while thrashing its horns. Two lamb mortalities occurred as a result of predator-control efforts. Sandoval (1979c) reported lamb mortality resulting from entanglement in a paddock fence.

Drowning is one of the most common types of accidental deaths recorded by McQuivey (1978). At least 15 sheep have drowned since 1962 in the Colorado River below Hoover Dam. Thirteen were males with heavy horns, one was a female, and the other was a lamb. Similar reports of drownings have been reported in Arizona (Welsh 1971) and California (Mensch 1969). Mensch (1969) found the remains of 34 bighorn trapped in a 3-m deep pothole in the Chocolate Mountains, California. He surmised that the sheep fell into the pothole when the water level was low and were unable to climb out.

The sheep population in Aravaipa Canyon, Arizona, experienced a 52% decline from 1988 to 1989, probably as a result of livestock-related viral diseases compounded by nutritional stress (Mouton et al. 1991). In 1991, survival of lambs in the Aravaipa population was higher than in 1989 or 1990 and the population appeared to be increasing (Mouton et al. 1991).

Cunningham and deVos (1992) monitored 49 adult sheep in the Black Mountains, Arizona, adjacent to Hoover Dam to document mortality factors. Of the 12 mortalities recorded, 6 were caused by collisions with automobiles, 2 were from coyote kills, 1 was killed by a mountain lion, and 3 were due to unknown causes. Female sheep with home ranges that crossed Highway 93 had more than a 24% chance of being killed while crossing the highway.

SURVIVAL

McQuivey (1978) reasoned that lamb survival, rather than adult mortality, is the most important factor in determining population trends for desert bighorn sheep. He computed the mean survival rate of lambs, based on observed lamb:female ratios, into a lamb percentage figure to obtain actual recruitment needed to replace natural mortality. His data show that a lamb: female ratio of 26:100 is the minimum fall survival rate required to maintain a stable population.

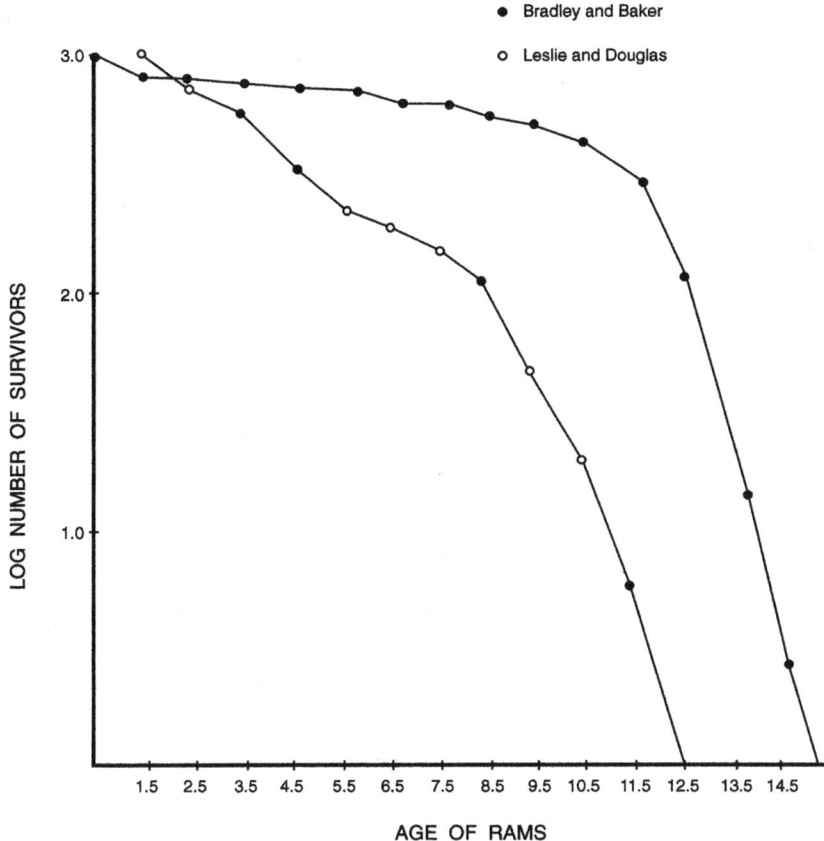

Figure 4.6. Survivorship curves of desert bighorn rams (not corrected for population expansion; data from Bradley and Baker 1967a, Leslie and Douglas 1979).

Survival patterns for desert bighorn sheep vary (fig. 4.6). Data presented by Bradley and Baker (1967a) show negligible mortality among males between two and nine years of age. This age interval represents the majority of individuals in the population; consequently, it is their mortality rates that characterize the mortality pattern for the entire population. In contrast, the data from Leslie and Douglas (1979) follow a quadratic equation. These data were not corrected for population expansion; therefore, the age 0–1 interval was not represented, resulting in an artificially high origin of the survivorship curve. However, once yearling status was reached, mortality was relatively constant among all other age groups.

The recruitment dynamics of a bighorn population in southern California were influenced by precipitation during November, January, and February

(Wehausen et al. 1987). Rising population density also was implicated as causing a slow decline in recruitment (Wehausen et al. 1987).

Factors Influencing Dynamics

Since 1900, desert bighorn populations have declined significantly and their distribution has been reduced (Russo 1956, Buechner 1960, J. A. Bailey 1980, M. C. Hansen 1982*b*). This decline has been attributed to human-induced factors such as disease, overgrazing of livestock (Gallizioli 1977) and feral burros (Seegmiller and Ohmart 1981), road construction (Leslie and Douglas 1979, Krausman et al. 1989), housing developments (Gionfriddo and Krausman 1986), canals (Krausman 1985), fire suppression (Etchberger et al. 1989), and recreation activities (L. K. Harris 1992). Habitat of desert bighorn sheep is naturally fragmented because precipitous mountains used by sheep are surrounded by large, relatively flat areas rarely used by sheep (Bleich et al. 1990*b*). Desert bighorn sheep are considered to be of unique aesthetic and recreational value and have stimulated the formation of numerous special-interest organizations dedicated to their conservation (i.e., Arizona Desert Bighorn Sheep Society, The Desert Bighorn Council, The Fraternity of the Desert Bighorn, and Society for the Conservation of Bighorn Sheep). Management of desert bighorn has been directed at describing habitat characteristics, identifying limiting factors for populations, and reintroducing sheep into historic areas.

Resource managers are concerned with factors limiting populations because empirical evidence suggests that many populations having fewer than 50 individuals may be susceptible to extinction (Berger 1990, Krausman et al. 1993*a*). The nature of mechanisms regulating the sizes of animal populations is a central problem in modern ecology. Authorities on population dynamics and their ideas have become so polarized that new students are introduced to the problem as a controversy (Horn 1971).

For many years it was assumed that epizootics, famine, and climatic factors terminated the rises in population size and precipitated the often spectacular crashes. However, it has become apparent that none of these mechanisms explain some observed phenomena of population mechanisms, suggesting that factors intrinsic to a given population are involved in their regulation. In other words, density-dependent mechanisms have evolved within individual animals to regulate population growth and to curtail it short of destruction of the habitat. This does not imply that all populations are self-regulatory— merely that within any species one can expect to find a population whose rate of increase is slowed down and possibly stopped because of the way its members interact with one another.

There is increasing evidence that density-dependent control of a population can result from behavioral interactions of individuals, rather than from negative response of the environment. This form of social regulation of animal populations may be thought of as resulting from numbers of interactions between individuals. This behavior can have the effect of keeping a population below the level permitted by available resources. Such population controls have been found mostly among higher animals with well-developed social behavior (McLaren 1971:1–21).

Wynne-Edwards (1963) stated that the impact of stress on the individuals concerned, arising from competition and acting through the pituitary-adrenal system, influences both physiological and behavioral responses. More recently, Christian and Davis (1971) presented data that indicated the existence of endocrine feedback mechanisms that can regulate and limit population growth in response to an increase in overall "social pressure." This pressure, in turn, is a function of increased numbers and aggressive behavior.

Recent studies suggest that wild sheep populations may be limited by interactions between individuals. Moser (1962) stated that excessive mating by males may cause sterility and that physical exhaustion resulting from the strenuous activity of the rut may also have a detrimental effect on females. L. O. Wilson (1968) suggested that the low lamb:female ratios that characterize bighorn sheep populations could be caused partially by large males warding off younger males from estrous females, but doing little or no actual breeding themselves. Working with Asiatic mouflon sheep *(Ovis orientalis),* Valdez and Alamia (1977) found that despite the seemingly random interactions of males and females, there was a definite rank system involved in these encounters. Older males had a more refined courting technique than their younger competitors. The behavioral mechanisms by which older males are dominant over younger males may have evolved to protect the females from unnecessary stress, conserving the energy of both sexes during the mating season.

Bighorn sheep populations are characterized by a predominance of old individuals, low recruitment, and failure to disperse into contiguous, suitable habitat. In the San Andres Mountains, New Mexico, the disappearance of 60% of the lambs corresponded with the peak in rutting activities. During this period, numerous observations were documented of lone or congregated lambs. In addition, sheep hunters reported lone lambs that walked up to them, bleating constantly, that had to be discouraged from following the hunters. The skeletal remains of lambs found in the field exhibit articulated bones, indicating that predation was not the causative factor (Sandoval 1979*a*).

Sandoval (1979a) reported that as many as three males attempted to mount unreceptive females at once, extended chases were a common occurrence, and males were overtly aggressive with females during the breeding season. These males were usually "socially immature" males approximately two to four years of age. Apparently, these males had not mastered the courtship rituals and were rough and unpolished in their breeding techniques. Geist (1971) noted that when a male was alone with an estrous female, his courtship became cautious; furthermore, if a male courted roughly, the female withdrew and attempted to escape. Therefore, rough courtship decreases a male's chance to sire offspring and, hence, the male selects against his own genotype. In bighorn populations, the larger-horned males generally will do most of the mating and are preferred by females, as demonstrated by Geist (1971).

It is well known that stress causes illness in animals (Farb 1963, Dasmann 1964). Shipping fever (myxovirus parainfluenza-3), heart disease, and liver disorders are types of illness most commonly attributed to stress syndrome (C. G. Hansen and Deming 1980). Cowan (1974) found that even mild harassment in domestic sheep increased fighting, hypersexual mounting, and depressed milk production. Vincent (1974) has shown that stress imposed on adult female sheep during late pregnancy led to serious losses of adult and newborn animals. It has been shown that "loneliness" in sheep caused excitement (Reid and Miles 1962). Geist (1971) reported that when a female bighorn withdraws for lambing, the yearlings are left on their own and unless they can attach themselves to a barren female, they may wander aimlessly. Wishart (1958) found a number of yearling sheep remains in localities not normally visited by sheep.

Excessive stimulation of the endocrine system, particularly the adrenal gland, can be induced when lambs become separated and are unable to attach themselves to adult sheep. To produce sufficient hormones to regulate body chemistry, the adrenal cortex enlarges under stress conditions. However, a point is reached beyond which further increased production is impossible, thus causing a breakdown in the endocrine mechanism, with death resulting (DeForge 1976, 1980). Therefore, excessive stimulation of the endocrine system might increase lamb mortality by lowering the lambs' resistance to parasitism, infection, shock disease, and predation.

Aside from environmental factors, including drought and resultant food and water shortages, it is possible that when a bighorn population reaches a high density with a corresponding high male:female ratio, a population regulatory mechanism may be triggered resulting from the interactions of individuals. Excessive harassment of females by socially immature males during the rut may disrupt family groups and separate females and lambs. Lambs may then be subjected to stress syndrome, which lowers their resistance to disease, predation, shock disease, and infections.

HUMAN DISTURBANCE

With few exceptions, desert bighorn are intolerant of human activities. However, quantitative data are lacking, and development of tolerance levels of desert bighorn relative to human activity requires further work to establish dependable criteria.

Desert bighorn are a species that have great difficulty adjusting to human encroachment. Behavior of bighorn is extremely rigid and ritualized (Geist 1971). Behavior patterns such as periodic range shifts are passed from adults to young animals; these shifts play an important role in population survival beyond the obvious advantage of distributing their impact on the vegetation portion of the habitat. The available literature (Geist 1971, Krausman 1993a, Krausman et al. 1995) indicates that bighorn do not adjust well to disruptions of these patterns.

Disturbances, whether directed toward bighorn or not, have been observed to cause reactions adverse to population welfare. The point at which harm results is not clear, but bighorn have abandoned historic ranges when human activity increased suddenly over a few years' time (Krausman 1993a, Krausman et al. 1995). While addressing the development of a ski resort in the San Bernadino National Forest, California, Light and Weaver (1973) studied bighorn behavioral responses to humans in numerous environmental situations. They found that human visitation created a spatial displacement effect on bighorn. Bighorn and their sign were absent in a line-of-sight pattern from the center of the human influence, that is, bighorn adapted to this human influence by using habitat out of sight. Light and Weaver (1973) concluded that bighorn (1) maintain their area of distribution as a living tradition and rarely depart from it; (2) fail to extend their range despite ample opportunity; (3) reduce use of historical habitat where human use is greater than 500 visitor-days/year; (4) avoid historical range when visitor-days per year reach 500 to 900; (5) may reduce numbers as a result of insufficient forage, increased predation, increased disease, and external harassment; and (6) curtail movements, resulting in reduced gene flow and gene pool size, which may ultimately affect the future existence of a bighorn population.

Holl and Bleich (1983) reported on the effects of humans by documenting the reaction of bighorn to the presence of the investigators in the San Gabriel Mountains, California. At a mean distance of 645 m, bighorn ($n = 302$) usually were not concerned with the investigators' presence. At a mean distance of

167 m, sheep ($n = 78$) usually exhibited a curious or concerned reaction. The maximum distance at which one of these reactions occurred was 1,500 m. However, the mean distance at which flight occurred was 440 m. Similar bighorn-human interactions have been documented in Utah (King and Workman 1982, Bates and Workman 1983), Nevada (Leslie 1977, McQuivey 1978), California (Kovach 1979, DeForge 1980, K. S. Hamilton et al. 1982), Arizona (Purdy and Shaw 1981, Seegmiller and Ohmart 1981), and New Mexico (Sandoval 1979a, Watts 1979, Bavin 1982, Elenowitz 1983). The influence of humans on bighorn sheep is not always immediate. The bighorn sheep in Pusch Ridge Wilderness near Tucson, Arizona, declined from a huntable population to being nearly extinct over the course of 15 years. The sheep were not able to withstand the increasing influence of humans (e.g., recreation, housing, fire suppression; Krausman 1993b, Krausman et al. 1995).

No detrimental effects of riverboats on desert bighorn sheep were observed in Cataract Canyon, Utah (Stanger et al. 1986). For the sheep observed, 58% showed no response, 39% seemed aware of riverboat activity, and only 3% changed their behavior in response to riverboat activity (Stanger et al. 1986).

The effects of disturbance from various stimuli on desert bighorn sheep have been measured with remote, heart-rate telemetry (Bunch et al. 1989, Krausman et al. 1993c, Krausman et al. 1998). In Utah, captive sheep were least disturbed by a motorcycle being driven past the pen. A human running past or standing in the pen had the most effect on heart rate (Bunch et al. 1989). Simulated noise representing military jets had little influence on the heart rate of captive desert bighorn (Weisenberger et al. 1996). Jets flying over sheep also produced a limited increase in heart rates of sheep, lasting less than three minutes (Krausman et al. 1993c, Krausman et al. 1998).

COMPETITION

Exploitation competition, or amensalism, arises when two species use common resources that are in short supply or, if the resources are not in short supply, competition arises when the animals seeking those resources nevertheless harm one another in the process (Pianka 1978). Exploitation competition may result in reduced population quality of one or both species through increased mortality, increased dispersal, or decreased reproduction.

Overgrazing by domestic livestock may result in competition for forage, water, and space. Competition is considered one of the most important factors in the historic decline of populations of desert bighorn in western North America. Overgrazing and its contribution to degraded bighorn ranges are difficult to document because much of the damage occurred long ago. The effects of diseases, human encroachment, and poaching can be documented more convincingly than the detrimental effects of overgrazing by livestock (Gallizioli 1977).

Overgrazing by livestock results in large-scale changes in plant composition and density, reduction of important plant species used by bighorn, and the permanent reduction in the overall carrying capacity for bighorn sheep populations. In addition to reducing bighorn numbers, the conversion of primarily grassland communities into predominantly homogenous shrub communities provides better habitat for increasing deer populations.

Spatial and forage competition between cattle and desert bighorn continues to be a major problem resulting in limited sheep populations. Overgrazed ranges also may be the major obstacle to the reintroduction of desert bighorn to historic habitats (Gallizioli 1977, Sandoval 1979b). McQuivey (1978) compared currently occupied bighorn ranges in Nevada that are grazed by cattle versus those that are not, regardless of differences in forage types and water availability. He found that grazed areas supported significantly lower bighorn sheep densities than ungrazed areas. Grazed areas supported an average of 2.3 bighorn/km^2, whereas densities in ungrazed areas averaged 6.6 bighorn/km^2.

Microscopic analysis of fecal samples from desert bighorn and cattle in the Big Hatchet Mountains, New Mexico, revealed 12 of 18 major forage species common to both bighorn and cattle. Winter fat *(Ceratoides lanata)* comprised an average of 17.9 and 15.4% of the annual diet of bighorn and cattle, respectively (Bavin 1982). Bavin concluded that cattle could be serious competitors with bighorn sheep for available forage, especially during years of limited forage production. A 32% dietary overlap (the percentage use in common of shared forage plants) between desert bighorn and cattle in the Peloncillo Mountains, New Mexico, was documented by Elenowitz (1983).

L. O. Wilson (1975) reasoned that spatial competition resulted in the displacement of a herd of desert bighorn in Utah. A recognizable group of bighorn occupying an ungrazed study area had been under observation for five years. A herd of 30 cattle was introduced into the area for two weeks. The bighorn sheep left the area and did not return for eight months. Similar patterns of livestock and bighorn interactions have been reported by F. L. Jones (1980) and Steinkamp (1990). Steinkamp (1990) experimentally introduced cattle into bighorn core areas; the sheep responded by relocating.

In Aravaipa Canyon, Arizona, diet and spatial overlap were low between cattle and desert bighorn, primarily due to cattle preference for level slopes and bighorn use of steep slopes (Dodd and Brady 1986). However, plans were under consideration to redistribute cattle grazing to include areas with moderate slopes, which might have increased overlap and competition (Dodd and Brady 1986). Diets of desert bighorn and cattle in Carrizo Canyon, California, did not overlap because the populations were not sympatric; however, competition could arise if cattle were introduced to bighorn range (Cunningham and Ohmart 1986).

Historically, the most serious ecological competitors with desert bighorn have been domestic sheep and goats. These species have similar feeding habits, forage preferences, and affinities for rough topography, and they harbor parasites and other disease agents detrimental to bighorn. The extirpation of desert bighorn from numerous ranges in California (Weaver 1972), Arizona (Russo 1956, Gallizioli 1977), Nevada (McQuivey 1978, Kelly 1979), New Mexico (Gross 1960, Sandoval 1979*b*), Utah (L. O. Wilson 1968, Dean and Spillett 1976), western Texas (W. B. Davis and Taylor 1939, Kilpatric 1982), and northwestern Mexico (Mendoza 1976) has been attributed to competition for forage and space and to the transmission of diseases from domestic sheep and goats (Bunch et al., this volume).

Concurrent with introduction of the first domestic sheep in western North America, bighorn sheep died off on a large scale, ostensibly from psoroptic scabies (*P. ovis;* Buechner 1960). Mortalities caused by scabies reduced populations of bighorn sheep at the time domestic sheep were introduced at Greybrell River, Wyoming (Honess and Frost 1942, Honess and Winter 1956*b*), Rocky Mountain National Park, Colorado (Wright et al. 1933, Packard 1946), Sierra Nevada, California (F. L. Jones 1950), and the Owyhee River, Oregon (V. Bailey 1936). Psoroptic mites on desert bighorn sheep have been reported from the Desert National Wildlife Range, Nevada (Cater 1968, Decker 1970), the San Andres Mountains, New Mexico (R. E. Lange et al. 1980, Sandoval 1980), southeastern Utah (Irvine 1969), and western Arizona (deVos et al. 1980, Remington 1981).

The existence of free-ranging, exotic ungulates on occupied or potential bighorn habitat poses a serious threat to desert bighorn sheep survival. The feral burro *(Equus asinus),* aoudad *(Ammotragus lervia),* and Persian wild goat *(Capra aegagrus)* currently occupy historical bighorn habitat or are rapidly radiating into habitat vital to desert bighorn (Sandoval 1979*b*).

The habitat requirements of aoudad (Simpson and Krysl 1981) and Persian wild goats (Bavin 1975) are similar to those of desert bighorn. Seegmiller and Simpson (1979) reported that the initial competition between two ecologically similar species consists of (1) niche overlap and/or behavioral intolerance; (2) shortage of limited resources; and (3) reduced population fitness.

Hardin (1960) proposed the competitive exclusion principle, which states that two species cannot coexist on the same limited resources. Therefore, competition results in reduced population fitness for one or both species through decreased natality, decreased recruitment, increased mortality, or increased dispersal. The survivor is the species able to perpetuate itself under competitive interactions, while the other is excluded from the area of sympatry (Seegmiller and Simpson 1979).

The aoudad and Persian wild goat would probably dominate any competitive interaction with desert bighorn. Their evolutionary history and demonstrated rates of colonization indicate that these exotic species have adapted to a variety of habitats and vegetation (Barrett 1967, Bavin 1975). They have a higher reproductive potential, with an annual rate of increase of approximately 75%, and may be able to survive on the fewer species and lower-quality forage that results from exploitation competition and deteriorated range conditions (Ogren 1965, Bavin 1975, Seegmiller and Simpson 1979, Dickinson and Simpson 1980). Seegmiller and Simpson (1979) stated that desert bighorn would have to either change their method of habitat and resource utilization or be excluded from the areas of sympatry. The adaptations necessary for these changes were probably beyond the desert bighorn's genetic capabilities, and it would be unlikely that aoudad-bighorn interactions would reach an equilibrium until desert bighorn became extinct in sympatric areas (Seegmiller and Ohmart 1979).

Competition between feral burros and desert bighorn has attracted much attention and has been a source of controversy. Many argue that feral burros are disrupting the desert ecosystem and propose strong control and elimination measures. Others consider feral burros to have significant aesthetic value and support their preservation.

Feral burros efficiently utilize coarse, low-quality forage, but are not primarily browsers by preference. They are opportunistic and prefer grasses when available (C. L. Douglas and Norment 1977). McMichael (1964), Walters and Hansen (1978), and Seegmiller and Ohmart (1981) substantiated an overlap of burro and bighorn sheep range and competition for available forage in the Black Mountains, Grand Canyon, and Bill Williams Mountains, Arizona, respectively. McMichael (1964) showed a 50 to 58% similarity index in burro and desert bighorn diets, Walters and Hansen (1978) indicated a 52% overlap, and Seegmiller and Ohmart (1981) reported a 64% overlap. Based on fecal analysis, R. M. Hansen and Martin (1973) found that grasses dominated the diet of feral burros in Grand Canyon, Arizona. Browse species dominated feral burro diets in Death Valley, California (B. Browning 1960, Moehlman 1974), and the Chemehuevi Mountains, California (Woodward and Ohmart 1976). C. L. Douglas and Norment (1977) demonstrated the impact of burros upon shrubby species in the Panamint Mountains, Death Valley, California. Of all shrubs, 46% exhibited some evidence of utilization and the survival of 12% was found to be threatened by severe browsing.

Geist (1985*b*) presented evidence that suggests "that current species in lower North America are there because of the absence of more capable competitors, and that most Siberian forms were excluded for hundreds of thousands of years." As competitors (i.e., burros) return, they can be expected to outcompete bighorn sheep for resources. J. A. Bailey (1980) argued that large mammals in North America may not be robust competitors.

Potential predators of desert bighorn include the mountain lion, jaguar *(Felis onca),* ocelot *(F. pardalis),* bobcat *(Lynx rufus),* coyote *(Canis latrans),* gray fox *(Urocyon cinereoargenteus),* and golden eagle (*Aquila chrysaetos;* Kelly 1980). All are capable of killing either young or adult sheep with varying degrees of success. Bobcats and coyotes are believed to be the most effective predators of lambs, particularly around waterholes (L. O. Wilson 1968) or in areas void of escape terrain (Watts 1979), whereas mountain lions are inclined to kill more adult sheep than other predators (Sandoval 1979a, Kelly 1980, Wehausen 1996). Most authors (Welsh 1971, McQuivey 1978, DeForge 1980) conclude that predation is rarely significant in affecting population survival. Free-ranging populations of bighorn sheep are well adapted for survival against the various predatory species encountered on their ranges. However, predation can be an important limiting factor. Mountain lions were the limiting factor for bighorn sheep in two mountain ranges in California (Wehausen 1996) and in New Mexico (Hoban 1990).

Special cases of transplanted sheep or small remnant populations may be exceptions also. When bighorn are transplanted or confined to an enclosure, the normal escape routes are unknown or unavailable to the population. The available data on various reintroduction efforts for desert bighorn strongly indicate that large predators, particularly mountain lions, frequently take advantage of the situation, resulting in serious consequences to transplanted or remnant populations.

In Texas, after a three-year struggle against repeated mountain lion predation in the Black Gap Wildlife Management Area, the entire bighorn reintroduction effort was abandoned. The few remaining bighorn in the Black Gap enclosure were captured and moved to the Chilicote Ranch, where adequate protection from predators could be provided by large-scale and long-term predator control programs by private land owners (Kilpatric 1982). Nevada encountered a severe setback in their attempts to introduce bighorn into the Mount Grant enclosure. The entire effort was destroyed over the five days following placement of the bighorn in the enclosure. Four mountain lions entered the enclosure and killed all the transplanted bighorn with the exception of a single yearling male (Broadbent 1969).

Mountain lion predation also has hampered reintroduction efforts in Zion National Park, Utah (McCutchen 1976), and has been documented as the major mortality factor in the remnant population in the San Andres Mountains, New Mexico (Hoban 1990). Bighorn lost to lion depredation in the San Andres Mountains between December 1980 and March 1982 represented 24% of the total known population and 36% of the radiocollared sheep (Munoz 1982). In the Big Hatchet Mountains, New Mexico, 34% of the sheep transplanted there were killed by lions within three months postrelease. Similar bighorn–mountain lion interactions involving reintroduction efforts have been reported in Colorado (Creeden and Schmidt 1983), Arizona (Remington 1983), and New Mexico (Elenowitz 1983).

LOSS OF HABITAT

Desert bighorn sheep historically existed in large numbers and were widely distributed throughout much of southwestern North America. The fact that they dispersed and maintained viability in this vast, arid region indicates that bighorn were successful colonizers (Geist 1971). After the arrival of Europeans, bighorn sheep suffered dramatic declines. The declines were attributed to human settlement disturbance, increasing encroachment and corresponding loss of habitat, competition with and diseases introduced by domestic livestock, and excessive illegal hunting. Bighorn populations have been reduced and have disappeared from numerous areas due to demands and abuses of the land (Valdez and Krausman, this volume).

Between 1948 and 1982, bighorn sheep have disappeared from 18 different ranges in California (Weaver 1982), primarily due to competitive land uses (DeForge et al. 1981). From 1960 to 1980 bighorn sheep have disappeared from 4 mountain ranges in Arizona and are declining in 14 others (Arizona Game and Fish Department 1980). In New Mexico, only two populations remained by 1950 (Sandoval 1979*b*); native desert bighorn have been completely extirpated from Texas (Winkler 1977).

Thus, desert bighorn have been forced into the most isolated portions of their former range. In many areas desert bighorn presently exist as isolated remnant populations of questionable viability, separated from former ranges by water impoundments, highways, fences, and residential communities. These developments are deleterious to desert bighorn in several ways. They directly or indirectly increase mortality, act to prevent use of water sources and overall habitat, erect barriers to local travel and movement corridors, and increase human activities beyond innate tolerance levels (Graf 1980).

Once extirpated, bighorn do not demonstrate the degree of adaptive plasticity that they historically exhibited. The colonizing abilities of bighorn have been suppressed because of human-related disturbances, competition with other ungulates, and their high degree of specialization (DeForge et al. 1979).

DISEASES

Desert bighorn are susceptible to, and have been negatively impacted by, numerous infectious diseases, particularly those diseases common to domestic livestock. Diseases were largely responsible for the extirpation of bighorn sheep throughout much of North America (Bunch et al., this volume). Diseases that have influenced desert bighorn include chronic sinusitis (Bunch 1979, Bunch and Webb 1979), psoroptic scabies (deVos et al. 1980, Sandoval 1980, Weaver 1982, Welsh and Bunch 1982, Thorne and Walthall 1983, R. K. Clark et al. 1988, Hoban 1990, R. K. Clark and Jessup 1992), and pneumonia (Russo 1956, Buechner 1960, Spraker 1977, Hailey et al. 1972, R. E. Taylor 1976, Feurenstein et al. 1980, DeForge and Scott 1982, Foreyt and Jessup 1982, Elenowitz 1983).

Problems associated with surveying desert bighorn for the presence of diseases are presented by Wehausen (1987). Small sample sizes may lead to inaccurate conclusions about disease prevalence, and Wehausen (1987) outlines an alternative classification scheme.

Human-induced injury to desert bighorn is associated with improperly fitted radiocollars (Bleich et al. 1990*c*). When radiocollars are fitted too loosely, serious injury to the osseous and dermal tissues may alter foraging behavior and decrease fitness (Bleich et al. 1990*c*). Radiocollars should be attached snugly and high on the neck to prevent injury. However, care must be taken when working with subadult animals because the neck will grow and injury could result from an overly tight radiocollar (Bleich et al. 1990*c*).

BEHAVIOR

Bighorn sheep are highly social animals that use the same range yearly. Home ranges are inherited by young sheep from their parents through learning behavior. Unlike many species, bighorn, especially females, do not appear to explore and colonize new territory, but rather seem content to live where their ancestors have (Geist 1971).

Bighorn sheep are localized into distinct herds, with little apparent movement and dispersal into potentially suitable habitat. The failure of bighorn to colonize suitable, contiguous habitat could be attributed to the complex mechanisms that dictate their behavior. The areas of highest sheep concentration are usually the home ranges of the female-juvenile segment of the population. These home ranges encompass lambing grounds, mating grounds, and summer and winter grounds, which frequently overlap each other. Mature males readily accept females as leaders when in mixed groups. Also, males are seldom observed assuming the role of group leaders when accompanying females. Mature males disperse from female-juvenile groups as mating activities diminish.

Desert sheep traditionally have been managed on a case-by-case basis, usually within a mountain range. However, the flats between mountains may act as important corridors for sheep to gain access to other ranges for lambing and foraging (Bleich et al. 1990b, 1996; Krausman 1997). Bleich et al. (1990b) proposed a model for the conservation of desert bighorn sheep and Schwartz et al. (1986) were among the first to suggest a management strategy on a landscape basis for desert bighorn. Overall, they suggested that management needs to "seriously consider intermountain travel corridors for sheep, taking steps to minimize potential barriers." Bleich et al. (1990b) concluded that we still have the raw materials; what is needed is a commitment to protect and manage them properly. Only with the recognition that stewardship responsibilities extend beyond areas of "traditional" habitat and "viable" populations will we be assured the long-term stability of desert-dwelling mountain sheep and other fragile species that inhabit naturally fragmented habitats.

five

ADAPTIVE STRATEGIES IN AMERICAN MOUNTAIN SHEEP

Effects of Climate, Latitude and Altitude, Ice Age Evolution, and Neonatal Security

Valerius Geist

Introduction

Behavior allows individuals to adjust to their environment and thereby maximize inclusive fitness. Indeed, much of what we observe in mountain sheep can be related directly to this supreme goal. However, adaptations do not exist in isolation, rather each is part of a larger, integrated system called an adaptive strategy or syndrome. Each adaptive strategy is aimed at a specific objective and is realized by appropriate tactics, which, in turn, depend on tactical requirements or habitat factors. In this chapter, I examine several such syndromes in mountain sheep and attempt to revise our understanding of their evolution. I use the term mountain sheep to denote American and East Siberian sheep of the subgenus *Pachyceros*.

Some Strategic Expectations about Mountain Sheep

To maximize inclusive fitness, an individual must continually monitor its environment and adjust to various options. It must maintain security to minimize the risk from predators, but also from parasites, accidents, and dangerous climatic factors. An individual must gather the best diet that security permits, minimize waste of scarce nutrients and energy (Zipf's law of least effort; Zipf 1949), and store energy adequate for superior reproduction. Finally, an individual must strive to maintain its body in sound, functional order; restore its body after suffering the costly stresses and injuries of reproduction; and prepare it again for the next round of reproduction. To do this, an individual must maintain access to scarce resources essential to reproduction and deny these resources to competitors, be they conspecific or not. Therefore, to maximize inclusive fitness an individual must be able to compete successfully for material assets essential to superior reproduction and it must compete for gametes from sexual partners that are preferably its superior. Alliances should be made if they help to increase fitness, preferably with others of the same genetic background (i.e., kin). However, an individual cannot dally indecisively, lest others act to its disadvantage.

The foregoing deductions suggest that an organism's success depends on its ability to evaluate situations and then make strategic and tactical choices with reference to numerous objectives, each of which contributes to the goal of maximizing fitness. Clearly, in more varied social, ecological, and physical environments, more options need to be recognized and evaluated, and individuals therefore must possess more cognitive and motor abilities to survive— let alone to prosper and reproduce.

At least three major factors have impinged on mountain sheep in these regards. (1) Landscapes with sharply delineated seasons, such as are found at high latitudes and altitudes, demand greater cognitive ability than environments with minor seasonal variations. (2) Great landscape changes, such as the transitions from glacial to interglacial conditions (Pielou 1991) and vice versa, also require greater cognitive ability. For instance, the change from widespread grassland to forest when moving from glacial to interglacial climates shaped the dispersal adaptations of mountain sheep (Geist 1967, 1971). (3) Finally, the late Pleistocene Rancholabrean fauna of North America, with its great diversity of large predators (Kurten and Anderson 1980), had an evolutionary impact on bighorn sheep. In contrast to the Eurasian Pleistocene fauna, the Rancholabrean fauna appears to have been predator limited (Geist 1998). This is suggested by the relatively primitive feeding organs of the Rancholabrean herbivores, their large bodies, their large hornlike organs (where present), their extreme security adaptations, the great specialization of the predators, and the severe wear and tear on the predators' bodies (Van Valdenburgh and Hertel 1993).

These three factors predict a large brain size in mountain sheep *(Ovis canadensis),* because brain size reflects the range of adaptive demands imposed on individuals by past and current environments (Geist 1978:188–189). For instance, Pees and Hemmer (1980) found that cold-adapted argalis *(Ovis ammon)* have relatively larger brains than do warm-climate urials *(Ovis gmelinii orientalis)*. This finding is in line with expectations that mammals from lower latitudes have smaller brain sizes relative to forms from higher latitudes (see Geist 1978:188–189). Also, Groves (1989) found that island mouflons from Corsica, Sardinia, and Cyprus have relatively smaller brain sizes than do mainland mouflons and urials, with the relatively smallest brain size being found in domestic sheep; Grove's findings for goats *(Capra)* were similar. Domestication has apparently reduced brain size, which a return to the feral state on islands failed to reverse completely. Presumably, domestication reduces the adaptive demands on an individual. Wishart and Brochu (1982) measured the brain case volumes of bighorn older than five years, reporting 234.5 ± 13.2 cc (SD, $n = 10$) for rams from southern Alberta and 223.2 ± 11.4 cc ($n = 14$) for rams from northern Alberta. For southern ewes, the average was 216.6 ± 13.6 cc ($n = 22$) and for northern ewes it was 213.9 ± 14.8 cc ($n = 23$). These data show that northern bighorn sheep have the same absolute brain size as argalis (Pees and Hemmer 1980), which have a body mass that is twice as large.

Through their thorough habituation to unusual but harmless stimuli and their sophisticated spatial evaluation of danger, bighorn sheep reflect their significant association with Rancholabrean predators. That is, they act much like the surviving Old American species and thus quite different from the easily spooked European mouflons, for instance. These Old American species (i.e., coyote, *Canis latrans;* black bear, *Ursus americanus;* puma, *Puma concolor;* white-tailed deer, *Odocoileus virginianus;* black-tailed deer, *Odocoileus hemionus;* pronghorn, *Antilocapra americana;* and peccary, *Tayassu tajacu*) are surprisingly capable of adjusting to human presence. This contrasts to either the East Siberian species that spread postglacially in North America as benefactors of megafaunal extinctions (e.g., grey wolf, *Canis lupus;* grizzly bear, *Ursus arctos;* wolverine, *Gulo gulo;* caribou, *Rangifer tarandus;* and moose, *Alces alces*) or introduced European species such as the wild boar *(Sus scrofa)*. The Eurasian species do not form inconspicuous, suburban populations; the Old American species do. Also, while predator eradication campaigns by livestock interests in the western United States eliminated wolves and grizzly bears in the early decades of this century, the Old American predators (i.e., coyote, puma, and black bear) generally survived.

Ice Age Mammal Characteristics

Ice Age mammals, of which North American mountain sheep are a good example, are relatively new forms of mammalian "design." During the cooling climates of the late Pliocene and the subsequent glacial ages, cold-adapted forms evolved by dispersing from the tropics to highly seasonal environments at high latitudes and altitudes (Geist 1971, 1978, 1987a). In the process they evolved from primitive, efficiency-selected, tropical ancestors into ornate, "grotesque" giants characterized by abundant "luxury tissues" (i.e., horn, fat, hair, brain). Mammals were transformed from life in the biologically harsh, but climatically benign, tropics to the biologically benign, but climatically harsh, environments of the cold regions. In these cold areas organisms must be ecological generalists and opportunists, not the specialists and resolute competitors we expect tropical forms to be.

Ice Age mammals had little ability to colonize lower latitudes for three major reasons. (1) As ecological generalists, they could not compete against a multitude of densely packed, ecological specialist species. (2) Because Ice Age mammals were adapted to low predator density and diversity, they could not be expected to readily deal with multiple predator species. (3) Ice Age mammals also could not deal with the myriad of pathogens and parasites found in warm climates. The immune system of Ice Age mammals is adapted to cold, dry environments where few parasites and pathogens can survive.

However, there is one exception: Ice Age species could colonize low latitudes if these, fortuitously, become empty of their megafauna. This happened in North America at the end of the Pleistocene when the Rancholabrean fauna collapsed (see Guthrie 1984). This allowed large, East Siberian mammals (including humans) to colonize the near-vacant continent, beginning about 12,000 years ago (Hoffecker et al. 1993).

Earlier in the Pleistocene, several Siberian species immigrated into North America, including short-faced bears *(Arctodus/Tremarctotherium),* lions *(Panthera leo),* elephants *(Mammuthus;* Kurten and Anderson 1980), bison *(Bison* spp.; Guthrie 1989), primitive musk oxen *(Praeovibos,* McDonald et al. 1991), and mountain sheep *(Ovis* spp.; Geist 1985b). Mountain sheep remained rare, but expanded in numbers at the height of the last glaciation about 20,000 years ago (Martin and Gilbert 1978, X. Wang 1984). When the Rancholabrean fauna was abundant, sheep made little progress. However, mountain sheep did spread from southern California, where they have apparently existed for some 350,000 years or more, into Wyoming by the last interglacial period about 110,000 years ago (Martin and Gilbert 1978, X. Wang 1984). Bighorn then dispersed across western North America in the late glacial period (20,000–12,000 B.P.), becoming large bodied in the process (A. H. Harris and Mundel 1974).

Such growth is expected of phenotypes during colonization in response to abundant food (Geist 1971, 1978, 1989; Shackleton 1973). Limb bones of late Pleistocene bighorn from Trap Cave, Wyoming, averaged 6.3 and 9.6% longer

Table 5.1. Limb-bone measurements of late Pleistocene and Recent bighorn, giant sheep, and ibex.

	A	B	C	D	E	F[a]
Humerus (H)	246.6 (15)	246	260	225.8 (5)	224	246
Radius (R)	257.6 (28)	254	302	232.6 (5)	225	212
Metacarpus (M1)	206.5 (30)	221	242	189.8 (5)	177	153
Femur (F)	299.9 (15)	315	320	272.7 (5)	245	262
Tibia (T)	350.2 (17)	350	355	323.2 (6)	300	311
Metatarsus (M2)	231.3 (33)	234	241	209.2 (6)	190	163
Ratios						
H/M1	1.19	1.11	1.07	1.21	1.27	1.61
R/M1	1.25	1.15	1.25	1.28	1.27	1.38
F/M2	1.29	1.35	1.33	1.32	1.29	1.61
T/M2	1.51	1.49	1.47	1.55	1.58	1.91

Note: Sample sizes are shown in parentheses.
[a] A = samples of late Pleistocene bighorns from Wyoming's Trap Cave (X. Wang 1984); B = measurements from a 5.5-year-old male *Ovis ammon polii* from Afghanistan, courtesy of the late John Boone; C = measurements from an old High Altai argali *(O. a. ammon)*, courtesy of the late John Boone; D = samples of extant bighorns *(O. c. canadensis)* from Wyoming (X. Wang 1984); E = measurements from a desert bighorn ram *(O. c. nelsoni)*, courtesy of the late John Boone; F = measurements from an old ibex *(Capra sibirica)*, courtesy of the late John Boone.

in linear dimensions compared to extant Montana and Wyoming bighorn, respectively (X. Wang 1984). Skull dimensions of the Trap Cave bighorn, however, averaged 12–51% larger than those of Alberta bighorn rams, but fell well short of the dimensions of Tibetan argali rams *(Ovis ammon hodgsoni)*, except for horn core diameters and circumferences (Geist 1991a). The Trap Cave Pleistocene bighorn were probably 20–40% heavier than extant Alberta bighorn (table 5.1), though other Pleistocene populations, such as the Lake Bonneville specimens, fell largely within the range of extant bighorn (Rutter et al. 1972).

Late Pleistocene sheep colonized western North America during early deglaciation, while the Rancholabrean fauna was still intact; however, mountain sheep did not become numerous (see McDonald 1978). Bighorn were in peak abundance in the Trap Cave deposits during cold phases. Two peaks, one at about 18,000 B.P. and another at 14,000 B.P., coincided respectively with the glacial maximum and the severe, Older Dryas cold phase. Such evidence confirms that mountain sheep are glacier followers (Geist 1971).

The bighorn survived the severe megafaunal extinctions. Mountain goats *(Oreamnos)* also survived in the northern Rocky Mountains, but disappeared from Mexico and almost all of the western United States. Bighorn occupied a vast region from north of the Peace River in northern British Columbia to north-central Mexico and the southern tip of Baja California. McDonald (1978) noted a parallel expansion of the southern bison *(Bison antiquus)* during the late glacial period. However, whereas southern bison probably met and hybridized with the long-horned Siberian bison *(Bison occidentalis)* that were moving southward from the Alaska-Beringian refugium (Guthrie 1989), the bighorn sheep from the south did not meet the thinhorn sheep from the north. Therefore, both species survive to this day. Bighorn and thinhorn sheep diversified morphologically by region (Cowan 1940), but this may be due to epigenetic adjustment and may be of little taxonomic significance (Geist 1978, 1989; F. C. James 1983).

Late glacial immigrants to North America from East Siberia are adapted primarily to East Siberia and not to North America. These species entered North America only with the decline of the Rancholabrean fauna. We expect East Siberian species to have relatively poor adaptations to North American landscapes and the native biota. This may, however, be less true of bighorn, despite adaptive deficits (J. A. Bailey 1980), because they have been in North America for more than 350,000 years (Geist 1985*b*). We can never be certain which features of the species of East Siberian origin are adapted to North American environments, which features reflect East Siberian conditions, and which features reflect ecological circumstances no longer extant, such as existed during the last glaciation. Therefore, North America's large mammal fauna is a unique source for the study of adaptations in process. The lessons from evolution, such as the susceptibility of North American mammals to diseases of East Siberian faunal elements, need to be respected in the management and conservation of North America's wildlife (Kistner et al. 1975; Welsh and Bunch 1983; Geist 1985*c*, 1988; Foreyt et al. 1994).

European Colonization

Superimposed on the Pleistocene factors that affected native sheep are the impacts caused by the colonization of North America by Europeans. On one hand, the colonizers inadvertently generated "wilderness" for over two centuries, when their diseases swept away most native people. This essentially removed the hand of native agriculturists and subsistence hunters from North America's landscapes and biota (Geist 1996). On the other hand, the expansion of European settlement and their livestock inflicted new pathogens and parasites on bighorn sheep. This settlement also brought American species of Siberian origin, which were in the process of adapting to southern latitudes, into contact with related Old World species that were already adapted to lower, warmer latitudes. This had consequences for North America's disease-prone mountain sheep (Foreyt and Jessup 1982, Goodson 1982, Jessup et al. 1984, Spraker et al. 1984, Clark et al. 1985, Onderka and Wishart 1988, Onderka et al. 1988, Foreyt et al. 1994). In addition, due to their incomplete adaptations to deserts (J. A. Bailey 1980), American desert bighorn are endangered by competition with exotics that are highly adapted to deserts, such as the North African aoudad (*Ammotragus lervia;* see Simpson et al. 1978).

For Native American people, the tragic consequences of encountering the diseases of Europeans are well known (Cook 1973, Baruzzi et al. 1977, Lightman 1979, Neel 1979, Joralemon 1982, Dobyns 1983). Black (1993) has added a new perspective on the cause of this demise that is crucially important to wildlife conservation. He argues that individual immunity is not particularly weak in Native Americans; however, their close genetic relationship to one another and a paucity of different alleles is the root cause of their susceptibility to diseases. When individuals in a population have a similar genetic composition, they are likely to fall prey to viral adaptations that overcome immune system responses in their relatives. The closer the recipient is in genetic composition to an infected individual, the more likely the recipient is to be infected by viral agents already adapted to overcome its immune system. Consequently, mammalian populations that arose from a small founder population during the rapid colonization of a continent are likely to succumb to disease epidemics. Black's (1993) hypothesis bodes ill for native mammals of Siberian origin, such as elk *(Cervus elaphus)* and moose *(A. alces),* for relict populations, and for reintroduced populations based on a small founding stock of closely related individuals.

The foregoing discussion generates expectations about the adaptations of mountain sheep. They should have (1) an ability to deal with the great landscape and habitat changes between glacial and interglacial conditions; (2) an ability to deal with ecological diversity and climatic extremes brought about by their adaptations to seasonality; (3) a deficit in adaptation to habitats found at lower latitudes; (4) an inability to deal with competition from southern exotics that have greater genetic heterogeneity, superior immune system responses, and superior adaptations for consuming toxic, sclerotic, and fibrous vegetation; and (5) superior behavioral adaptations due to the changes in predator faunas between the late Pleistocene and the Post-Glacial, with man and wolves emerging as major predators in the American fauna during the Recent.

Antipredator Adaptations

Bighorn sheep experienced a change in predator fauna 12,000–8,000 years ago, when the Rancholabrean fauna collapsed and Siberian immigrants entered North America. The native North American fauna had been dominated by a diversity of large predator species including the huge, predacious, short-faced bears *(Arctodus simus)*; large lions *(P. leo)*; big, cheetahlike cats *(Felis trumani)*; mountain lions *(P. concolor)*; large, dire wolves *(Canis diurus)*; large-bodied coyotes *(C. latrans)*; powerful, stocky, saber-toothed cats *(Smilodon)*; and long-legged, more fleet-footed scimitar cats *(Homeotherium)*. Whether *Homeotherium* were important predators of mountain sheep is uncertain.

However, mountain sheep remains from the last interglacial and glacial ages are found together with those of the short-faced bear, lion *(P. leo)*, and American cheetah (Martin and Gilbert 1978). These late Pleistocene bighorn (table 5.1) were as large as Pamir argali *(Ovis ammon polii)*, but had slightly shorter metapodia. In the relative dimensions of their limb bones, they stood between those of argali and current bighorn. Compared to current bighorn, late Pleistocene bighorn associated with the Rancholabrean fauna had a more slender body shape. Late Pleistocene bighorn also had relatively longer legs and were probably more cursorial than their descendants today, just as bison of the late Pleistocene had a more cursorial body shape than current bison (Smiley 1978, Guthrie 1989). It would be most useful to study muscle insertions on limb bones of late Pleistocene sheep to see if the limbs were muscled for power or for speed. The implication noted by Martin and Gilbert (1978), that relatively long-legged sheep were found with long-legged predators appears to be valid. Like bison, bighorn apparently experienced a change in body shape with the change in predator fauna after the late Pleistocene megafaunal extinctions (Guthrie 1989), though bighorn changed less profoundly. Both species became smaller and their legs became geared for power rather than for speed.

The major change in predation from the late Pleistocene to the Recent was in the rise of gray wolf *(Canis lupus)* and human populations. Bighorn sheep were regionally important prey for native people. For bison, the switch from a predator fauna dominated by large cats and predacious bears to one dominated by wolves and human hunters apparently had a severe evolutionary impact (Guthrie 1989, Geist 1996). In less than 3,000 years, this switch in predators changed the big, long-horned, highly cursorial bison that probably *confronted* predators to the small, short-horned, less fleet-footed, relatively shy modern form (M. Wilson 1980). Guthrie (1989) attributes this change solely to wolf predation. However, humans armed with throwing spears or bows and arrows would have been an even greater factor in selecting for a small, shy bison because hunters would have rapidly killed off pugnacious bison that confronted them (Geist 1993).

Studies of heart rates in instrumented, free-living bighorn sheep show that sheep reacted sensitively to wolflike predators (R. A. MacArthur et al. 1979, Stemp 1983). Coyotes elicited higher heart rates than did domestic dogs, and both merited higher heart rates than did disturbances by people, vehicles, or aircraft. It may be that in the absence of true goats and under intense wolf predation, both East Siberian and American sheep occupied the ibex niche (see Geist 1971).

Observations of sheep exposed to hunting in wilderness areas indicate that sheep withdraw to localities distant from where they have been severely disturbed and may not return for weeks (Geist 1971:87). Where hunted sheep cannot escape disturbance by humans, they may opt to stay in a given locality but maintain long distances between themselves and the disturbances. However, hunted sheep may also dash into forests. In either case, there is fragmentation of sheep bands, probably because in steep, broken cliffs and in cover it is difficult to maintain visual contact with other individuals. Although I have not studied systematically the spatial distribution of ram bands suddenly subjected to severe harassment, the distribution of Stone's sheep rams subjected to three years of hunting in my study area differed greatly from the distribution I had known earlier. The rams became scattered widely in tall cliffs, in pairs or as singles, while previous foraging areas were devoid of all sheep. Such grazing areas were vacated by ram bands for only three to five days following nonhunting human disturbance.

Mountain sheep living in smooth, coverless terrain with good visibility use grouping as an antipredator strategy, as do many other ungulates living in open landscapes. That is, sheep form "selfish herds" (W. D. Hamilton 1971, Treisman 1975). As Dehn's (1986) calculations show, the most important factor protecting members of the selfish herd is the dilution effect. Although individuals may share vigilance costs by one scanning for predators while others feed, with a frequent change of roles, such scanning is of little use at night. Moreover, sounds made by an approaching predator may be drowned by common herd noises such as the sounds of feeding, ruminating, shaking, scratching, and the frequent vocalizations commonly found in herding species. Nevertheless, in small groups, vigilance appears to be important because individual sheep do spend more time looking around as herd size decreases (Horejsi 1976, Berger 1979*b*). Also, compared to carnivores, herbivores need much more time to scan for and select digestible food. Herbivores have less time to scan for predators than predators have to scan for prey. Consequently, by grazing alone and scanning frequently for predators, individual herbivores may run short in their food intake. Therefore it is beneficial to form a group and share scanning time with others.

The primary strategy of mountain sheep to escape predators is to place obstacles in their path. This demands that sheep stay close to appropriate obstacles and dash to these when predators approach. The most common obstacles chosen by sheep are steep, sharply dissected cliffs, where gravity and footing are in the sheep's favor and where they may dodge quickly out of the predators' sight. Ragged cliffs permit hiding, as does shrubbery. Whether sheep hide deliberately in forest is contentious; there is no evidence of them hiding in places in cliffs, such as in crevices, in caves, or behind rocks. I have seen Stone's sheep emerge from caves in the morning during midwinter cold snaps with extremely low nighttime temperatures. However, I have never seen mountain sheep flee into caves, despite opportunities to do so.

Prior to estrus, female mountain sheep normally escape into cliffs when pursued by rams. A result of this behavior is that during the rut American mountain sheep rams can guard only one female at a time, but cannot guard harems. In Marco Polo sheep *(O. a. polii)*, which dwell primarily in rolling, open terrain with unobstructed vision, rams do herd harems during the rut (R. Petocz, International Union for the Conservation of Nature, pers. comm.).

Antipredator Strategy and Its Adaptive Repercussions

An antipredator strategy of placing obstacles in the path of pursuing predators has profound consequences for the adaptations of American mountain sheep. This strategy ties sheep to escape terrain, thereby fixing their grazing areas to narrow bands around such terrain. The finding by Stemp (1983) that heart rate in bighorn rises with distance from cliffs supports this view. The use of narrow grazing bands should make the annual reproductive output of mountain sheep a function of the annual fluctuations in available food supply. A synchrony of horn growth in rams and lamb production in females (Bunnell 1978) supports this view. The antipredator strategy also shapes the morphology and escape behavior of mountain sheep, making both remarkably similar to that of ibex *(Capra sibirica)*. However, bighorn have not reached the limb proportions of ibex and are anatomically intermediate between argalis and ibex (table 5.1). The need for visual obstacles in the environment prevents harem formation, thus forcing serial polygyny. Finally, this antipredator strategy imprisons and concentrates sheep during deep snow into small feeding areas around cliffs, which sets the stage for males competing for food with the females they bred and the offsprings they sired (Geist and Petocz 1977).

Cliff Escape Strategy as an Upper Limit to Body Size

The above mode of escaping predators appears to set an upper limit on the body size of mountain sheep. In cold climates with deep snow this strategy also places constraints on fecundity (after Geist 1981): because mountain sheep currently produce one lamb weighing about 4 kg at birth, what changes would be needed to produce two lambs of the same birth weight? Clearly, to support double the gestation and lactation costs, females would need to be twice the *metabolic* size. This translates into an absolute increase in body mass of 2.540 times and a corresponding increase in the linear body dimensions of 1.359 (the cube root of 2.540). I refer to a sheep so transformed as a "megasheep."

A megasheep that requires twice the food intake of a mountain sheep requires double the grazing area. To double the grazing area requires that the home range increase linearly by a factor of 1.420. If, to minimize grazing effort, a megasheep were to take as many steps during grazing as a mountain sheep, then its legs need to be increased not by a factor of 1.359, but by 1.420.

Also, to forage successfully, a megasheep must spend 50% of its grazing time at distances 1.420 times beyond the *safe* distance from the cliffs. Worse still, the maximum effective escape distance from the cliffs is *less* for the megasheep than for the mountain sheep. At twice the body mass, the power of the megasheep's muscles increases only as the square of linear dimension (about 1.840 times) to push 2.540 times the body weight; there is shortfall in power of nearly 30%. In short, the ability of the megasheep to accelerate is considerably impaired not only due to reduced power but also because its legs are *relatively* longer, thus increasing the length of the lever on which the muscles must act. Furthermore, the megasheep could not jump as high or as far as a normal sheep. Therefore, the antipredator strategy of escaping into cliffs, effective for American mountain sheep at their current size, is impaired at double the *metabolic* body mass. A sheep much larger than bighorn cannot rely on sprinting across open terrain to the safety of cliffs.

What would be the escape strategy of a sheep that has double the metabolic mass of American mountain sheep and that gives birth to two young weighing 4 kg each at birth? Fortunately, the answer is at hand because these specifications fit the giant sheep or argalis *(O. ammon)*. Argalis escape wolves by running at high speed with great endurance across open terrain. They are cursors (Gambaryan 1974) or, more precisely, they are long-legged cursors

Table 5.2. Dimensions of male Dall's sheep *(Ovis dalli)*, Altai argalis *(O. ammon ammon)*, and the hypothetical megasheep.

	DALL'S RAM	ALTAI RAM	MEGARAM
Body weight (kg)	90	200–235	227
Length of skull (mm)	280	350–385	378
Female skull length (mm)	255	305–337	346
Shoulder height (cm)	92	125	128
Weight of largest horns (kg)	8.6–9.1	23.2	21.7–23.0
Record horn length (cm)	123–127	160–179	167–173
Birth weight (kg)	3.2–4.1	4.0	4.0

Source: Geist (1981).

(table 5.1), that is, they are adapted to run across partially obstructed terrain. Long-legged cursors such as argali, pronghorn, or dama gazelle *(Gazella dama)* run with elevated heads that allow them to scan their flight path. Short-legged cursors such as saiga antelope *(Saiga tatarica)*, addax *(Addax addax)*, or buffalo *(Bison bison)* run with lowered heads over level, even terrain. However, though argalis are cursors they also still climb cliffs, and their limb proportions and morphology fall short of such highly evolved cursors as pronghorns (see X. Wang 1984).

There is a close fit between the body and horn sizes of a "megaram" and a Siberian argali ram (table 5.2), which suggests that mountain sheep are the product of opposing selective forces. Large body size selects for large neonatal size, which is essential for neonatal survival in cold climates. Such climates select for neonates large enough to survive hypothermia (Geist 1971:284–287). However, large body size reduces the efficiency of rapid escape into cliffs by decreasing the power-to-mass ratio, thus reducing acceleration and forcing individuals to forage beyond safe distances from cliffs. Mountain sheep have difficulty achieving survivable size for neonates, as indicated by their long gestation period compared to that of Old World sheep (Geist 1971).

However, selection for large body size in sheep is fostered once they adopt the antipredator strategy of cursors. This, in turn, encourages twinning and thus higher reproduction, an advantage in times of unstable climates. Conversely, a shift to warmer climates, and thus a reduction in the dangers of hypothermia to neonates, allows mountain sheep to shrink *adult* body size. That, in turn, allows niche expansion into marginal environments, which may explain the origin of desert bighorn and other small-bodied sheep that inhabit desert environments.

Latitudinal and Altitudinal Changes in Body Size

Average body size in mountain sheep appears to be a function of the *annual plant productivity pulse,* which predicts an initial increase in body size with latitude that is followed by a decline (Geist 1987*b*). Contrary to Bergmann's rule, large mammals do not continue to increase in size with latitude. The inflection point toward smaller body size for large, northern mammals appears to lie between 60 and 65°N latitude. However, because mountain sheep live at high altitudes, we expect the inflection point in body size to lie well to the south of 60°N.

Because every 100-m increase in altitude lowers annual temperatures on average by 0.6–0.7°c, equivalent temperatures to those between 60 and 65°N latitude are reached on sheep ranges between 1,200 and 2,000 m elevation at 50°N latitude. This is where we expect sheep to reach their largest average size. Indeed, the largest bighorn sheep are found in the Rocky Mountains between 49 and 51°N latitude, that is, in southern Alberta and British Columbia. Mountain sheep decline in body size to the north and to the south.

Because the decline of average annual temperatures accelerates with increasing latitude, we expect the decline in body size above 65°N latitude to be more rapid than below 65°N. The decline in average annual temperature per degree of latitude is about 10 times greater at 60°N latitude than at 20°N latitude. Indeed, caribou, for which there is adequate data to check this hypothesis, decline rapidly in size above 65°N latitude (see Geist 1987*b*). There is currently little adequate data for American mountain sheep to test this hypothesis. Nevertheless, a crude test is possible: according to Cowan (1940), the basilar length of Hensel of mature bighorn rams at latitudes 48–50°N is about 285 mm; for desert forms from California, Nevada, Arizona, and Mexico it is about 261 mm; for Baja California rams the value is 271 mm. This is a decrease of 5–8% in linear dimension over about 20° of latitude, or a 0.25–0.40% decrease in linear dimensions per degree of latitude.

The test is complicated to the north because here the species *O. canadensis* gives way to *O. dalli,* which has somewhat different skull proportions. Ignoring these differences, we note that for Alaska Dall's rams Cowan (1940) reported a basilar length of Hensel of about 246 mm. This does not include the small sheep from the Brooks Range and British Mountains living at 69°N. Consequently, the test is conservative. The decline in linear dimensions is at least 13% over about 17° latitude, or a 0.76% decrease in linear dimensions per degree of latitude.

Considering an average annual temperature of 26.2°c at the equator, annual temperatures approximating those at 60–65°N, and thus favoring maximum body size in mountain sheep, would be reached at about 4,500 m elevation. Therefore, south of 50°N we expect bighorn to increase in body size with elevation up to the level characterized by about -5°c average annual temperature, approximating 65°N, and to decrease in body size with higher elevation. At latitudes higher than 52°N, the body size of mountain sheep can only decrease with elevation.

Antipredator Strategy and Horn Size

Antipredator strategy has not only had an effect on neonatal size (Geist 1981), but also on horn size *through* neonatal size (Geist 1986, 1991*b*), so that increases in the size of hornlike organs parallel increases in neonatal size. In cursors, neonatal size must be maximized to (1) have highly developed young at birth; and (2) allow the neonate to rapidly grow to survivable size by beginning its postnatal growth at a relatively large body mass. In addition, females must produce either a very high flow or rich milk. This, however, requires females to have a superior ability to allot nutrients and energy to reproduction rather than to maintenance and growth. In bovids, as in cervids, there is a positive correlation between horn size and the percentage of milk solids produced (Geist 1991*b*).

In mountain sheep rams, superior ability at foraging, surviving, and sparing nutrients from maintenance and body growth is reflected in the size of horns. This should lead to males advertising with their horns during courtship, as Geist (1971) found. Also, females should select large-horned rams over small-horned rams for mating. Geist (1971) reported that ewes accepted more mounts by large-horned than by small-horned rams, and Hogg's (1987) work confirmed this.

This suggests that in species with harems and serial polygyny, horns should reflect heavy predation by enlarging in size. In males, horns reflect their ability to spare resources from maintenance. Unlike the thinhorn and snow sheep of relatively recent East Siberian origin, during their Pleistocene history in North America bighorn were subjected to the Rancholabrean predator fauna. This fauna was characterized by many large-bodied, fleet-footed predators (Kurten and Anderson 1980). It is noteworthy that in this fauna the extinct American stag moose *(Cervalces scotti)*, southern bison *(Bison latifrons)*, shrub-oxen *(Euceratherium)*, long-legged woodland muskox *(Bootherium bombifrons;* see McDonald 1984), and American mammoths evolved very large antlers, horns, or tusks. These species were relatively large bodied, as is expected if culling predation (during which prey are run down) is severe; they show evidence of cursorial adaptations in elongated limbs, or, as in the case of long-horned bison, limbs muscled for speed, not power (Smiley 1978).

Late Pleistocene bighorn were large and had *relatively* longer legs than today's bighorn. Today's large horns are an evolutionary relict from the late Pleistocene, when bighorn ran quickly away from more numerous predators. This is evidence for the hypothesis that bighorn were exposed to heavy predation by cursorial predators. The large horns of bighorn may thus be more than weapons, shields, and rank symbols for male engagements (Geist 1966b,c). They may have also evolved under sexual selection.

Ecological Segregation of Sexes

Not all sheep are equally capable of escaping from predators. Pregnant females and small lambs are expected to be more vulnerable to predation than adult rams, which suggests that rams could use areas that females and lambs could not. This would fit with the growth objectives of rams because they must maximize growth of body and horns to succeed in the rut and thus must maximize forage intake. This is particularly necessary in late spring and summer. However, having a minimum of energy expenditures in winter is an asset because it permits quicker recovery and a longer annual period of growth. Because rams are very capable runners and give no parental care, they can roam further and go into foraging areas that are less secure than those used by females. This hypothesis was developed in some detail for elk (Geist 1982). Also, the data of Geist and Petocz (1977), Morgantini and Hudson (1981) and especially Shank (1982) support this interpretation.

Note that ewes, who are bound closer to escape terrain than rams (Geist and Petocz 1977), are expected to graze more intensely on the relatively smaller grazing areas that are close to escape terrain. Females should, therefore, overgraze their foraging areas. However, females are expected to take in relatively less food than rams, because to *optimize* birth weight during gestation females must *not* maximize food intake. Otherwise ewes might grow a fetus too large to pass through the birth canal intact, if at all (Geist 1971). This condition (dystocia) may damage the lamb and/or the mother, or may cause the female to desert the young after an exceptionally painful birth. Thus lambs that are too large or too small are at risk, and the objective of females must be to bear lambs of *optimum* size (Geist 1971:286). Females can therefore compromise food in favor of security, whereas rams may compromise security in favor of food. Consequently, ewes and lambs are expected to be more closely associated with escape terrain than rams and to have smaller home ranges. For example, adult Alberta rams weigh on average 102 kg and ewes weigh 72 kg (D. A. Blood et al. 1970). That is, rams are 1.42 times heavier or 1.26 times larger in metabolic size and thus require home ranges *at least* 1.26

times larger in area or 1.12 longer in linear dimensions than those of ewes. No test of this hypothesis is currently available.

Sexual dimorphism is extreme in older mountain sheep; the only overlap in body size is found between adult females and yearling males just before the males' first rutting season. These two classes engage frequently in agonistic interactions during the preceding summer (Geist 1971). As a rule of thumb in all ungulates, females match in external appearance the class of males they are most likely to confront aggressively (Geist and Bayer 1988).

Adaptive Male Altruism

After the rut, old rams that have bred females ought to abandon the small, crowded wintering areas used by the females to maximize the survival of their prenatal offspring and their mates. This condition can only arise where the sheep are imprisoned in midwinter by deep snow on tiny feeding areas close to escape terrain. Data show that the older the rams, the later they appear on female ranges and the earlier they leave (Geist 1971:173). That is, old rams spend less time with females than do young rams. Also, the least spatial overlap between rams and ewes is found after the rutting season in midwinter (Geist and Petocz 1977). Adaptive altruism is only possible in a geographically closed system of resource availability, where the gain of one individual is the loss of another.

Home Range Traditions as Adaptation

Mountain sheep habitat is permanent (i.e., it changes little, if at all, from year to year) and may be distributed widely in small patches. Mountain sheep require habitat to have adequate escape terrain. Sheep migrate predictably between these habitat patches.

How does a mountain sheep assimilate the best patches into a home range quickly, cheaply, and safely? It appears that individual sheep incorporate the home range knowledge of older sheep by following the most successful ones (Geist 1967, 1971). Young rams tend to follow old, large-horned rams, whereas young ewes follow lamb-leading ewes. These are respectively excellent choices. In the case of rams, the size of horns is a function of the quality of their food and longevity. That is, horn size is a reflection of the quality of the home range occupied by the ram: the home range contained superior forage, superior security terrain, or both. Therefore, a young ram choosing to follow a large-horned ram maximizes the quality of the home range it inherits.

Striking out on its own is not a winning strategy. By dispersing to unknown terrain, a young ram is likely to expose itself to large areas of unsuitable habitat where, in addition to inefficiently allocating its resources, it risks death due to accidents and inadequate knowledge of the area. The ability to maintain home range knowledge as a tradition is an adaptation to *interglacial* conditions, in which open sheep habitat is at its minimum and at the highest elevations.

The above is a useful paradigm to explain the reluctance of relict and some introduced sheep populations to pioneer new areas (Fairaizel 1980). However, observations on reintroduced populations indicate that when bighorn sheep express high-quality, or dispersal, phenotypes (see Geist 1971:303–308, 1978: 116–144; Shackleton 1973; Horejsi 1976), they are prone to disperse. This was found by Butts (1980), who studied a reintroduced bighorn population that was characterized by early maturation, high fecundity, and extensive horn growth. These sheep did what bighorn normally are reluctant to do (and which may elevate their heart rate; R. A. MacArthur et al. 1979), namely cross strange, wooded gorges and canyons. Even bighorn, who are normally most conservative in their movements, when expressing a "dispersal phenotype," may become less sensitive to punishing stimuli (see Geist 1978:140) and disperse. The theory of epigenetic phenotype response to material resource abundance, which was first developed in detail during the study of mountain sheep (Geist 1971, Shackleton 1973), is useful in explaining differences in the behavior, anatomy, health, and taxonomy of vertebrates, including humans (Geist 1978:116–144, 1989).

six

DISEASES OF NORTH AMERICAN WILD SHEEP

Thomas D. Bunch, Walter M. Boyce,
Charles P. Hibler, William R. Lance, Terry R. Spraker,
and Elizabeth S. Williams

Introduction

Disease has had a significant impact on populations of North American wild sheep since Europeans colonized the mountainous regions of western North American. Prior to the arrival of Europeans, wild sheep were more numerous than today and were distributed over mountains or rocky outcrops that gave them protection from the open range. Competition from domesticated ungulates, loss of habitat, and disease reduced wild sheep numbers estimated in the millions to present-day estimates (Valdez and Krausman, this volume). This decline happened quickly in the last half of the nineteenth century. Although the Texas bighorn *(Ovis canadensis texiana)* and the Black Hills bighorn *(O. c. auduboni)* are considered by some experts to be invalid subspecies (Valdez 1982), they are now, nevertheless, extinct.

This chapter describes the major diseases and parasites that affect North American wild sheep. Individual sections cover pasteurellosis, verminous pneumonia, ectoparasites, psoroptic scabies, chronic sinusitis, gastrointestinal parasites, bluetongue, paratuberculosis (Johne's disease), contagious ecthyma (sore mouth), and mandibular osteomyelitis (lumpy jaw). A section is also devoted to "capture myopathy." Although capture myopathy is a metabolic-disorder syndrome rather than a disease, it leads to losses of wild sheep in capture programs.

This chapter represents the efforts of several scientists who have invested considerable time and effort in studying diseases that affect the well-being of North American wild sheep. It is hoped that wildlife biologists and managers, veterinarians, wildlife students, and interested people in general who are concerned about diseases of North American wild sheep will find this chapter a helpful and useful reference.

Pasteurellosis

North American wild sheep are susceptible to many diseases, the most important of which is bronchopneumonia in free-ranging and captive animals (Potts 1937, Marsh 1938, Cowan 1944, Buechner 1960, Pillmore 1961:83–97, Post 1962). Although several types of pneumonia infect wild sheep, the bacterial species *Pasteurella multocida* and *P. hemolytica* are usually a part of the infectious process. Three types of pneumonia involving *Pasteurella* are found in populations of North American wild sheep: (1) the verminous pneumonia of lambs; (2) the peracute fibrinous pneumonia of captive sheep; and (3) the acute to chronic bronchopneumonia of captive or free-ranging sheep. Because *Pasteurella* has such a significant impact on populations of wild sheep, it will be discussed at greater length and in more technical terms than the other diseases.

Recently captured wild sheep that are kept in an enclosure are the animals most susceptible to acute fibrinous pneumonia. Although the sheep appear to be adapting well to the enclosure, some type of additional stress, such as a storm, sudden change in the weather, or visitation of the pens by people, cause the sheep to die. Some do not show any clinical signs of pneumonia before death. Others will be moderately to severely depressed with droopy ears and lowered head, have frequent paroxysms of coughing, and lose their fear of humans. Sheep showing these clinical signs usually die within four to eight hours.

The postmortem lesions of sheep with the peracute form of pasteurellosis include a reddened (congested) nasal cavity and trachea, acute fibrinous pneumonia, moderate to severe hemorrhage of the epicardium, and congestion with petechia of the subcutaneous tissues. The sheep are usually in excellent body condition. Fibrinous pneumonia is characterized by the ventral and anterior or the posterior aspects of the diaphragmatic lobes being dark red and firm. These consolidated areas are often covered by a thin coat of fibrin. The cut surface of these consolidated areas is dark red; exudate (fluid) can be expressed from the cut bronchiolus and, occasionally, small areas of necrosis are evident.

Histologically, the trachea, bronchi, and bronchiolus usually contain neutrophils and are surrounded by a few lymphoid cells, macrophages, and plasma cells. The alveoli are dilated and filled with neutrophils, edema, macrophages, and fibrin. Occasionally, focal areas of necrosis of alveolar walls are evident. Thrombosed vessels rarely are found. The cortex of the adrenal gland is hyperplastic and congested. In most cases that involve Rocky Mountain bighorn sheep *(Ovis canadensis canadensis)*, a suppurative process occurs in the bronchiolus and alveoli within the lungworm nodules, and few to moderate numbers of first-stage lungworm larvae may be present within the ventral aspects of the apical and cardiac lobes. In other cases, lungworm larvae are not found within the anterior aspects of the lungs.

The gross lesions of the next type of pneumonia, the acute to chronic bronchopneumonia, are observed in captive sheep, but are more common in free-ranging animals. This pneumonia is similar to the peracute fibrinous pneumonia of captive sheep; however, one of the main differences in the acute to chronic bronchopneumonia is the increased amount of fibrin on the pleural surface. Occasionally, fibrinous pericarditis and epicarditis are observed. Pyothorax (i.e., pus in the thoracic cavity) is sometimes found. The sheep are usually in excellent to good body condition, although a few will be in poor body condition.

The histologic lesions demonstrate the more chronic nature of this pneumonic process. The condition is characterized by mild to severe hyperplasia of tracheal, bronchial, and bronchiolar epithelium; mild to extensive fibrosis surrounding bronchiolus and alveolar ducts; more infiltration of macrophages; areas of atelectasis; and areas of fibrin and edema within alveoli. Often there is a mild to severe pleuritis and, frequently, varying degrees of organization (i.e., granulation tissue) of the fibrin. Bacterial colonies and, on occasion, mycotic agents are found through histopathology.

Bacteria isolated from these types of pneumonia include *P. multocida, P. hemolytica, Corynebacterium pyogenes, Neisseria* spp., *Staphylococcus* spp., *Streptococcus* spp. *Mycoplasma* and parainfluenza virus type-3 (PI_3) have not been isolated from cases of acute to chronic bronchopneumonia.

The sequence of events leading to pasteurellosis, whether the peracute form or the acute to chronic form, can be triggered by varying degrees of stress. Many factors can cause stress and, generally, a single factor cannot be incriminated. Some examples of stress include overcrowding, poor nutrition, loss of escape cover, high lungworm burdens, harassment by dogs, encroachment by humans and their domestic animals, heavy snowfall, increased noise, and atmospheric dust (i.e., from blasting, dusty roads, or increased use of a road).

Increased stress initiates a mild to severe adrenal cortical hyperplasia, probably resulting in higher levels of circulating steroids, which predisposes the animal to infection. Steroids have many physiological actions. They affect protein and carbohydrate metabolism. Also, because of their action in the liver with regard to glycogen and catabolic action of proteins, steroids cause an increase in blood glucose. These steroids also lead to reduction in size of the thymus, spleen, and other lymphoid tissue, and high dosages are believed to cause necrosis of lymphoid tissue, thereby lowering the animal's resistance to infection. Steroids can interfere with growth, retard healing, and mitigate the inflammatory process.

After a die-off has begun, whether in the wild or in captivity, treatment usually is of little value. Occasionally a few captive sheep can be saved, but many of these become "pulmonary cripples." The main objective of treatment is to prevent a die-off from occurring. This is accomplished by reduction or removal of the assumed stressor(s). For captive sheep, the most important management objectives are to put the animals in large enclosures, keep people and dogs away, and give the animals an abundance of escape cover, good water, and food. Escape cover can be provided by placing walls or objects within the pens for the sheep to hide behind, thus giving them a feeling of security. This factor is very important and will markedly reduce the overall stress on the animals.

If a die-off is occurring in the wild, the most important form of treatment is to determine the stress factor(s) that triggered the die-off or the initial source that led to the *Pasteurella*-associated die-off. It must be emphasized that there is usually not just one factor involved, but a combination of factors. To save a few animals, or to prevent a future die-off, these factors should be curtailed. For example, if a herd contracts pneumonia and the predominant factors are crowding and poor range conditions, these factors should be corrected by recognition of the problem and removal of the excess sheep prior to the onset of the disease. If the pneumonia is thought to be related to commingling with domestic or exotic wild sheep, then corrective action should be taken to prevent further contact. Attempted removal of sheep after onset of the disease will only hasten the die-off. Treatment of sick sheep with antibiotics and anthelmintics usually does little to no good and under certain circumstances may facilitate the transmission of an infectious agent by concentrating the sheep, thereby increasing the morbidity and mortality of the animals.

A major concern in protecting populations of North American wild sheep from pasteurellosis is the potential transmission of *Pasteurella* spp. from domestic and exotic wild sheep (Foreyt 1989, Callan et al. 1991, Foreyt et al. 1994). Vaccination of North American wild sheep against cytotoxic strains of *Pasteurella* has not proven to be an effective measure to prevent the spread of *Pasteurella* (Foreyt and Silflow 1996). *Pasteurella* has been a serious mortality factor in the desert bighorn *(Ovis canadensis nelsoni),* causing the decimation of entire populations. A recent example is the widespread decline of a well-established herd of Nelson's bighorn sheep in the Canyonlands National Park complex east of the Colorado River. The Utah Division of Wildlife Resources counted 259 sheep ($N = 500$) in the North San Juan District in the late 1970s prior to the die-off. During the early 1980s the population steadily declined, and by 1988 only 6 animals could be located. The Utah Division of Wildlife Resources documented the decline in annual censuses. Commercial rafters reported dead sheep along the Colorado River during the early 1980s and alerted the National Park Service and the Utah Division of Wildlife Resources of the ensuing problem. A National Park Service employee spent most of a summer searching for surviving sheep and located only one ram. The animal was captured and sent to Utah State University, where it died within two days. The cause of death was chronic pneumonia. At the time of capture, the ram was extremely emaciated and had lost 33% of its normal body weight (J. Karpowitz, Utah Division of Wildlife Resources, Price, pers. comm.).

Until more is known about interspecies transmission of *Pasteurella,* it is absolutely critical that land managers and biologists avoid circumstances that allow domestic sheep and exotic wild sheep (e.g., mouflon, red sheep, and various exotic wild sheep hybrids) to commingle on ranges that harbor viable populations of North American wild sheep.

Verminous Pneumonia

Verminous pneumonia is undoubtedly a very serious parasitic disease in North American wild sheep and has had a major impact on several populations of Rocky Mountain bighorn sheep. Transplacental transmission of the lungworm *Protostrongylus stilesi* predisposes sheep to a fatal bacterial pneumonia. Overwhelming numbers of transplacentally transmitted lungworm larvae provide an opportunity for invasion by species of *Pasteurella.* Verminous pneumonia generally is acute in newborn lambs, especially if they are born with large numbers of infective lungworm larvae, and death will occur within four to six weeks. Frequently this is known as "summer lamb mortality" because the extremely young lambs simply disappear. In populations in which verminous pneumonia is severe, 75–95% of the lambs will die during this acute phase; however, in most populations, mortalities continue for two to three months after the onset of clinical signs.

Verminous pneumonia also infects adult sheep compromised by lungworm. Here, too, the disease manifests as pasteurellosis, but among adult animals death generally follows some form of stress, such as overpopulation, competition, inadequate nutrition, inclement weather, or harassment by humans. Any one or a combination of the above factors can provide sufficient stress to precipitate extensive mortality, even an "all-age die-off." For a more extensive treatise on verminous pneumonia see Buechner (1960), Forrester (1971), Uhazy et al. (1971), and Spraker (1979). An excellent bibliography has been published by Post (1971).

The clinical signs of verminous pneumonia in lambs are frequent paroxysms of coughing and a rough, yellow, shaggy coat. Affected lambs are small in stature and have lower body weights compared to offspring in healthy populations. Compromised lambs seldom frolic and generally lag behind the herd during any activity. When sheep succumb to verminous pneumonia as a direct result of some form of stress, all age groups are affected. In this situation, the primary clinical signs are frequent paroxysms of coughing and a nasal discharge.

The two species of lungworm most commonly found in wild sheep of North America are *Protostrongylus rushi* and *P. stilesi*. Although both species are important, *P. stilesi* is the most important primarily because of its ability to cross the placenta (Hibler et al. 1972, 1974; Spraker 1979). In adult males and females, *P. stilesi* resides in the lung tissue, generally in the dorsal tips of the diaphragmatic lobes. Female nematodes produce a thin-walled egg that is deposited in the tissue. Development of the egg is quite rapid, and a first-stage larva develops within a week. Upon hatching, first-stage larva make their way into the air passages, eventually migrating up the bronchial tree and into the mouth. Once in the mouth, larvae are swallowed and pass unharmed through the gastrointestinal tract to be expelled with the fecal material. Lungworm larval development cannot begin until the larva penetrates a land snail. A number of species in the genera *Vallonia, Pupilla, Gastrocopta, Pupoides, Vertigo,* and *Euconulus* serve as intermediate hosts for the development of *Protostrongylus* larvae. Several of these genera are ubiquitous throughout wild sheep habitats in North America (Spraker 1979). Development from the first-stage larva to the infective form (third-stage larva) requires 45–60 days, depending upon environmental conditions. Infected snails are accidentally ingested by wild sheep while they graze or forage in areas of decayed vegetation. If infected snails are ingested by an uninfected sheep, or a sheep with a very low level of lungworm infection, the infective form makes its way to the lungs (by some means presently unknown but presumably via the circulatory or lymphatic system) and begins development to the adult stage. Development to the adult stage generally requires about 20 days. The adults mate, deposit eggs, and the first-stage larvae hatch, thus commencing the cycle again. First-stage larvae generally can be found in fecal pellets between 28 and 35 days after ingestion of infected snails.

This basic type of life cycle is found in both *P. stilesi* and *P. rushi*. However, sometime during its evolution, *P. stilesi* developed the capability of transplacental transmission to ensure its perpetuation. Personnel with the Wild Animal Disease Center at Colorado State University have shown that if wild sheep are infected with *P. stilesi,* they are immune to a higher level of infection (C. P. Hibler and T. R. Spraker, pers. comm.). Infective larvae accidentally ingested by infected sheep become stored in the lungs and are transmitted to the fetus during pregnancy. Examination of pregnant ewes has revealed that undeveloped infective larvae are stored in most lobes of the lungs until the last six weeks of pregnancy; thereafter the larvae migrate across the placenta and remain in the fetal liver until birth. At birth, infective larvae enter the neonatal lamb lung and commence development, reaching sexual maturity at about 20 days of the lamb's life. The numbers of infective larvae that can be found in lungs of adult ewes, fetal livers, or neonatal lambs range from 1 to 1,500 (Spraker 1979). Lambs born into sheep populations in which verminous pneumonia is responsible for severe lamb mortality frequently are infected with between 100 and 500 larvae. The exact number necessary to predispose the lamb to a fatal pneumonia is not known, but 100 is more than adequate.

Readers interested in a more extensive treatise on the biologic cycle of lungworm should consult the annotated bibliography by Post (1971) and especially the extensive articles written by Forrester (1971), Hibler et al. (1972, 1974), Monson and Post (1972), R. E. Lange (1973), and Latson (1977).

The pathology of verminous pneumonia in wild sheep is an extremely complex subject and will only be given cursory coverage here. Readers interested in more extensive coverage on the subject should consult Spraker (1979). In summary, at birth the infective larvae leave the lamb's liver and enter the neonatal lung. Their presence is indicated by small foci of hemorrhage distributed throughout all lobes of the lungs. Shortly after arrival in the lungs, the larval stages begin a migration toward the dorsal aspects of the diaphragmatic lobes of the lungs. When lambs are 3.5–4.0 weeks of age, the great majority of the lungworms have migrated to the dorsal tips of the diaphragmatic lobes and the lesions indicative of infection consist of red-gray nodules surrounded by a zone of hyperemia. As the lungworm in the nodule matures, the entire nodule becomes surrounded and infiltrated by lymphoreticular cells. When the egg is deposited, an intense, mononuclear, inflammatory cellular reaction develops; as more eggs are deposited, the more intense the reaction becomes. Histologic evidence strongly suggests that the inflammatory cellular reaction is directed almost exclusively against the egg and the first-stage larva that subsequently develops. As a result of this, the lamb becomes sensitized (similar to any other allergy) to the egg and first-stage larva. Lambs from populations severely affected by verminous pneumonia often have 30,000,000 or more first-stage larvae in their lungs. As these massive numbers of first-stage larvae begin to make their way up the bronchial tree in sensitized lambs, some are aspirated into the anterior and ventral lobes of the lungs. These larvae immediately are surrounded by inflammatory cells, forming a nodule or granuloma. Because the parasites continue to produce, the granulomas increase in number and coalesce, which quickly depletes the immune system in the young animal. Bacteria summarily invade the compromised tissue. The species primarily associated with verminous pneumonia are *P. multocida* and *P. hemolytica,* but *Corynebacterium* spp., *Streptococcus* spp., *Staphylococcus* spp., and *Neisseria* spp., all of which are pathogenic secondary or tertiary invaders, are also present. Shortly thereafter (at about six weeks of age) a suppurative bronchopneumonia develops. The anteroventral portions of the lungs become consolidated (i.e., liverlike in color and consistency). The cut surface of this consolidated area develops a red-gray "cobblestone" appearance due to the presence of small, gray, clover-leaf-shaped, raised areas surrounded by a depressed red zone. Within two to five days following the consolidation, the lamb succumbs to "verminous pneumonia."

Verminous pneumonia becomes a mortality factor in a population of bighorn sheep when overpopulation and crowding results from mismanagement, limited habitat, loss of habitat due to competition with other wild and domestic ruminants, human activity (including housing subdivisions), increased recreational use on sheep habitat, and other forms of harassment. Their gregarious nature and marked tendency to develop patterns of habitat use make wild sheep their own worst enemy. As populations increase, sheep either refuse or are usually slow to migrate into new (or unused) habitat, even though the habitat may be quite suitable and have all the attributes (e.g., food, water, and escape cover) necessary for the health and well-being of the population. This promotes and perpetuates infection with lungworm and ensures the expiration of the population. Given all of the necessities for health and well-being, wild sheep tend to frequent preferred lambing, resting, and bedding areas and thereby produce an ideal habitat for snails (i.e., dense undergrowth of plants and moistened microenvironments). The added ingredient of feces promotes lush, green vegetation, and sheep find fewer reasons to leave bedding areas to forage for food. Thus they infect the habitat, reinfect themselves, transmit the larvae to the young, and begin again the verminous pneumonia cycle.

A typical example is the Pikes Peak herd in Colorado (Spraker 1979).

When the population had reached a maximum number of animals (approximately 1,000) in 1969–1970, large bands of sheep used preferred bedding areas night after night. These areas were approximately the size of a football field and contained a considerable amount of fecal matter, which resulted in a more lush and palatable vegetation. Bedding areas situated on southeastern-facing slopes tended to be warmer in winter and greened-up earlier in summer, which also provided an ideal environment for snails. Sheep left the winter areas between mid-April and mid-May and entered the lambing areas. They generally left these lambing areas around mid-July for the higher elevations (i.e., summer range). By mid-August to early September they moved back into the winter ranges in time to become infected by accidentally ingesting snails that were infected on these ranges the previous winter. The end result was that ewes began to store infective lungworm larvae in numbers sufficient to predispose their lambs to a fatal pneumonia by the time the lambs were four to six weeks of age. In 1970 the population suffered a 95% lamb mortality, and mortality continued at this rate until 1975. Due to the lack of lamb recruitment and old-age attrition, the population stagnated and began to decrease in numbers. By 1975, only approximately 162 sheep from the 1,000 estimated in 1970 remained. Although Pikes Peak is used here as the example, wild sheep populations throughout North America (with the exception of the desert bighorn sheep) crash in a similar manner. More often than not, this is the result of mismanagement—too many sheep on too small a range. Desert bighorn sheep populations often crash due to outbreaks of pasteurellosis, but the cause is usually not related to lungworms.

Stress is another prominent problem that is infinitely more subtle and almost impossible to measure. As indicated earlier, stress can be caused by a number of factors. Some of these are easy to measure, whereas others are less tangible. Too many sheep in a given habitat can constitute a form of stress. Irrespective of the cause, the end result is bronchopneumonia, due to a combination of lungworm and bacteria.

A presumptive diagnosis of verminous pneumonia can be made by observing the clinical signs of pneumonia in lambs aged four to eight weeks; however, final diagnosis necessitates obtaining fecal samples or initiating an animal collection. Fecal samples should be obtained and evaluated by the Baermann technique when lambs are four to eight weeks of age (Spraker 1979). A lungworm larval output of 500–10,000 larvae/gram is indicative of the disease. As a last resort, animal collections of those showing clinical signs should be considered.

Investigators should be cautioned that using fecal examination of adult ewes as an indicator of the lungworm burden in the population often is misleading. For example, by virtue of immunity to a higher level of infection, adult females may be passing between 100 and 500 larvae/gram; yet these animals could have stored a considerable number of third-stage larvae to pass transplacentally to the fetus. As the population number increases, more and more infective larvae are stored for transplacental transmission. However, because of their previous infection, adults would not be producing larger numbers of first-stage larvae. Diagnosis is best accomplished by using a combination of clinical signs, lungworm larval output in young lambs, and observation of the population and their habitat.

Based upon our experience, a number of compounds can be used successfully for the treatment of bighorn sheep for this disease. The key is to break the chain at its weakest link, which currently is the destruction of infective larvae stored in the ewes' lungs. Animals in a population in which verminous pneumonia is a problem should be treated with cambendazole, whereas animals in a population that are destined for transplant should be treated with fenbendazole. Cambendazole is effective only against the stored infective larvae and shows very low efficacy against adult nematodes, but fenbendazole has high efficacy against adult nematodes. When treatment of a population is contemplated, the animals must first be accustomed to feeding on apple pomace. Once they are readily eating the pomace, cambendazole can be mixed at the rate of 2.8 mL per 2.2 kg of pomace. Fortunately, adult sheep will only eat about 6.6 kg of pomace and will do so in a period of approximately 30 minutes. They will then leave the bait station and not return until later in the day, eating an additional 4.4 kg (Spraker 1979). As might be anticipated, yearlings and lambs eat a proportionately smaller amount of pomace. Once animals have been allowed to treat themselves with drugs mixed in pomace, the remainder should be picked up to prevent a second dose. If transplanting sheep from a population infected with lungworm into a clean area, such as historic range, is contemplated, then upon capture the animals should be drenched with fenbendazole at a rate of 10 mL per 110 kg of animal. Fortunately, both cambendazole and fenbendazole are extremely safe compounds; however, cambendazole is used at a dose level such that any extra drug may cause toxicity.

Control of verminous pneumonia in a bighorn sheep population is best accomplished by proper management techniques. Generally, this means trapping and transplanting excess animals into their historic range because the current policy on hunting favors an extremely limited, ram-only hunt. Wild sheep from a growing, thriving population should never be transplanted into a remnant population with confirmed verminous pneumonia. If transplants are made into remnant populations with verminous pneumonia, the population will generally increase for a few years and then return to its previous remnant status. Remnant populations with confirmed verminous pneumonia should be treated first for lungworm before being bolstered with additional sheep.

Ectoparasites

Arthropod parasites of wild sheep in North America may be important as direct agents of disease, as vectors, or as pathogens. The scabies mite (*Psoroptes* spp.) and the sheep nasal botfly *(Oestrus ovis)* are important ectoparasites of wild sheep in selected populations and are described in detail below. At least three genera of ticks infest wild sheep: *Dermacentor, Ixodes,* and *Otobius*. *Dermacentor hunteri,* a tick restricted to desert bighorn sheep, has been implicated as the primary vector of anaplasmosis (Crosbie et al. 1997). *Ixodes* spp. have been recovered from bighorn sheep infected with the protozoan *Babesia* spp. and are the likely vector of this pathogen (Goff et al. 1993, Kjemptrup et al. 1995). *Otobius megnini* is occasionally found in the ears of wild sheep, but has not been associated with clinical disease.

Psoroptic Scabies

Infestations with ectoparasitic mites in the genus *Psoroptes* appear to be widespread in wild sheep throughout the western United States. Most infestations are confined to the ears and are likely to be of limited clinical significance. Rocky Mountain and desert bighorn sheep have had serious *Psoroptes* infestations in the ears and on the body; it is becoming increasingly clear that psoroptic scabies is not confined to any particular geographic region or subspecies. However, it appears that Dall's sheep *(Ovis dalli dalli)* have not been exposed to the mite (Boyce and Zarnke 1996).

Although *Psoroptes* mites also infest a variety of other potentially sympatric domestic and wild ungulates such as domestic sheep and goats, cattle, mule deer *(Odocoileus hemionus),* and elk *(Cervus elaphus),* it is not clear whether mites are transmitted between these hosts. Recent evidence from New Mexico suggests that one interbreeding population of mites may actually be infesting sympatric bighorn sheep and mule deer (Boyce et al. 1990, Boyce and Brown 1991), and free-ranging elk and deer have also been infested with mites in an area known to contain infested Rocky Mountain bighorn sheep in Idaho. However, another study in California (R. S. Singer et al. 1997) indicated that mites were not shared between sympatric bighorn sheep, cattle, and deer.

Because of uncertainties about host specificity, the literature regarding the identification and taxonomy of the genus *Psoroptes* is confusing, and a variety of different specific epithets (e.g., *P. ovis, P. equi, P. cervinus*) may have been used to describe the same species of mite. Until further information is available, one should consider that infestations may be transmitted between host species and refer to infestations as being caused by *Psoroptes* spp., rather than attempting to make a species-level identification.

The issue of whether *Psoroptes* spp. mites were introduced into bighorn populations by domestic sheep and cattle is of considerable interest. Although early observers in the 1800s were convinced that bighorn sheep acquired scabies from domestic sheep, there is no way to definitively address this question and their conclusions must be regarded as speculative. Epizootics were noted in the late 1880s and early 1900s in Colorado, California, Wyoming, Montana, and Oregon; in recent years bighorn scabies has also been diagnosed in Nevada, Arizona, New Mexico, Idaho, and Utah (Hornaday 1901, Seton 1929, Wright et al. 1933, F. L. Jones 1950, V. Bailey 1936, Cater 1968, Decker 1970, deVos et al. 1980, R. E. Lange et al. 1980, Foreyt et al. 1985, Boyce et al. 1990, Hoban 1990, Muschenheim et al. 1990, Mazet et al. 1992). At this point in time, it is not clear whether the apparent upsurge in documented accounts of scabies in populations of wild sheep in the western United States is due to an actual increase in the incidence of infestations or to increased effort being devoted to assessing the infestation status in wild sheep populations. However, it is clear that scabies can be a serious disease, as evidenced by the loss of more than 90% of bighorn sheep in the San Andres National Wildlife Refuge during an epizootic that began in 1979 (R. E. Lange et al. 1980, Hoban 1990).

Diagnosis of scabies infestations can be accomplished by examining ear swabs and skin scrapings and by serologic testing. Boyce et al. (1991a,b) recently developed a sensitive and specific serodiagnostic assay (enzyme-linked immunosorbent assay, ELISA) that has been used to retrospectively determine the distribution and prevalence of infestations in populations of bighorn sheep in California (Mazet et al. 1992). The assay also has been used prospectively to assess infestation status of animals being relocated from one site to another in California, Oregon, and Idaho. Although treatment of bighorn scabies is not likely to be seriously considered as a management tool except under extraordinary circumstances, ivermectin has been shown to be effective against scabies at high dosages (Kinzer et al. 1983). Sustained-release implants containing ivermectin have been developed that could potentially be delivered to free-ranging animals via a remote delivery system (Boyce et al. 1992); however, the possibility of reinfection following treatment due to the presence of an infested reservoir host (e.g., deer) may completely preclude treatment as a management option.

Chronic Sinusitis

Chronic sinusitis is an infectious, noncontagious disease that primarily affects desert bighorn sheep. One case of chronic sinusitis has been documented in Rocky Mountain bighorn sheep (Turner 1982). Although the causative agent has not been unequivocally established, the disease is hypothesized to be initiated primarily from bacterial invasion of dying or dead nasal botfly larvae *(O. ovis)* entrapped within the paranasal sinuses (Bunch et al. 1978b,c; Paul and Bunch 1978). Turner (1982) reported a case of chronic sinusitis in a Rocky Mountain bighorn ram that he attributed to the disruption and subsequent infection of the tooth arcade. Because desert bighorn sheep have a high incidence of dental anomalies, he further suggested that an abscessed tooth or teeth could lead to chronic sinusitis in desert bighorn sheep; however, there was no such association in the studies of Bunch et al. (1978b,c). The highest prevalence of tooth anomalies in North American wild sheep is found in Dall's sheep (Glaze et al. 1982). As of yet, there has not been a reported case of chronic sinusitis in Dall's sheep. More than likely, there are several factors or agents that contribute to chronic sinusitis in desert bighorn sheep; however, the nasal botfly should still be considered to be the primary causative agent.

Chronic sinusitis infections usually progress to a pyogenic osteomyelitis, osteonecrosis, and osteolysis within the skull. Chronic sinusitis–induced lesions are extremely diverse and range from various degrees of spongy, porous bone within the matrix of the frontal and cornual sinuses to extensive lysis of the frontal bone, horn core, and sheath (Bunch et al. 1979, Turner 1982). In rams, infection often is localized in the middle to posterior region of the cornual sinus and results in extreme chafing of the horn sheath and eventual breakage of the horn. Infections centered at the base of the cornual sinus or within the frontal sinus generally erode the orbital region of the skull and cribriform plate, resulting in blindness and the brain becoming abscessed. Manifestations of chronic sinusitis in ewes are usually blindness, meningitis, and encephalitis. An oil-like or sometimes mucoid to slimy, purulent exudate often drains from fistulas of the frontal bone and horn sheath. *Corynebacterium pyogenes* is commonly found in the exudate. Chronic sinusitis often is associated with progressive debilitation, and the infected animal may lose one-half of its normal body weight before dying.

The life cycle of the nasal botfly has been thoroughly described in domestic

Table 6.1. Survey of chronic sinusitis in bighorn sheep endemic to Arizona, California, Nevada, New Mexico, and Utah.

LOCALITY/SUBSPECIES	EWE			RAM			LOCATIONS WHERE SKULLS WERE EXAMINED
	NEG.	POS.	FREQ.	NEG.	POS.	FREQ.	
Arizona							
O. c. nelsoni and O. c. mexicana	12	10	(0.45)	38	13	(0.25)	Arizona Game and Fish Dep., at Phoenix and at Yuma; Univ. Arizona, Tucson desert bighorn study site; Quartzsite
California							
O. c. californiana, O. c. nelsoni, and O. c. cremnobates	59	17	(0.22)	120	44	(0.27)	Univ. California, Berkeley and Riverside; Anza-Borrego Desert State Park; Joshua Tree National Monument; Twenty-nine Palms, Death Valley National Monument, Death Valley
Nevada							
O. c. nelsoni	100	19	(0.16)	124	20	(0.14)	Univ. Nevada, Las Vegas; Desert National Wildlife Range, Corn Creek
New Mexico							
O. c. canadensis and O. c. mexicana	12	0	(0.00)	29	1	(0.03)	New Mexico Game and Fish Dep., Sante Fe
Utah							
O. c. nelsoni				10	2	(0.17)	Utah Division of Wildlife Resources, Regional Office, Price, Utah; Arches and Canyonlands National Park, Moab
Total	183	46	(0.20)	321	80	(0.20)	

Note: Frequency of sinusitus is shown in parentheses.

sheep (Cobbett and Mitchell 1941, Capelle 1966, Cole 1969, Krull 1969, J. W. Davis and Anderson 1971). The gravid female fly deposits larvae near the external nares from early summer to autumn in cooler, temperate regions and during most of the year in warm, desert areas. The larvae migrate up the nasal passages to the paranasal sinuses, where they undergo two successive molts within 2–10 months. The mature larvae return to the nostrils and are forcibly expelled by sneezing. They then undergo a pupation period of 4–6 weeks. The mature female will live from 2 to 30 days after mating and will deposit fewer than 500 larvae.

The nasal botfly aggravates its host as it attempts to deposit larvae. Deposited larvae irritate the nasal epithelium and cause a copious mucous production, which causes sheep to sneeze frequently, shake their heads (often violently), grate their teeth, and (often) refrain from eating. Sheep that cannot expel mucus suffocate or die from abscesses induced by dead larvae.

Domestic sheep and goats are the common host for the larval stage of the nasal botfly, although the flies have occasionally been observed to parasitize Rocky Mountain bighorn sheep, white-tailed deer, and various species of African antelopes (Coney 1950, Capelle 1966, Horak and Butt 1977). Dogs and humans, when they associate regularly with infected sheep, are aberrant hosts (M. T. James 1947).

Chronic sinusitis has been diagnosed in populations of Rocky Mountain bighorn, California bighorn, and desert bighorn that are native to Arizona, California, Nevada, New Mexico, and Utah (Bunch et al. 1978*b,c*; Bunch and Webb 1979; Bunch 1980; Bunch and Allen 1981; Welsh and Bunch 1983). Examination of 630 bighorn sheep skulls showed that 20% had been affected by chronic sinusitis at the time of death (table 6.1). In many cases the necrosis was the primary cause of death. The highest prevalence of chronic sinusitis in ewes (0.45) was observed in Arizona and the highest prevalence in rams (0.27) was observed in California. Chronic sinusitis has been diagnosed in live bighorn sheep from Arizona, California, and Utah. Although Rocky Mountain bighorn sheep with chronic sinusitis have been reported in New Mexico, the disease has not been observed in captive or free-ranging Rocky Mountain bighorn sheep in Colorado.

Cobbett and Mitchell (1941) found the heaviest larval infections in heads of horned sheep whose frontal and cornual sinuses are especially large. Consequently, the severity of sinusitis in the bighorn may be aggravated by the extensive pneumation of their paranasal sinuses. The higher prevalence of chronic sinusitis in desert bighorn could be attributed to the xeric conditions, which favor the fly, and limited water resources, which predispose sheep to continued fly strikes. Flies do not have to travel long distances in search of sheep, but need only to wait at water holes or watering areas.

Chronic sinusitis affects bighorn sheep ranging in age from 1 to 18 years. Most sheep skulls with evidence of lesions characteristic of chronic sinusitis are 6 to 8 years of age at the time of death. Chronic sinusitis has not been observed in lambs.

Treatment of captive animals for chronic sinusitis has not been very successful. Furthermore, by the time of clinical diagnosis, the infection is usually so extensive that treatment (i.e., pressure flushing the infected area with antiseptic and antibacterial agents) is of little benefit (Paul and Bunch 1978). Treated sheep generally redevelop clinical signs of chronic sinusitis within 2–12 months.

Gastrointestinal Parasites

Wild and domestic ruminants throughout the world are parasitized by a myriad of gastrointestinal parasites, including protozoa, cestodes, trematodes, and nematodes. Domestic sheep are notorious for their number and variety of gastrointestinal parasites, and wild sheep are no exception. It would be unusual to find a wild sheep that was not parasitized. Although the presence of parasites should not be construed to indicate that animals are suffering detrimental effects, high levels of gastrointestinal parasitism are excellent indicators that an animal population has been subject to crowding, competition, and inadequate nutrition.

Clinical parasitism (i.e., disease) generally results from a combination of several factors including age, nutritional status, crowding, environmental conditions, and interspecific competition. Young animals, due primarily to their lack of exposure to gastrointestinal nematodes, are more apt to develop clinical parasitism when exposed to high levels of infection. Older animals are more likely to have been exposed to the parasites and, thus, to have developed some immunity. Despite previous experience with gastrointestinal parasites, undernourished animals are more prone to disease because they are not in sufficiently good body condition to develop an immune response to the parasites. Overpopulation, or reduced habitat due to interspecific competition, is conducive to high levels of gastrointestinal parasitism. When young, susceptible animals are present, the crowding effect predisposes them to levels of parasitism capable of causing disease. Environmental conditions suitable to the development of the free-living larval stages must be present for clinical parasitism to develop. The eggs or larval stages of most parasites require a certain amount of warmth and moisture. If these conditions are met, the parasite can survive. However, the suitability of these conditions determines the number of parasites that are successful. Interspecific competition, at least in part, can be considered to be a predisposing factor to the nutritional status of the host animal. Use of wild ruminant range by parasitized domestic ruminants leads to competition for nutrition and space requirements, resulting in clinical parasitism.

The wild sheep of North America occupy habitats with environmental conditions generally too extreme to support high levels of gastrointestinal parasites. There has never been a report on clinical gastrointestinal parasitism in wild sheep. However, this does not imply that clinical parasitism does not or will not arise when the factors necessary for perpetuation of the parasite species are adequate. Sheep are notoriously gregarious and tend to be creatures of habit. They prefer the same range and frequent the same resting and bedding areas. This, of course, promotes infection with those species of parasites having a direct life cycle. Moreover, species with an indirect life cycle, such as the cestodes of sheep, would be found more commonly in repeatedly used areas because intermediate hosts (i.e., arthropods) would have a better chance of becoming infected and subsequently eaten in resting or bedding areas.

The clinical signs of severe gastrointestinal parasitism generally are loss of body condition, rough coat, diarrhea, and anemia. However, infections with bacterial or viral agents often may be responsible for similar clinical signs. Therefore, if gastrointestinal parasitism is to be blamed as the cause of the morbidity, it must be confirmed by the recovery of parasites from the gastrointestinal contents in numbers sufficient to have caused the observed signs and lesions.

As might be anticipated, a tremendous number of genera and species of gastrointestinal parasites have been reported in various species and subspecies of North American wild sheep, and the great majority of the parasites are thought to be of domestic animal origin. Readers interested specifically in surveys of parasites in wild sheep are urged to read articles listed in the annotated bibliography by Post (1971). Excellent surveys have been conducted by R. W. Allen (1955, 1960, 1961, 1962, 1964), Honess and Winter (1956a), Becklund and Singer (1967), Uhazy and Holmes (1971), and Kistner et al. (1977).

The protozoans that affect wild sheep have not been studied as thoroughly as the nematodes and cestodes. A species of *Sarcocystis* has been reported as well as eight species of coccidians in the genus *Eimeria*, including *E. ahsata, E. arloingi, E. crandallis, E. faurei, E. granulosa, E. intricata, E. ninakohlyakimovae,* and *E. parva.*

Only a few species of adult cestodes parasitize ruminants, and all of these have been reported from wild sheep. These include *Moniezia benedeni, M.*

expansa, and *Thysanosoma actinioides.* One tapeworm of wild sheep, *Wyominia tetoni,* is probably the only species of wild sheep origin; the others are of domestic animal origin. Many of the reports of *T. actinioides* in wild sheep are probably actually *W. tetoni.* The only larval tapeworm reported from wild sheep is *Taenia hydatigena,* a common parasite of dogs, coyotes *(Canis latrans),* and wolves *(Canis lupus).*

A myriad of gastrointestinal nematodes have been reported from wild sheep. Like the protozoans and cestodes, nematodes originated in domestic animals. The nematodes reported from the stomach are *Haemonchus contortus, Marshallagia marshalli, Pseudostertagia bullosa,* and five species of *Ostertagia,* including *O. circumcincta, O. lyrata, O. occidentalis, O. ostertagi,* and *O. trifurcata.* Nematode parasites reported from the small intestine are a *Trichostrongylus* spp., *Cooperia oncophora, C. surnabada,* and 11 species of *Nematodirus,* including *N. abnormalis, N. archari, N. davtiani, N. dogieli, N. filicollis, N. helvetianus, N. lanceolatus, N. maculosus, N. odocoilei, N. oiratianus,* and *N. spathiger.* The nematodes reported from the large intestine include the pinworms, *Skrjabinema ovis, Trichuris ovis,* and *T. discolor.*

The above parasites have not been reported in all species and subspecies of North American wild sheep. However, as additional surveys are conducted, their discovered distribution widens because the great majority of these parasites have a broad host range and infect a variety of domestic and wild ruminants.

Quantitative or qualitative diagnosis of gastrointestinal parasitism is truly an art and should be undertaken by qualified parasitologists. Before attempting postmortem recovery of helminth parasites, investigators are urged to read articles by the following parasitologists: Allen (1955, 1960, 1961, 1962, 1964), Honess and Winter (1956a), Becklund and Singer (1967), Uhazy and Holmes (1971), and Kistner et al. (1977). These methods consist essentially of washing stomach and intestinal contents through a series of graded screens, followed by total or aliquot counts and subsequent identification of the various species recovered.

Using fecal examination to determine the burden of gastrointestinal parasitism in animals is a subjective technique, but often the only means of assessing the parasite status of a population. When used in conjunction with data on population size, body condition, forage quality, competition, and stress levels, the technique is useful for recommending management practices. A number of techniques are available and may be found in textbooks on medical and veterinary parasitology (e.g., Soulsby 1982, Georgi and Georgi 1990).

The best and most effective means of minimizing the effects of gastrointestinal parasitism is to obviate the cause, which probably will necessitate a change in management practices. At the risk of sounding redundant, it cannot be overemphasized that clinical disease due to parasitism is an indicator of another, much more major problem with the population and the habitat.

Bluetongue

Bluetongue is an acute viral disease of wild and domestic sheep, cattle, mule deer, white-tailed deer, and pronghorn antelope *(Antilocapra americana)*. The bluetongue virus is classified as an arbovirus in the family Reoviridae and is transmitted by a small, biting gnat, *Culicoides variipennis* (Luedke et al. 1967). Populations of the *Culicoides* gnat increase throughout the summer months, especially when conditions are conducive to large populations (e.g., warm and moist). Bluetongue usually arises in the late summer and fall and is more prevalent during a long, Indian summer.

To date, bluetongue has been diagnosed only twice in North American wild sheep. The first case was suspected by Robinson et al. (1967). They reported that bluetongue could have been responsible for losses of desert bighorn sheep in western Texas. One sheep was observed sick; the clinical signs shown by this animal were weakness and staggering. The primary postmortem finding was pneumonia (Robinson et al. 1967).

The second case report of bluetongue was observed in two captive Rocky Mountain bighorn sheep in Fort Collins, Colorado, in the summer of 1973. Domestic sheep were within 1.6 km of the 15 captive bighorn sheep, and cattle surrounded the facility. Two of the 15 sheep died due to bluetongue. The clinical signs of these two sheep were mild depression and a bloody diarrhea. The gross lesions included hemorrhages on the epicardium, endocardium, adventitia of the pulmonary artery and aorta, serosa of the rumen, small and large intestine, and a severe hemorrhagic colitis (especially of the spiral colon). Also, there were small petechia and ulcerations of the lower portions of the small intestine. A mild degree of edema was noted on the lungs. Histologically, the main lesions included hemorrhages within the epicardium, endocardium, myocardium, and adventitia of the pulmonary artery and aorta; mild to moderate hemorrhage and edema within the long parenchyma; moderate to severe hemorrhagic colitis; mild generalized hyalinization of muscular walls of the arterioles; and moderate endothelial-cell swelling of these affected arterioles. Bluetongue virus was isolated from one of these bighorn sheep. Unfortunately, the virus was not serologically typed.

The overall impact of bluetongue on wild sheep populations is unknown (Jessup et al. 1984). Robinson et al. (1967) believes that bluetongue was responsible for reduction of desert bighorn sheep, but there is really no proof for their supposition. The only statement that can be made in regards to the effects of bluetongue on populations of North American wild sheep is that Rocky Mountain and desert bighorn sheep are susceptible to the bluetongue virus and that bluetongue can be fatal. Stressed bighorn sheep probably are more susceptible; it is unknown whether domestic sheep and cattle can be carriers (Luedke et al. 1967, Metcalf 1977).

Paratuberculosis (Johne's Disease)

Paratuberculosis caused by *Mycobacterium paratuberculosis* is a specific, infectious, enteritic disease of domestic and wild ruminants. In bighorn sheep the disease is characterized by gradual loss of weight to a state of emaciation, intermittent or constant diarrhea in the later stages of disease, submandibular edema (i.e., bottle jaw), and an incubation period that may last several years, during which time the individual may excrete infectious bacteria into the environment. Clinical signs are usually seen in animals older than two years of age, though occasionally yearlings are affected. The known distribution of paratuberculosis in free-ranging wild sheep in North America is limited to a herd in Colorado (Williams et al. 1979) and a case in Wyoming (Williams 1982).

Paratuberculosis generally is acquired by ingestion of bacteria shed in feces; however, congenital infection of domestic calves and lambs born to clinical and subclinically affected dams has been documented and may be common (Doyle 1956, Lawrence 1956, McQueen and Russell 1979, Barker et al. 1993). The infecting organism has been isolated from the uterus of a bighorn ewe, thus suggesting that intrauterine transfer may also occur in wild sheep. *Mycobacterium paratuberculosis* has been recovered from milk (Doyle 1956), suggesting transmammary transmission, and from semen (Lawrence 1956), indicating the possibility of venereal transmission; however, these are not likely to be significant modes of transmission in wild sheep.

Mycobacterium paratuberculosis is extremely resistant to fluctuations in environmental conditions and under some circumstances may remain viable at least one year after being shed in feces (D. C. Blood and Henderson 1974). The organism is difficult to culture in the laboratory; special, enriched medium is required, and 6–14 weeks may lapse before growth is observed. Newer detection tests have reduced the time required for recognition of the organism in feces (Collins et al. 1990, Challans et al. 1994, Cousins et al. 1995).

Gross lesions in animals with paratuberculosis are variable but generally characterized by mild thickening of lower portions of the small intestine, markedly enlarged mesenteric lymph nodes, fluid ingesta throughout the intestinal tract, and emaciation. Submandibular and subserosal edema are commonly present. Microscopically, intestines and mesenteric lymph nodes are infiltrated with large numbers of inflammatory cells (i.e., macrophages and giant cells), often to the point of functionally destroying the organs. Bacteria may be numerous within the inflammatory cells (Williams et al. 1979, 1983).

Diagnosis of paratuberculosis is difficult, especially in subclinically affected sheep, and culture of *M. paratuberculosis* from tissues or feces is necessary to confirm the diagnosis. A number of serologic tests have been used for the diagnosis of paratuberculosis in domestic species; however, they are unreliable in bighorn sheep especially in the subclinical stage of infection (Williams et al. 1985). Recently developed serologic tests that may be useful in bighorn sheep include the ELISA (Collins et al. 1991, Burnside and Rowley 1994, Sweeney et al. 1995) and radioimmunoassay (Riemann et al. 1979). Agar-gel immunodiffusion tests have been used in domestic species with variable success (Dubash et al. 1996). A test of cell-mediated immunity, the lymphocyte blastogenesis test, is effective for diagnosing paratuberculosis in bighorn sheep and other wild ruminants, but is not very practical for free-ranging sheep (Williams et al. 1985). Additional, newly developed tests for domestic species have not yet been tested on wild ruminants (Stabel 1996).

Paratuberculosis is fatal in animals that develop clinical disease and treatment is not effective. Mortality rates due to paratuberculosis in bighorn sheep are not known, but have been estimated at approximately 5% in one infected herd. In domestic sheep, mortality rates vary from 1 to 25% per year (Sigurdsson 1954).

Paratuberculosis is very difficult to control. Several factors contribute to this: many animals may be subclinically affected and shed organisms into the environment, the organism is quite resistant to fluctuations in environmental conditions, suitable diagnostic tests are not widely available, and congenital infections are possible. Management of infected wild sheep herds by removal of clinical and some subclinical animals is possible, but will not free the herd of paratuberculosis. Depopulation of infected herds of European mouflon *(Ovis musimon)* and aoudad *(Ammotragus lervia)* has been used to control paratuberculosis in a zoo situation (Boever and Peters 1977).

Impact of paratuberculosis on wild populations is not known and is quite difficult to assess. Decreased productivity as an early effect of the disease may reduce the quality of affected herds. Subclinical and clinically affected individuals excreting *M. paratuberculosis* pose a threat to susceptible conspecifics and other species of wild and domestic animals. Paratuberculosis has been diagnosed in Rocky Mountain goats *(Oreamnos americanus)* associated with an affected bighorn sheep population (Williams et al. 1979). *Mycobacterium paratuberculosis* cultured from bighorn sheep will infect elk, mule deer, bighorn-mouflon hybrids, and domestic sheep (Williams 1981, Williams et al. 1983).

Contagious Ecthyma (Sore Mouth)

Contagious ecthyma, also known as sore mouth, scabby mouth, or orf, is a recently described disease of bighorn sheep. Early accounts of diseases of bighorn sheep (Seton 1929, Honess and Winter 1956*a*) do not describe contagious ecthyma. It was first documented in the herds of Banff National Park, Alberta, in the early 1950s (Connell 1954). Since 1954 contagious ecthyma has been described in the wild sheep of Alaska, northern California, southern Colorado, and southern New Mexico (Samuel et al. 1975, Blaisdell 1976, Neiland 1978). It is now considered to be ubiquitous to the herds and/or populations of wild sheep endemic to the western United States.

It is not known how contagious ecthyma is spread in wild sheep. Canadian researchers believe that it may be spread through artificial salt blocks. Colorado studies could not transmit contagious ecthyma to susceptible sheep through a highly contaminated salt block (R. E. Lange, pers. obs.). The contagious ecthyma virus is highly resistant to environmental extremes and may live up to 20 years in shed scab material. Once a salting area, feed ground, or bedding area is contaminated, elimination of the disease from a population is unlikely. Contagious ecthyma is observed only sporadically in a given population. For clinical disease to develop, other factors such as stress must alter the bighorn sheep's resistance to this environmentally constant virus.

Contagious ecthyma is a viral disease produced by a large, environmentally resistant virus that results in lesions limited, with few exceptions, to external epithelial surfaces (Livingston and Hardy 1960). The clinical disease is characterized by large (up to 4 cm), painful, pruritic, proliferative cauliflower-like growths over the muzzle, oral cavity, nose, and external genitalia. Externally exposed lesions become covered with a hard, dark, dried crust that, when removed, causes hemorrhage of the underlying lesion. Secondary bacterial infections exacerbating the condition is a common sequel. Within the oral cavity, the virus produces proliferative lesions at the gum line of the teeth in addition to ulceration of the hard palate and tongue. Contagious ecthyma lesions on exposed eyelids may form crusts over the eyes, thus producing impaired vision or clinical blindness.

The impact of contagious ecthyma on populations of wild sheep varies depending on other preexisting or complicating factors. In healthy adult sheep, contagious ecthyma generally is a transitory, self-limiting condition that resolves within 21 to 28 days. In nursing lambs, oral lesions can be of sufficient severity to impair feeding and may produce secondary, metastic, udder lesions on the lactating ewe. Painful udder lesions may prevent a ewe from allowing her lamb to nurse, leading to nutritional stress of the lamb. Colostral immunity is not protective and lambs younger than two weeks of age may develop active lesions. Where other preexisting detrimental factors are present, contagious ecthyma may function as a contributory mortality agent in concert with other factors, such as bacterial or verminous pneumonia.

All documented reports of contagious ecthyma in free-ranging wild sheep are associated with sheep being artificially concentrated (i.e., at feeding and mineral-lick sites, in captive enclosures, or within park boundaries) recently stressed by outside activities, captured, transported, or in populations experiencing rapid growth. The first epizootic of contagious ecthyma in Colorado occurred in a herd that was at a high population peak, following two years of severe drought conditions, which concentrated the animals even more. The 1979 epizootic in New Mexico occurred following capture, pesticide treatment, and confinement (R. E. Lange, pers. obs.).

Die-offs of certain bighorn sheep herds in Colorado have been preceded by outbreaks of contagious ecthyma some 6–10 weeks prior to mortality due to other factors such as pneumonia. Although the evidence is circumstantial, contagious ecthyma in populations of bighorn sheep may be related to an environmentally ubiquitous agent combined with a subacute or chronic stress condition.

The presence of contagious ecthyma in a population may be determined by characteristic, active, clinical lesions; presence of characteristic scars on the oral and nasal mucocutaneous function; virus identification; or serology. When contagious ecthyma is suspected, crusty scab material should be collected from the infected site, promptly frozen, and submitted to a qualified laboratory for virus identification by electron microscopy or confirmation through exposure of susceptible domestic sheep to scab material, with subsequent development of clinical contagious ecthyma. When active lesions are not visible, the presence of scars resulting from recent infection may aid in diagnosis. Wild sheep that have recovered from moderate to severe oral and nasal lesions often have residual, irregular, white to gray (i.e., depigmented) areas on the characteristically solid black nasal and oral mucocutaneous junction. These areas persist for less than six months following clinical disease. Serum samples may be submitted for complement fixation or indirect fluorescent antibody procedures to determine recent exposure to the virus, although the antibody to this virus may subside nine months following clinical disease.

Wild sheep transplanted from known infected herds should be thoroughly examined by experienced, wildlife-disease professionals to insure they are free of active or developing lesions. Contagious ecthyma is a true zoonosis and produces clinical disease in humans. Biologists have contracted the disease from contact with infected bighorn sheep (Jessup et al. 1991). Care should be exercised when handling wild sheep with active lesions.

Mandibular Osteomyelitis (Lumpy Jaw)

Mandibular osteomyelitis is commonly encountered in some populations of mountain sheep (Murie 1944, E. L. Johnson 1957, Allred and Bradley 1965, Hoefs and Cowan 1979, Glaze et al. 1981). Gross lesions are characteristic and include distortion and proliferation of the jaw bones, damage to teeth, abundant scar tissue, and, in some cases, tracts draining through the skin.

Inciting agents of these physiological changes are bacteria that have become established in the bones of the oral cavity. *Actinomyces bovis,* which is responsible for lumpy jaw in domestic cattle, has seldom been recovered from active lesions in wild sheep. However, numerous other bacteria such as *Actinomyces pyogenes, Corynebacterium pyogenes, Staphylococcus aureus,* and *Streptococcus* spp. are involved in mandibular osteomyelitis in wild sheep (Glaze et al. 1982, Palmer 1993).

Pathogenesis of lumpy jaw is well understood in domestic animals and is no doubt similar for wild sheep. Mandibular osteomyelitis in cattle generally results from extension of infection from the gums and tooth alveoli to the jaw bones. Bacteria associated with the infection generally are normal residents of the oral cavity that gain entrance to the tissue following injury. Coarse forage and plant awns have been implicated often in the pathogenesis of this syndrome. Infection may spread via the blood to other organs.

Little information is available on the prevalence of mandibular osteomyelitis in wild sheep populations. A study of desert bighorn skulls by Allred and Bradley (1965) showed that osteomyelitis was extremely common; however, the severity of disease, as judged by mandibular or maxilla distortion, was variable. Lesions tended to be more severe in older animals and the mandible was most affected. Prevalence of mandibular osteomyelitis ranges from 11 to 67% in some Dall's sheep populations (Hoefs and Cowan 1979, Glaze et al. 1982, Bunch et al. 1984). The syndrome has been identified in Stone's sheep in northern British Columbia (Sheldon 1930), Rocky Mountain bighorn sheep from Montana and Alberta (Cowan 1944, Green 1949, Couey 1950), and Dall's sheep in southwestern Yukon (Glaze et al. 1982).

Effects of mandibular osteomyelitis on sheep populations are difficult to assess. In areas where forage is predominantly coarse, lumpy jaw syndrome would be expected to be relatively prevalent (Glaze et al. 1982). Allred and

Bradley (1965) hypothesized that osteomyelitis and dental problems should be considered an important factor influencing the life expectancy of desert bighorn sheep. Affected individuals are unable to efficiently consume forage and in serious cases may lose body condition. These animals are also predisposed to predation (Murie 1944) and starvation. This syndrome, which affects sheep in widely differing habitats, may have population implications and warrants continued study.

Capture Myopathy

Capture myopathy is a syndrome observed in wild sheep often following capture, chemical immobilization, excessive exercise, or stress. Clinical signs of the condition appear from several hours to two weeks later and are characterized by sudden death, hyperthermia, depression, renal failure, shock, and disease of the muscular tissue.

Capture myopathy has been described in many free-ranging wild ruminants in North America. The condition is well documented in bighorn sheep, white-tailed deer, elk, antelope, moose, and Rocky Mountain goats (Spraker 1975, Wobeser et al. 1976, Chalmers and Barrett 1977, Haigh et al. 1977, Lewis et al. 1977).

The clinical signs of capture myopathy can be categorized into three different syndromes; however, the pathogenesis of each is similar. These syndromes include the (1) sudden death syndrome; (2) ataxic and myoglobinuric syndrome; and (3) ruptured muscle syndrome.

The sudden death syndrome is characterized by severe depression beginning 30–60 minutes after capture. Sheep lose fear of humans and become reluctant or unable to rise, the eyelids are usually partially closed, and the pupils are slightly dilated. Many of the animals are hyperthermic. The isoenzyme levels within the serum, including glutamic oxalacetic transaminase (GOT), creatine pyrokinase (CPK), and lactate dehydrogenase (LDH), will be markedly elevated. Postmortem findings are usually meager in these sheep and include a mild to diffuse congestion of the viscera, pulmonary edema, and congestion. The skeletal muscles are uniformly dark red; however, upon histopathologic examination, a mild to moderate degree of acute granular necrosis of myocytes is usually found in the large muscles of the hind legs. The muscles are markedly swollen, with loss of striations and with fragmentation and cleavage of the myofibrils. The nuclei in the sarcolemma are pyknotic. Loss of striations in the myofibrils can be confirmed with a phosphotungspic acid-hematoxylin stain. Edema and infiltration of inflammatory cells is not seen in cases of sudden death.

The second condition, the ataxic and myoglobinuric syndrome, usually is not observed until 8–12 hours after capture. Animals with this condition usually have experienced additional and prolonged stress, such as capture followed by transportation. The clinical signs in these sheep are characterized by mild to marked posterior ataxia, mild to severe torticollis, and myoglobinuria evidenced as a brown (i.e., coffee-colored) urine. Sheep with mild signs of ataxic myoglobinuric syndrome usually are not depressed; however, sheep with moderate to severe signs may be markedly depressed. If the signs are mild, animals often will recover if the stress is removed. If the clinical signs are moderately severe, such as posterior ataxia (a tendency to favor one or both hind legs), about 50% will die. If sheep have severe signs, such as severe torticollis of the neck or dragging one or both hind legs, there will be 100% mortality. Serum isoenzyme levels (i.e., GOT, CPK, and LDH) are markedly elevated. In mild cases, the blood urea nitrogen is normal to slightly elevated, but in the moderate to severe cases the blood urea nitrogen will be highly elevated due to renal failure. Death with the ataxic form of capture myopathy usually occurs within two to five days postcapture.

Postmortem examination of animals with ataxic and myoglobinuric syndrome will reveal that the major lesions are within the kidneys and skeletal muscles. The kidneys are usually dark reddish brown, swollen, and firm. The bladder is empty or contains a small amount of dark brown urine. Gross lesions in the skeletal muscles are usually found in the muscles of the cervical region, lumbar areas and the flexor, and extensors of the hock and stifle joints. These are characterized as multifocal, pale, soft, dry areas. The lesions are usually bilateral, but not necessarily symmetrical, and often are so subtle they are difficult to see in animals that die two to three days postcapture. However, because the condition is progressive in animals dying four to six days postcapture, the lesions are much more obvious. If animals with the ataxic syndrome are forced to move or run, muscle rupture frequently occurs within the compromised areas.

The most remarkable histologic lesions are found within the skeletal musculature and the parenchyma of the kidney. Lesions in skeletal musculature are the same type described for the sudden death syndrome: acute granular necrosis of myocytes. However, the lesion is more extensive and severe. If animals survive for four to five days postcapture, macrophages begin to infiltrate the area for the purpose of eliminating the necrotic sarcoplasm. Nuclei surviving in the sarcolemma begin to proliferate, attempting to repair the damaged muscle. The renal lesions are characterized by marked congestion of vessels and glomeruli, degenerated and necrotic tubular epithelial cells, and protein (myoglobin) casts within the degenerated tubules. Congestion and edema of the lungs, or congestion of the viscera, is usually not evident in such cases.

The third syndrome, the ruptured muscle syndrome, is characterized by a rupture of one or both of the gastrocnemius muscles. This condition is usually noticed from 12 to 36 hours following capture. Generally animals in this category will have ruptured muscles within 4–6 hours of transportation. Sheep with this syndrome do not show prior evidence of capture myopathy, and usually the affected sheep are only slightly depressed. The most evident clinical sign is an extreme weakness in the hind quarters and a hyperflexion of the hock; the animals seem incapable of keeping the hind legs functioning properly. Although they can still walk or run, they do so with great difficulty. Animals with ruptured muscles usually die within 5–6 days, but death can occur as late as 3–4 weeks postcapture. With treatment, about 50% will survive for 6–12 weeks and a small number of these will recover. In those recovering, an extensive amount of scar tissue develops in the ruptured gastrocnemius musculature. Serum isoenzyme levels are extremely elevated, but the blood urea nitrogen levels are usually within normal limits.

The predominant gross lesions are extensive hemorrhage within the subcutaneous lesions of the hind legs. Small to large, multifocal, pale, soft lesions are found disseminated throughout the muscles of the forelegs, hind legs, diaphragm, cervical region, and intercostals. These pale lesions are accentuated by small, white foci distributed regularly along individual muscle bundles. The lesions are bilateral, but not symmetrical. Occasionally, the common digital extensor muscle of the foreleg is entirely white. Many of the muscles containing the pale lesions also have slight to extensive ruptured areas. The specific muscles usually found ruptured include (beginning with the most common): gastrocnemius, subscapularis, middle and deep gluteal, semitendinosus, and cervical muscles. Most of the muscles rupture unilaterally, except the gastrocnemius, which is usually ruptured bilaterally. The remaining organs in these animals generally are unremarkable. Histopathologically, the muscle lesions are characterized by acute granular necrosis of myocytes and an extensive proliferation of nuclei in the sarcolemma, regeneration of myocytes, moderate macrophage infiltration, and fibrosis. Lesions in the other tissues are mild.

The pathogenesis of capture myopathy is believed to be related to the exhaustion of several normal physiological mechanisms within the body. The two primary physiological processes that occur during capture are exhaustion of the voluntary and autonomic muscles. The autonomic muscles, innervated by the sympathetic nervous system, function to supply energy, remove waste products, and maintain a cellular-oxygen tension and blood pressure that is compatible with life. The voluntary muscles function as a result of conscious thought and are involved in movement such as walking, trotting, or running to escape danger. With prolonged stimulation these physiological processes become refractory and, when complicated with hyperthermia, result in tissue hypoxia and stagnation of blood in the capillaries. Continued stimulation and exhaustion of the sympathetic nervous system lead to decreased tissue perfusion, vasculature collapse, and shock, which, in turn, result in tissue hypoxia and cellular necrosis, especially of the kidney tubular epithelium and skeletal muscle. Potassium can easily leak out of the damaged cells, causing ventricular fibrillation of the myocardium, which is already sensitized with epinephrine. Prolonged muscular exertion, especially when accompanied by heat, can quickly lead to anaerobic metabolism and a buildup of lactate in the muscles, causing cellular death, acidosis, and eventually death of the animal.

Animals with capture myopathy should be treated. The purpose of the treatment should be to correct the hyperthermia (if present), restore normal blood pressure by giving fluids, and correct the lactic acidosis by giving sodium bicarbonate. Measurement of the blood pH prior to the initiation of any treatment is the best approach to determine how much fluid and what pH to administer; however, generally this is not feasible. The treatment of choice is 500–750 mL of 5% sodium bicarbonate (pH = 8.0) administered intravenously. If the animal has renal failure (i.e., the ataxic myoglobinuric syndrome), fluids and diuretics also should be given with the sodium bicarbonate. When a sheep has a ruptured gastrocnemius muscle, the leg or legs may be casted and the animal kept in a small, confined area and provided with readily accessible food and water.

The most important "treatment" of capture myopathy is prevention. This is best done by reduction of stress on the animals. Capture of animals should be done on days that are neither extremely hot nor cold. Few, experienced personnel should be used on all trapping and transportation endeavors. Moreover, the number of visitors (the public or other professionals not participating in the operation) should be kept to a minimum because extra personnel lead only to confusion, resulting in more stress on the captured animals. The length of time that an animal is pursued (as in helicopter chases) or handled (as in a drop net or enclosure) should be kept to a minimum (i.e., < 30 min.). If animals are tranquilized, the correct dosages should be used. When animals are transported, the vehicle or compartment should be well constructed, easily darkened, and of a size sufficient to accommodate four to five animals without crowding. The compartment should be constructed to facilitate adequate flow of air, but with controls so that overheating or overcooling of the animals can be prevented. Care should be taken to prevent carbon monoxide poisoning during transport. Animals in late pregnancy should not be trapped, because the physiological processes that function to cause capture myopathy can also cause death of the fetus. In general, young animals are more resistant to capture myopathy than are older animals.

It must be emphasized that capture myopathy is a very important cause of death in captured animals and can make the entire transplant project a total failure. It can be prevented by reducing the stress of capture and transportation, and sometimes it can be treated, but prevention is far more successful.

seven

MANAGEMENT OF BIGHORN SHEEP

Charles L. Douglas and David M. Leslie Jr.

Introduction

Relative to the management of other North American ungulates, management of bighorn sheep *(Ovis canadensis)* is in its infancy. Bighorn management has progressed rapidly since 1975, and enough good and bad experience has accrued to assist in selecting management guidelines. All states in the western United States have been involved in hands-on management, including transplant efforts and habitat improvements. Hunting of mature rams is allowed in only a limited number of populations; the yearly harvest of rams is small and the demand for bighorn tags is enormous. In addition to a lottery for selecting winners of bighorn tags, western states auction one to two tags per year to the highest bidder. Revenue from these tags is used exclusively for bighorn management and has permitted opportunities for management projects by state fish and game agencies that otherwise would not have been possible. Substantial progress has been made in understanding the biology of bighorn sheep and techniques for managing them, but much remains to be accomplished.

State and federal agencies have gained experience in conducting bighorn surveys, trapping and transplanting animals, and improving water availability. Current management strategies have proven to be relatively effective, and any improvement in methodology probably will produce only small advances. In this chapter, we examine the philosophy of bighorn management, rather than provide a cookbook of management techniques. We also identify aspects of bighorn management that we believe warrant a review of philosophical underpinnings.

The worldwide catastrophic loss of wildlife habitat, species extinctions, and escalating numbers of threatened and endangered species are commonly known. Desert bighorn sheep *(O. canadensis mexicana, O. c. nelsoni, O. c. cremnobates,* and *O. c. weemsi)* are not endangered species, although in March 1998 the northern peninsular bighorn *(O. c. cremnobates)* was listed as endangered under the Endangered Species Act of 1973. Intensive management over the past 20 years by state and federal agencies has generally led to increased numbers in many herds and maintenance of numbers in others (Valdez and Krausman, this volume). However, despite heroic efforts by state wildlife agencies to save them, some herds have been decimated by diseases.

Although there are thousands of bighorn in North America, they suffer from the same problems that threatened and endangered species suffer: reduced and fragmented habitat, loss of movement corridors, loss of fitness, diseases, competition, predation, and human impacts. State status reports in *Transactions of the Desert Bighorn Council* (1975–1995) suggest that numbers of bighorn have increased in most western states since 1975, largely as a result of translocations. However, it is difficult to evaluate population estimates made 20 years ago because census methods have improved and early estimates were, by necessity, often only best guesses. Statewide population estimates of wildlife always will be guess work. Management successes appear to be mixed; some states have been able to dramatically increase their numbers of sheep (e.g., Arizona), whereas other states' management actions may be staying even with, but not necessarily ahead of, habitat losses, competition, predation, and diseases. States having lower densities of humans and more federal lands (e.g., Nevada) have been more successful in increasing their bighorn herds. This decade and the next will be pivotal in determining whether bighorn sheep and their habitat will be protected successfully or degraded further by fragmentation. With increasing human populations and accelerated development, the next decade will be critical for ensuring long-term conservation programs for bighorn populations.

Relatively few transplanted bighorn populations have flourished and grown into large herds (Rowland and Schmidt 1981). This should sound a warning that it is time to examine our philosophical frame of reference, our methods, our successes and failures, and what we can learn from research on species extinctions and conservation efforts elsewhere. Wildlife managers have learned an enormous amount about bighorn ecology and management since 1975, but we have only begun to ask appropriate questions about ecological, behavioral, genetic, and demographic interrelationships.

Wildlife managers should frequently examine their basic premises about management strategies and not be lulled into complacency because some tasks have been accomplished for bighorn. It is critical to determine whether management actions have actually been successful and to share this information. We hope to stimulate an awareness that bighorn management should be more than just "moving animals around" and of how bighorn management philosophy could be improved.

Background

Bighorn sheep inhabit mountain ranges in the western United States. Journal accounts of explorers and settlers of this region indicate that bighorn were once more widely distributed and numerous than they are presently (Buechner 1960). Although this perception is based on limited data and considerable conjecture, it almost certainly is true. Bighorn sheep are relatively numerous in the western states, consisting of several thousand animals in some states (Valdez and Krausman, this volume). Large numbers of bighorn populations suggest that all is well with bighorn sheep, however, these estimates do not address the critical issue of herd sizes and isolation. For example, California has 81 populations of bighorn and a statewide total of about 4,600 animals; only one population has more than 300 animals and only 15 populations have more than 100 animals (Torres et al. 1994). California has identified 10 presumed metapopulations (Torres et al. 1994). New Mexico has 6 populations and an estimated statewide population of only 230 animals; all herds have fewer than 100 animals (Fisher 1992). The Utah Department of Wildlife Resources and Bureau of Land Management biologists counted 1,401 bighorn sheep on a statewide survey in 1994; 4 populations had more than 100 individuals (McKee et al. 1995). In Arizona, more than 60 populations of sheep inhabit deserts, but fewer than 7 have more than 100 individuals (Krausman and Leopold 1986). Population geneticists generally regard an effective population size (i.e., number of breeding individuals) of 500 animals as necessary for long-term survival (however, see Lande 1995). To achieve an effective population size of 500, a total population of more than 2,000 animals might be required (Brussard and Gilpin 1989). A minimum viable population (MVP) of desert bighorn has been assumed to be about 100 $\pm$ 20 animals (for a panel discussion on MVP see Krausman et al. 1992). Some population geneticists now think long-term genetic risks to small, isolated populations probably have been substantially underestimated in most conservation programs (Lande 1995). Based on results of more recent work, Lande (1995) has become less optimistic about the potential for small populations to retain adaptive potential. He concludes that the effective population size may need to be on the order of 5,000, rather than 500, to ensure long-term viability.

There are recent examples of habitat loss and fragmentation and loss of local populations due to livestock competition, predation, and diseases. A major loss occurred in New Mexico when 85% of the San Andres herd (formerly about 200 animals) died from a virulent outbreak of contagious ecthyma caused by scabies mites (*Psoroptes* spp.; Sandoval 1980). Herds have been lost in other states following construction of roads through their habitat or an outbreak of disease. Other populations have maintained low numbers of animals, with birth and death rates being about equal. Although bighorn sheep are not federally listed as threatened or endangered, they should be considered a species of concern because they are basically a montane wilderness species that is being surrounded and impacted by civilization (Krausman 1993*a*).

Large vertebrates have relatively high probabilities of being negatively affected by inbreeding and loss of ability to adapt to the environment, both of which are consequences of increased genetic drift and loss of heterozygosity (Gilpin and Soule 1986, Lacy 1987). By contrast, the high reproductive rate of small vertebrates may permit these species to survive frequent fragmentation and local extinctions. Large vertebrates have low reproductive rates and may persist at low population levels for generations, buffered by longevity and adaptations to short-term environmental changes (Gilpin and Soule 1986). Habitat fragmentation and local extinctions of patch populations have profound negative consequences on the effective number of individuals and produce effective population sizes that are orders of magnitude smaller than the total census count (Maruyama and Kimura 1980, Rolstad 1991, Lande 1995).

Colonization is the establishment of animals in unoccupied habitat, whether by natural or artificial means. Bighorn sheep are not noted for being colonizers, although their ancestors must have been reasonably good at colonizing, as is reflected in historic bighorn sheep distributions. Studies of island biogeography, colonization experiments, and efforts to introduce beneficial species to a habitat all provide important conceptual information pertinent to bighorn sheep (R. H. MacArthur and Wilson 1967). Species having high fecundity are better colonizers than those having low fecundity. Bighorn sheep have low fecundity, producing one lamb per birth. In high-quality populations, ewes may breed as yearlings and give birth at two years of age; in most populations, ewes begin parturition when they are older than three years of age. Generalists usually are more successful colonizers than specialists (R. H. MacArthur and Wilson 1967). Bighorn sheep are facultative generalists in diet selection, selecting fewer plant species during the peak of the growing season and a larger number of species during the remainder of the year (Ginnett and Douglas 1982). Because of their ruminant digestive system, bighorn must select the most nutritious parts of plants.

Ungulates, such as bighorn sheep and deer (*Odocoileus* spp.), have large home ranges and require relatively large areas of habitat. Wildlife preserves and national and state parks are too limited in numbers and size to provide adequate long-term protection for bighorn sheep. Even our largest parks do not, by themselves, provide adequate space. For example, Death Valley National Park comprises about 1,400,000 hectares and has a population of about 300 desert bighorn sheep, which move in and out of the park on the eastern, western, and southern boundaries. Bighorn from Yosemite National Park winter at lower elevations on the eastern side of the Sierra Nevada on national forest lands. Boundaries of parks may run along topographic features, such as the ridge of the Panamint Mountains in Death Valley, which, until the area became a National Park in 1994, effectively left half of the mountain range outside its boundaries. These arbitrary boundaries are meaningless to wildlife. Effective management of desert bighorn and other ungulates must be conducted on a regional landscape basis, despite different management strategies and mandates of individual state and federal agencies. Agencies must develop joint management plans and strategies to assure long-term perpetuation of bighorn sheep. This will not be easy, but it is possible. If the overriding concern is long-term perpetuation of bighorn sheep rather than politics, necessary management choices will be clear.

Habitat Considerations

Bighorn sheep occupy habitat islands within mountain ranges, and herds frequently are small and spotty in distribution. Spotty distributions may occur naturally because of habitat attributes, habitat fragmentation, losses associated with disease, or as the result of transplantation. Habitat fragmentation has two components: reduction in total habitat area, which affects population sizes and thus extinction rates, and subdivision of habitat into disjunct patches, which affects dispersal and immigration rates (Wilcove et al. 1986). Habitat degradation without subdivision serves to reduce an area's carrying capacity, thereby affecting population size. Bighorn habitat contains important attributes, such as different facings and degrees of slope, rocky and nonrocky terrain, plateaus, drainages, vegetation types, and water.

The amount of habitat animals require for their seasonal needs is determined by body size. The size of mammalian home ranges are related to body mass raised to the 1.4 power (Swihart et al. 1988). With the exception of grizzly bears *(Ursus arctos)*, bighorn sheep and other native ungulates are the largest land mammals in the western United States and thus require the largest home ranges. Except in oceans or arctic tundra, suitable habitat for a given species is not continuous over large areas; most habitats are naturally arranged in disjunct patches within a landscape matrix. Small mammals (e.g., rodents) may spend their entire lives within an area of several hectares, but an area that size would represent only a small part of the home range of a bighorn sheep. Size, distribution, and quality of habitat patches determines how many animals of a given species can be supported.

Distance between patches and an animal's mobility and propensity for interpatch movements determines whether adjacent habitat patches can be used. Interpatch distances can easily be too large for an animal to traverse. Fences and interstate highways may serve as physical barriers in areas where interpatch movements might otherwise be possible. Habitat fragmentation negatively affects animals because it involves division of habitat patches into smaller parts that may not be sufficient to meet their needs. Animals may move between patches if adequate dispersal corridors (Noss 1987, Quinn and Hastings 1987, Simberloff and Cox 1987, W. E. Hudson 1991) are present and interpatch distances are not excessive. To be successful, corridors must provide many of the requirements of the species under consideration. Movement through corridors could be relatively rapid or might involve a substantial time period, depending on the corridor's configuration and size of the animals involved. Research on corridors is more advanced in Australia than it is in the United States due to conservation efforts to salvage hundreds of threatened and endangered species and their habitats (T. W. Clark and Seebeck 1990).

Escape terrain is one of the most critical patch attributes for ewes and they generally are found within 100–200 m of it (Leslie and Douglas 1979, D. W. Ebert and Douglas 1993). Escape terrain has slopes that are greater than 60% and rocky outcrops (Holl 1982, McCarty and Bailey 1994). The amount of escape terrain was an accurate predictor of ewe numbers in winter ranges of eight herds in the San Gabriel Mountains, California (Holl 1982). Regression models indicated that about 60 hectares of escape terrain were needed to support 10 ewes (Holl 1982). Distance to water is another critical patch attribute associated with escape terrain (Leslie 1978, Leslie and Douglas 1979, Monson and Sumner 1980). Escape terrain more than 3 km from water would likely be used only during winter or during periods when animals were free from water stress. Patches of escape terrain can be isolated and receive little or no use, unless animals can move into these patches by way of a ridgeline or corridor of relatively contiguous, steep, broken terrain. Seasonal movements of ewes usually are limited to lambing areas within escape terrain until lambs are mobile. Reproduction and population size ultimately depend on numbers and ages of ewes present and their physical condition. Availability of ewe habitat and free water is fundamental for continued existence of a herd within an area.

During most of the year, bighorn rams occupy terrain having slopes that are less than 60%, but are found with ewes in more rugged terrain during the rut. Seasonal movements of rams greatly exceed those of ewes (Leslie and Douglas 1979) and may allow exchange between subpopulations within a metapopulation or between mountain ranges (Witham and Smith 1979, Schwartz et al. 1986). Home range size and seasonal movements of rams increase each year until they are three to four years old, at which time there is an order of magnitude increase in home range size and distances traveled (Leslie and Douglas 1979). At about three to four years of age, young rams enter the rut and may move large distances to locate receptive ewes; such movements may lead to exploration of adjacent mountain ranges.

Intermountain movements of ewes are uncommon, but have been reported (Witham and Smith 1979, Bleich et al. 1990b, J. Jaeger 1994). Radio-collared ewes have been reported to move seasonally between mountain ranges in Arizona (Witham and Smith 1979), southern California (Bleich et al. 1990b), and Nevada (J. Jaeger 1994). Rams and ewes have been seen in Nevada ranges that do not support a resident population, which suggests that they were in transit between ranges (McQuivey 1978). Our understanding of whether intermountain movements occur in most herds of bighorn sheep is inadequate. The modest size of many, if not most, bighorn herds suggests that metapopulation structure is, or once was, operative (Gilpin and Hanski 1989, Bleich et al. 1990b). The level of interaction between bighorn in adjacent populations needs to be determined, because small populations will not persist unless they interact reproductively with other populations (Berger 1990, Bleich et al. 1990b). This determination is not easily accomplished; however, even a few animals moving between populations per generation can maintain genetic diversity. Prudent management dictates that bighorn movement studies and habitat assessments be conducted prior to proposed construction that could disrupt interherd movements. Livestock fencing is common in the West and can endanger bighorn or restrict their movements unless proper fence design is used (Helvie 1971). Mitigation measures have been proposed that help protect movement corridors between habitat patches (D. W. Ebert and Douglas 1993).

In western states, areas of high species diversity are found in riparian systems and habitat islands in mountain ranges with elevational and vegetational diversity (M. J. Scott et al. 1991). These same mountainous areas often support (or supported) populations of bighorn sheep. It could be argued that conservation of bighorn sheep habitat would protect many other species that rely on portions of bighorn habitat, even though their habitat requirements differ from those of bighorn sheep. It is economically prohibitive to study habitat requirements of every organism within bighorn habitat. However, conservation of bighorn habitat could serve less mobile organisms effectively.

Habitat Evaluation

Several approaches have been developed for evaluating the relative quality of bighorn habitat. Evaluation methods use topographic maps on which habitat areas are divided into 1-hectare cells (Bleich 1993, D. W. Ebert and Douglas 1993), 2.6-km^2 cells (C. G. Hansen 1980d, Armentrout and Brigham 1988), or 4-km^2 cells (Cunningham 1989, Cunningham et al. 1993). Each cell is assigned a score for habitat suitability based on the sum of scores for individual habitat factors. The effectiveness of this method can suffer from large cell size. For example, a road crossing a small corner of a cell results in the entire cell being scored lower than a cell without a road, even though the road may have no influence on most of the cell. Additionally, topographic and habitat variation within 2.6 km^2 of mountainous terrain can be enormous.

Holl (1982) used a pattern recognition system (PATREC) to evaluate bighorn habitat in the San Gabriel Mountains, California; that system used conditions associated with high and low population densities as standards of habitat quality to develop a habitat suitability index (HSI). More recently, geographic information system (GIS) software has been used for habitat evaluation and modeling. Evaluation begins by establishing a base elevation map from USGS digital elevation data (the same data used to create topographic maps) or by hand-digitizing from a topographic map. Each habitat attribute is prepared as a separate map layer; layers can be combined to create combination maps and the underlying data can be examined in many different ways. In raster-based GIS, images are divided into a grid of equal-sized square cells. Each cell is assigned a score for each habitat feature, such as a road, vegetation community, or slope. Feature scores for individual cells are summed for the HSI. Because remote-sensing data are recorded in raster format, a grid-based system is useful for processing satellite imagery. Vector-based GIS uses points, lines, and polygons to separate map segments. Both kinds of systems georeference imagery to the terrain it represents; therefore maps of habitat attributes, such as distances between places, correspond to measurements in the real world. Many researchers use cells of different sizes to measure HSI; Cunningham et al. (1993) used 4-km^2 cells, whereas D. W. Ebert and Douglas (1993), Bleich (1993), and Dunn (1996) used 1-hectare cells. T. S. Smith (1991) used a vector-based program to assess habitat. The use of GIS permits more refined spatial analyses of habitat attributes than has previously been possible. It is now possible to (1) quantify amounts of habitat attributes within patches, corridors, or landscapes; (2) analyze fractal dimensions of patches; (3) assess patch or attribute continuity within and between mountain ranges; (4) identify threats to habitat integrity; (5) assess deficiencies in habitat; (6) perform statistical tests of habitat data; and (7) simulate changes in habitat quality or attribute availability under different management scenarios.

Critical thresholds of habitat degradation and fragmentation are poorly understood. Livestock overgrazing in bighorn habitat or corridors is ultimately detrimental to bighorn in the area, because overgrazing reduces forage resources and range carrying capacity. Also, livestock fences serve as barriers to bighorn. Proximity of livestock and bighorn increases the potential for transmission of livestock diseases to bighorn. Serum samples from virtually all western bighorn herds that have been tested have more than one positive titer for livestock diseases (D. Jessup, pers. comm.), indicating past exposure of bighorn to disease organisms. A positive titer does not confirm that bighorn have had a given disease, but numerous losses of bighorn to common livestock diseases such as bluetongue, parainfluenza-3, and bovine respiratory syncytial disease should alert managers to potential dangers of disease transmission between livestock and bighorn herds. Domestic livestock and bighorn sheep should never be allowed to use the same habitats (Bunch et al., this volume).

Population Assessment

Determining the number of animals in a wildlife unit remains one of the most difficult and challenging tasks for wildlife researchers and managers. Western mountain ranges and wildlife management units usually are large and may encompass hundreds of square kilometers of terrain. Large, free-ranging ungulates are difficult and costly to census; at best, only a fraction of the population will be seen on a survey, and population trends must be evaluated from the number of males, females, and young animals observed.

Population estimates for large mammals based on mark-recapture calculations commonly have 95% confidence intervals (CIs) as large as ± 50% of the observed animals (Caughley 1977). The challenge is to evaluate what portion of the total population is represented by the sample (Caughley 1977). For example, if 100 animals were observed, some of which were marked, and the mark-recapture estimate was 100 ± 50 (95% CI), the population size probably lies between 50 and 150 animals. Because 100 animals were seen, the true population is probably between 100 and 150 animals (if no double counting occurred). Any number within the 95% CI has an equal probability of being the actual number of animals in the population. This creates a difficult situation for the wildlife manager because error associated with the estimate may be so large that small fluctuations in the population cannot be detected. By the time a large decrease is detected, the population may already be in jeopardy.

Long-term trend data are needed to help circumvent inaccuracies in all survey and/or census methodologies, but long-term data exist for only a few bighorn populations (Leslie and Douglas 1986). The worst error would be to assume the best-case scenario and determine a management action, such as removals for transplantation, on the assumption that there were actually, for example, 150 animals in the population. If 50 animals were removed and the actual starting population was only 100 animals, removal could seriously jeopardize survival of the population. Furthermore, removal could dramatically alter age structure, which would be expected to negatively affect the population's productivity and rate of recovery (Leslie 1980, Leslie and Douglas 1986, Stevens and Goodson 1993).

Transplantation

Since the 1960s, transplantation of bighorn sheep has been a favored management strategy to reestablish populations in historic but unoccupied habitat and to supplement dwindling populations. Transplant efforts have had various degrees of success and most have not achieved a population size of more than 100 animals, which has been proposed as the minimum number required to assure a high probability of survival for the next 100 years (Berger 1990). Some transplant efforts have failed entirely due to heavy predation or unknown causes (Rowland and Schmidt 1981). Berger (1990) examined historic records of bighorn herds in the western states for which there were population estimates for more than 70 years. He found that all herds with fewer than 50 animals became extinct within 70 years, whereas all herds with more than 100 individuals persisted for 70 years and were still extant in 1990. However, not all small herds go extinct (Krausman et al. 1993a, 1996), and those that persist probably interbreed with animals in adjacent populations as part of a metapopulation (Schwartz et al. 1986, Gilpin and Hanski 1989). Until recently, bighorn transplant groups in the western states have consisted of fewer than 12 to 30–40 individuals; the recommended transplant size is a minimum of 20 animals for direct release (Rowland and Schmidt 1981, L. O. Wilson and Douglas 1982). Of 13 transplant efforts reviewed by Rowland and Schmidt (1981), only 6 started with more than 20 animals. It is not surprising that small transplant groups frequently have had difficulty increasing in size. Colonizing animals usually have higher mortality than those in established populations (Ebenhard 1991), and loss of reproductive ewes from a small transplant group could have a major impact on population growth (Leslie 1980). Unquestionably, small, isolated populations are extremely vulnerable to extinction.

Addition of individuals to bighorn populations may be warranted when a population has sustained major losses and is languishing or when a small transplant group is not successfully recruiting young individuals. Populations that continually have small effective numbers of breeding individuals are especially susceptible to loss of genetic diversity by genetic drift or inbreeding (Ellstrand and Elam 1993, Fitzsimmons et al. 1997). In populations that undergo occasional fluctuations to small population size, loss of genetic diversity also can happen by chance; such fluctuations may occur during population bottlenecks or in founder and/or colonization events (Ellstrand and Elam 1993).

Transplanting bighorn into historic but unoccupied habitats is a worthy and useful management action if habitat conditions are suitable and the transplant group has a high potential for success. Care must be exercised that a transplant group does not become another relict population, as apparently has happened in a number of desert bighorn transplant efforts (Rowland and Schmidt 1981). To have a reasonable probability of long-term survival, a transplant group must expand either naturally or through additions of animals to attain more than 100 individuals (Berger 1990; but see Krausman 1993a, Goodson 1994). Although multiple reserves offer enhanced protection for wildlife populations, the creation of isolated populations of bighorn serves to increase management problems and to jeopardize long-term survival of the transplanted animals. If movement corridors exist between an established herd and a transplant group and the combined groups total more than 100 individuals, then gene flow between the herds and long-term persistence should be possible (Berger 1990).

Most wildlife managers would not consider removing 50% of a bighorn population for transplantation. However, there have been situations in which 25–30% of the animals counted on surveys have been removed for transplantation several years in succession (Leslie and Douglas 1986, Stevens and Goodson 1993), under the apparently erroneous assumption that population estimates, based on marked and unmarked individuals, were the real figures that management decisions should be based upon. Confidence intervals for population estimates have been downplayed, and the upper CI value has been used to substantiate the conservative nature of large removals from the herd. In one instance, a population estimate was unrealistically larger than the total survey count or previous estimates, had large CIs, and indicated explosive population growth during drought years when there was essentially no survival of young. Nevertheless, a large number of animals were removed for transplantation (Leslie and Douglas 1986), which resulted in lower reproductive rates and a greatly reduced herd size for several subsequent years. These management actions were based on faulty assumptions, political expediency, and refusal to use common sense.

Similarly, removing transplant animals from a small population (< 100 animals) is risky. Ungulate harvesting theory has two major strategies: the harvestable surplus model (HSM) and the partial compensation model (PCM). The HSM assumes that there is natural compensation for animals removed by hunting or mortality. Under this model, population density is not reduced below carrying capacity (K_1) if cropped at a level equal to the production of young or yearlings. Under the PCM, cropping to produce a sustained yield always reduces population density and maintains it below K_1 (Caughley 1977). Under either strategy, repeatedly harvesting more animals than are recruited into the population will ultimately drive the population to extinction. Removals for transplantation should be based on recruitment of source herds and numbers of animals seen from year to year on surveys, rather than on population estimates or assumptions about compensatory responses (Leslie 1980; Leslie and Douglas 1982, 1986; Stevens and Goodson 1993).

The size of the transplant group and the ratio of natality to death are directly related to the probability of success in colonizing a habitat patch (R. H. MacArthur and Wilson 1967, Ebenhard 1991). Age composition of the transplant group also is important, because it influences the group's reproductive value (R. H. MacArthur and Wilson 1967, Williamson and Charlesworth 1976). Leslie (1980) demonstrated that age composition of the transplant group was of major importance in establishing a herd of desert bighorn sheep. Removal of older age classes of ewes had the least effect on the parent population but, due to older ewes' limited reproductive capacity, was not optimum for the transplant group. In contrast, removal of young ewes had the greatest effect on the parent population but was optimum for the transplant group (Leslie 1980). Natural replacement of reproductive ewes removed from the parent population may require five to seven years (Leslie and Douglas 1986, Stevens and Goodson 1993).

Except in the highest-quality populations, reproduction is confined to individuals older than three years of age (Geist 1971, Monson and Sumner 1980). In high-quality populations, yearling ewes may conceive and reproduce as two year olds, and bighorn sheep produce only one lamb per year. High lamb mortality is common in desert herds, especially during dry years (C. L. Douglas and Leslie 1986, C. L. Douglas 1993). The generation time of bighorn is approximately five to six years.

Desert bighorn sheep live in a highly stochastic environment with highly variable precipitation and forage production between years. In years having more than average precipitation, forage production and lamb survival generally increase. The timing of precipitation is more critical for forage production than the total yearly amount received. In drought years, forage production and lamb survival are minimal (Leslie and Douglas 1982, C. L. Douglas and Leslie 1986). The closer the size of the population is to the seasonally decreasing carrying capacity, the more severely it will be affected by climatic vicissitudes (Macnab 1985). The River Mountains, Nevada, support a highly productive herd of desert bighorn sheep. Over the past 18 years, October–November lamb:ewe ratios averaged 40:100, but have ranged from 8:100 to 80:100. Summer is the period of major lamb mortality for desert bighorn sheep (Monson and Sumner 1980); those lambs surviving until autumn likely will be recruited into the population. Vagaries in precipitation lead to "boom or bust" population demography.

Mortality in desert bighorn tends to be relatively linear from two years of age to the last year of life (Leslie and Douglas 1979). Therefore, an animal surviving to two years of age has a good probability of living another 10–12 years. Survival or mortality in lambs is normally the driving force behind population dynamics. A transplantation of desert bighorn is affected by demographic and environmental stochasticity. In desert environments, management actions must be conservative, based on demographic and weather data, and should be guided by long-term conservation perspectives. Bighorn herds in more stable, northern environments, where lamb survival and population density are high, show more predictable results from management actions.

Active, hands-on bighorn management began in the southwestern deserts in the early 1970s. Thus, we did not know until relatively recently that high-quality bighorn herds (Geist 1971) are more resilient to transplantation and hunting removal than had been anticipated. Conversely, managers had not recognized until recently how fragile low-quality herds could be and how readily they could be lost. Unfortunately, there are far fewer high-quality herds than low-quality herds of desert bighorn sheep. Wildlife managers should not hesitate to examine and reevaluate their basic assumptions regarding a species, the way they are managing it, probable reasons for successes and failures, and ways of improving management practices. All management actions should be considered experimental (Macnab 1983) and should be monitored to evaluate why they succeeded or failed.

Transplant efforts are popular with hunters and conservationists and enjoyable for wildlife managers and volunteers; much public recognition ensues from efforts to reestablish bighorn populations. Conversely, there is no appreciable public-relations benefit for monitoring a transplant group, which may explain the inadequacy and short duration of most follow-up monitoring. Unfortunately, many groups that have been transplanted over the past 20 years may not survive because they will not achieve adequate numbers; inbreeding depression may result from low founding numbers; habitat may be inadequate in size, quality, or proximity to the nearest-neighbor patches; or livestock diseases may be transmitted to the group.

Hundreds of bighorn sheep have been trapped and relocated to unoccupied habitats in the western United States and Canada. Procedures and techniques have been outlined elsewhere (Trefethen 1975, L. O. Wilson and Douglas 1982, Krausman et al. 1984, Bailey 1990). Our intent here is to provide a condensed summary that highlights standing protocols. Procedures have been developed over the past 25 years by various state, federal, and provincial game and land management agencies. Considerable expertise exists among personnel of those agencies.

TRAPPING TECHNIQUES

Bighorn sheep have been captured with various trap designs or with drop nets by baiting with apple pulp and hay, by trapping at watering sites, by driving animals into nets, and by immobilizing sheep from helicopters with capture guns using nets or drugs (e.g., etorphine plus azaperone, etorphine plus xylazine). Safety of the animals and personnel must be the primary concern during all phases of capture, and an experienced crew will minimize injury. Attempting to handle more than 25 animals in a trap or net at one time increases the chance of injury and the likelihood of capture myopathy (Bunch et al., this volume). Excessive chasing or hazing during periods of high temperatures should also be avoided. Winter captures are common in northern states, but are less common in desert areas. It is mandatory to seek experienced help before attempting to use drugs because individual tolerance and darting challenges vary with each circumstance. Use of a net gun is becoming the capture method of choice in many western states because this method allows the selection of animals to be captured and is more expedient and cost effective than administering drugs. Mortality losses from net gun capture are about the same as from drop nets or corral traps.

HANDLING BIGHORN SHEEP

Trapped bighorn sheep can be immobilized without drugs by using the appropriate number of field personnel. Generally, two people are required to immobilize ewes and two to three people are required for rams depending on their size. Blindfolds and leg hobbles are universally successful in calming captured animals. The sheep's head should be held up and the animal should be kept in sternal recumbency, not lying on its side. Each captured animal should be checked for injury and signs of possible stress (e.g., labored breathing, excessive salivation, heart rate > 130 beats/minute, and nystagmus). Heat stress should be avoided by moving captured animals into shade and the transport truck as soon as possible. Water should be sprayed or poured on excessively hot animals. The presence of experienced personnel and a veterinarian enhances the success of a handling operation. Long-acting antibiotics (e.g., LA-200 or Flocillin) are usually administered at time of capture. All animals should be tagged or marked for future identification, and data (e.g., morphometric measurements and blood, feces, and nasal swabs) should be collected in a timely fashion. Trapping operations that minimize handling time decrease the risk of sheep's injury, stress, and death.

TRANSPORTATION

Enclosed truck beds or stock trailers with sufficient room and good ventilation are best for transporting bighorn sheep. Blindfolds and leg hobbles should be removed after sheep are in the transport vehicle. Animals held longer than 24 hours need water and food. Large rams should not be transported with others because they are prone to butting small animals and may not remain calm during transportation. Bedding materials should be provided to a comfortable depth (approx. 30 cm), but loose or dusty bedding materials that may blow around during transport should be avoided to keep the air clean. Animals captured with etorphine can exhibit an uncontrollable urge to eat, so use of bedding materials that afford some nourishment may be advisable. Individual immobilized bighorn can be placed in a heavy canvas bag with a drawstring closure and slung by helicopter to transport vehicles or temporary enclosures. On occasion bighorn have been transported inside a helicopter; this dangerous practice should be prohibited, however, for it only courts disaster.

RELEASE

Animals should be allowed to leave a transport vehicle themselves; however, coaxing a group to leave together may minimize group fragmentation immediately after release. Releasing relocated bighorn sheep into permanent or temporary enclosures to permit acclimation to the new surroundings was popular in the late 1960s and early 1970s, but most releases now are directly into the new habitat (Nevada, California, Arizona [Dodd 1983], and Idaho). Some enclosures did not lead to viable reintroduction because significant problems with disease and predation were often encountered. (An enclosure in Texas was an exception: over 14 years, 80 bighorn were raised from an original 6 sheep through extensive predator control and electrified fencing.)

Unless animals are equipped with radiocollars before release, it is impossible to monitor short-term success of a transplant effort or identify management actions that could enhance its long-term success.

SITE SELECTION

Transplant locations have been chosen because they were, or were suspected to be, former bighorn habitat or potential habitat between widely separated herds where a new group could help reestablish metapopulation movements. It is imperative to know whether a prospective relocation site supported bighorn sheep in the past. Also, knowledge of the history of livestock use of the area is important because of high probability of disease transmission (e.g., contagious ecthyma) to transplanted animals. The area should have adequate water (several sources < 8 km apart), escape terrain, lambing sites, and forage diversity. Sites should be free of high concentrations of native and feral ungulates (e.g., burros [Seegmiller and Ohmart 1981, C. L. Douglas and Leslie 1996] and horses) to minimize interspecific competition, especially around summer water sources.

AGE STRUCTURE OF THE TRANSPLANT GROUP

To maximize reproductive output, transplant animals should, theoretically, be as young as possible (Lenarz and Conley 1980). A transplant group of ewes from various age classes is recommended, although very old females should not be relocated because of their limited future reproductive output. Computer simulations indicate that removing a random age selection of females has a moderate effect on demographics of the parent population (Leslie 1980, Leslie and Douglas 1986), but provides mature ewes for young ewes to learn from at the new site. Rams younger than four years of age are preferred for relocation because they are easier to transport than mature rams and they tend to stay with ewe groups during initial colonization of the transplant site. Young rams are fully capable of fertilizing receptive ewes. Old rams may wander off after release, increasing the risk of predation and/or injury, and may be unable to find females during the breeding season. Desert bighorn sheep display little or no group integrity through time (Leslie and Douglas 1979, Chilelli and Krausman 1981), so selection of specific groups for transplantation is unnecessary. Lambs-of-the-year should be transplanted with their dams. Minimum group size for direct release is 20, with a ram:ewe ratio of 1:3–5. Optimally, however, initial releases should be supplemented with additional sheep from a nearby population (of the same subspecies) to minimize negative genetic founder effects and to enhance population growth.

MONITORING SUCCESS

Lamb:ewe ratios are the best indicator of population status and need to be evaluated each year after a transplant effort. Generally, lamb:ewe ratios of less than 25:100 during the autumn of successive years should be cause for concern. Weather and forage availability vary greatly throughout the range of desert bighorn sheep (C. L. Douglas and Leslie 1986), and one or two years of poor lamb production may not be indicative of the long-term potential for population growth of the transplant group. Nevertheless, regular monitoring of population demography is critical to evaluate the success of a transplant effort; too frequently, little or no monitoring is conducted.

PHILOSOPHICAL CHECKLIST

Managers should address several critical questions before transplanting bighorn sheep. Answers to the following questions are fundamental to the success or failure of transplant efforts and should receive more consideration than they have in the past. We hope these questions will challenge standing protocols and improve future management and/or research activities.

Is the physical habitat, water, and forage adequate to support a population of more than 100 animals? Assessing adequacy of physical habitat and water relationships can best be accomplished using GIS software to develop a habitat assessment model (T. S. Smith 1991, Bleich 1993, D. W. Ebert and Douglas 1993, J. Jaeger 1994, Dunn 1996). Use of GIS software allows detailed, quantitative evaluations of habitat variables and their relationships for hundreds of square kilometers of terrain. The GIS modeling of bighorn habitat permits testing and manipulation of habitat data and assessment of alternative scenarios useful for management. For capabilities and versatility in evaluation and modeling, no other method of habitat assessment approaches GIS. Alternatively, if a cell size greater than 1 hectare is used, HSI models (C. G. Hansen 1980*d*, Holl 1982, Armentrout and Brigham 1988) provide a relatively rapid, qualitative evaluation of habitat that may provide a useful first analysis.

Physical habitat probably is the least restrictive variable for rams (Bleich et al. 1997), but the amount of escape terrain may limit lambing areas and ewe numbers. Armentrout and Brigham (1988) estimated that 2 hectares of escape terrain is the minimum size that bighorn will use for lambing. Holl (1982) estimated that in the San Gabriel Mountains, California, a ewe requires about 6 hectares of escape terrain in winter habitat. Habitat quality determines the number of bighorn that can be supported; desert ranges in Nevada support bighorn densities from 10.4 to 19.9 individuals/km^2 (McQuivey 1978). Dunn (1991) estimated potential population size from amount of escape terrain and predicted that 50 desert bighorn, of both sexes, use 633–862 hectares of escape terrain (areas with > 60% slope) on their yearly range. Dunn (1996) compared mountain ranges in New Mexico that had (historic ranges) and had not (nonhistoric ranges) historically supported bighorn sheep; historic ranges contained an average of 204 km^2 of potential habitat and 20 km^2 of escape terrain, whereas nonhistoric ranges contained an average of 66 km^2 of potential habitat and 5 km^2 of escape terrain. Escape terrain in historic ranges was more contiguous, and historic ranges also had more available water sources (Dunn 1996).

A major deficiency of both GIS and HSI models arises when assigning a score to vegetation based on community classification. Community classifications are subjective and have limited discriminatory power in identifying bighorn habitat. Vegetation communities in the desert and arid lands are widespread and probably are a better reflection of elevation and aspect than they are of bighorn habitat. Composition, diversity, biomass, and condition of forage species are not addressed in assigning a score for plant communities. For example, it makes intuitive sense to assign a higher rating to a mixed-shrub community (e.g., *Larrea tridentata–Ambrosia dumosa–Lycium andersonii*) than to a dense blackbrush shrub (*Coleogyne* spp.) community because the latter usually has limited species diversity. A community rating does not give wildlife managers the information that they need; important forage species for bighorn could have been removed or so degraded by livestock in an overgrazed mixed-shrub community that the rating should be lower than that of the blackbrush community. There also may be differences in density or diversity of forage species between communities of the same type that would warrant divergent ratings.

It is less important to know the density and volume of forage species in early spring, when biomass is at its peak, than it is to know what species will be available for bighorn during the most stressful season of the year. If bighorn do not use a plant species for food or shade, its presence is inconsequential. If several species are used disproportionately during drought years, distribution and abundance of those species are important. Additional research is needed to address the issue of forage species diversity, seasonal biomass, nutrient content, and secondary compounds as related to bighorn habitat quality and forage selection. Potential or former habitat with a low diversity of forage species might show promise for restoration. Potential habitat having adequate plant diversity but situated more than 10 km from permanent water will probably be used by sheep on a irregular basis, if at all.

The only reliable way to obtain information on plant species diversity and relative biomass is by running plant transects. This is too time consumptive to be warranted for evaluating bighorn habitat on a statewide or landscape basis.

However, running plant transects would be warranted when evaluating potential transplant sites or when comparing ranges in which sheep are doing well to ranges in which they are doing poorly. It is noteworthy that Dunn (1996) successfully evaluated bighorn habitat and potential habitat throughout the state of New Mexico without incorporating vegetation scores in his HSI.

Does the area contain enough ewe habitat and escape terrain to support 60–70 ewes, which is approximately the number required to maintain a population of at least 100 animals? Ewe habitat usually contains the steepest and most broken terrain and has a high degree of unimpeded visibility. McCarty and Bailey (1994) have provided an excellent summary of habitat requirements for desert bighorn. In canyon-mesa terrain, ewes also use talus slopes and rocky rims of mesas (Bates 1982). Quantity of escape terrain is directly correlated with the number of ewes that will be found in a mountain range. The total hectares of escape terrain in a mountain range plus a 100-m buffer around it could be used to approximate core ewe habitat. This area divided by the mean ewe home range in hectares would give an approximation of the number of ewes the terrain (but not the vegetation) could support. Because ewes have overlapping home ranges, this approximation would be an underestimate. Excluding escape terrain more than 10 km from water from such calculations is prudent, because those areas would be used only infrequently. Mean ewe home ranges vary between mountain ranges, but home range calculations for bighorn in a nearby range having similar habitat conditions should give a close approximation of the space required.

What is the area's grazing history? How close is the nearest livestock grazing allotment, and what kinds and numbers of animals are grazed there? An area's grazing history must be known to help evaluate it as a transplant site. Areas formerly grazed by domestic sheep may not be safe for bighorn for more than four years following cessation of grazing because disease organisms such as footrot and sore mouth remain in the soil and can be transmitted to bighorn when conditions are favorable (Jessup 1985). The sore mouth virus can remain viable in soil for 10–20 years (Jessup 1985). The proximity of current livestock grazing is critical; both domestic sheep and cattle have diseases that can be transmitted to bighorn sheep via vectors, such as biting gnats. A buffer strip of more than 13.5 km between ranges used by bighorn and domestic sheep is recommended (Technical Staff Desert Bighorn Council 1990). Under no circumstances should bighorn and domestic sheep use the same habitat or be allowed to graze on lands immediately adjacent to each other. Experience throughout the western states has shown that when bighorn and domestic sheep share the same or immediately adjacent rangelands, it is only a matter of time before the bighorn population begins to die out (Technical Staff Desert Bighorn Council 1990).

Has the habitat been degraded by livestock? How is this degradation reflected in vegetation measurements and abundance of forage species? A lengthy grazing history in potential or existing bighorn habitat should be viewed with concern, because livestock may have degraded the habitat. Perennial bunchgrasses and other preferred species receive heavy grazing pressure and will be among the first species to exhibit signs of overgrazing. It cannot be assumed, without examination, that the range has recovered after 10 or 20 years of no livestock grazing. Relatively rapid recovery of vegetation could be expected in mesic environments at higher elevations and higher rainfall areas, but recovery in arid areas at lower elevations with little rainfall may take several decades or centuries (Lathrop and Rowlands 1983).

Are bighorn forage species so degraded that they should be reseeded or planted several years before release of a transplant group? What guideline for forage plant density or biomass should be used to determine the adequacy of forage for supporting a bighorn transplant effort? To our knowledge, no management has been undertaken to improve quantities or kinds of forage species in desert bighorn habitat. However, prescribed burns can be used to reduce dense vegetation and promote more open habitat. Habitat improvements generally have only addressed water developments, fence design, or domestic grazing. If an area of former or potential bighorn habitat is so severely degraded that restoration is necessary, reseeding or planting should be done with native species distributed in mixtures approximating the relative percentages found in prime bighorn habitat. Also, relative germination rates must be taken into account. Broadcast seeding or planting should be accomplished under moist conditions between late winter and early spring.

How far is the transplant site from the closest bighorn herd? Are movement corridors between the transplant area and closest herd suitable for bighorn movements? Are livestock, fences, or other barriers present in the intervening area? The greater the distance between habitat patches, the lower the probability that bighorn will traverse it. Compared to movements during the rest of the year, bighorn rams move relatively long distances during the rut (Geist 1971, Leslie and Douglas 1979). Transplanted animals probably will have had no contact with herds in adjacent mountain ranges; if adjacent herds are at distances greater than are covered during normal exploratory forays, they may not be located. Large distances between herds may serve to effectively isolate them, which could create additional management problems. Herds that are separated from other herds by distances of 48–60 km have a very low probability of interbreeding unless they are in the same mountain range. Bighorn rams can move more than 60 km on occasion, but conditions must be favorable for them to do so (Monson and Sumner 1980). Conditions that could promote such movements include: prior knowledge of adjacent herds, freedom from water stress (i.e., moist conditions, cool weather), and easily traversed movement corridors. Livestock grazing in movement corridors between habitat patches could expose bighorn to diseases of livestock. Also, the presence of livestock (or feral livestock) or livestock fences may stop movements of bighorn between habitat patches.

What human activities in and near the mountain range may negatively impact bighorn? Bighorn are impacted by many human activities including mining, using high-speed highways (Leslie and Douglas 1979, Cunningham et al. 1993), building housing developments (Krausman et al. 1979), hiking, camping, driving off-road vehicles (P. Jorgensen 1974, K. S. Hamilton et al. 1982), grazing domestic livestock (Dean and Spillett 1976), and diverting or modifying water sources (Leslie and Douglas 1980). Subdivision of bighorn habitat by roadways or housing developments presents serious management problems. Because bighorn are relatively intolerant of disturbance, any habitat intrusion such as a road has a zone of influence considerably larger than that of the road itself (Krausman et al. 1989, D. W. Ebert and Douglas 1993).

Are animals being released into habitat similar to that with which they are adapted? Although bighorn are not without adaptive capabilities, it is best to transplant them to habitat that is generally similar to habitat from which they were removed and to which they are adapted. Transplanted animals are stressed by capture, handling, and transportation (capture myopathy; Bunch et al., this volume). Releasing stressed animals into habitat dramatically different from their native habitat is certain to create additional stress. Stress in mammals is exhibited in various ways: reduced breeding, spontaneous abortion, and aberrant behavior. Bighorn that have lived with mountain lions *(Puma concolor)* and bobcats *(Lynx rufus)* in their habitat have developed effective predator-avoidance skills. These skills are less developed in bighorn that have not coexisted with larger predators; when transplanted into lion country, naive bighorn have experienced high rates of mortality (McCutchen 1982). Predator control may be required before transplanting bighorn into a new habitat, but this should be evaluated on a case-by-case basis.

Are environmental conditions in the new habitat similar to those to which the animals are adapted? Mating and reproduction of ungulates are synchronized with normal climatic conditions of the region in which they live. Females must obtain adequate nutrition and reach a certain level of body condition before the breeding season, otherwise they may not ovulate or conceive. Underweight females that conceive generally give birth to underweight young, which have a lowered probability of survival. Through adaptation, perhaps in response to photoperiod, reproduction is synchronized to coincide with the most favorable time of year for obtaining relatively high levels of nutrition. Ewes need much higher nutritional intake during the last three months of gestation to support rapid growth of the fetus and permit nursing and weaning of the neonate. Survival of lambs appears to be associated with forage availability, as indexed by precipitation, during gestation (C. L. Douglas and Leslie 1986, C. L. Douglas 1993). In the desert, the breeding and reproductive season is attenuated; lambs may be born anytime from late January through June (Leslie and Douglas 1979, Turner and Hansen 1980, Krausman et al. 1989). In northern areas, breeding and reproduction is tightly restricted in time, with lambs being born from mid-May to mid-June in Idaho and Oregon and mid-May to mid-July in Wyoming (Turner and Hansen 1980).

Bighorn from desert ranges of southern Nevada have been transplanted to northern sites in the state and to Colorado, Utah, and Texas. Desert bighorn usually are transplanted in the autumn, often after ewes are pregnant. If ewes lamb between March and April, which are peak lambing months in the desert, they will be out of synchrony if transplanted to northern areas, where May and June are peak lambing months (Turner and Hansen 1980). Desert lambs born in northern areas in March or April would have a low probability of survival. Giving birth out of synchrony with climatic regimes of the transplant area also places high levels of nutritional and physiological stress on ewes. Thus, not only are lambs almost certain to be lost, but health of the ewes can be compromised. In subsequent years, ewes adapt and lamb more in synchrony with normal expectations for the given latitude.

Although the Nevada bighorn transplanted to Colorado appear to be doing well, we regard relocating animals from southern to northern latitudes and hundreds of kilometers out of their normal range as ill advised. If bighorn were so rare that the only choice was to transplant them to areas having dramatically different climates, then such transplants could be justified. However, it seems sensible and more humane to keep transplant groups in the same general climatic regime as their parent population. Genetic adaptations of bighorn appear to be clinal in distribution; thus, it would be prudent to obtain transplant stock from the nearest-neighbor herd to the site for new herd establishment (Ramey 1993).

When will follow-up releases be planned to ensure the population reaches more than 100 animals? If the transplant group is released in a mountain range that may be isolated from nearby bighorn herds, it is critical for long-term survival that the transplant group be large enough to have acceptably low levels of genetic drift and inbreeding and to deal with climatic vicissitudes as they affect herd productivity. This suggests that the herd size should reach a minimum of 100–125 individuals (Geist 1975, Berger 1990) and should be managed to remain at that level or larger. If a transplant group has poor reproductive success for several years following release, it would be prudent to supplement the herd rather than to wait for a turn of events. If the transplant group is truly isolated from other herds, managers must simulate immigration and release new animals into the herd every four to five years. If the transplant group is near adjacent herds and there is evidence of movements between them, growth in size of the transplant group is somewhat less critical. Nevertheless, any evidence of poor reproductive success (if one assumes favorable weather conditions and good range conditions) warrants consideration of adding supplementary animals.

Future Obligations

Bighorn research and management have flourished since 1975. An unprecedented amount of funding has been, and is being, spent for research and management of bighorn sheep by federal and state agencies. Despite all that has been learned and accomplished, researchers and managers are just now beginning to address critical questions about habitat requirements, methods of habitat evaluation, population dynamics of bighorn, relationship of domestic livestock diseases to bighorn, metapopulation dynamics, and population genetics.

Bighorn management has been conducted with enthusiasm and good intentions, but with a small amount of the scientific information or research design that was needed. Only a few transplant efforts have had adequate follow-up monitoring. Reasons for successes or failures are not well understood, because transplant efforts were not approached in an experimental manner. The oldest transplant groups have only about 22 years of history; some have flourished, but most have persisted as relatively small herds. It is too early to ascribe success to the majority of transplant efforts. We naturally want transplant efforts to be successful. Therefore, if a transplant group persists for several years and young animals are observed, we may wrongfully considered it a success—even though there may only be about the same number of animals as originally released (Rowland and Schmidt 1981).

Hundreds of animals have been captured and handled for transplantation without obtaining blood samples, nasal swabs, parasites, and body measurements; data from these samples and measurements could have provided an invaluable database and could have greatly enhanced our knowledge. It is wasteful to capture and handle animals without obtaining as much information as can be obtained from them in a timely manner. Body temperature and respiration rate should be monitored while obtaining samples so that corrective actions can be taken if the animal becomes stressed.

Until quite recently, habitat evaluations have been qualitative, subjective, and marginally adequate, at best, for predicting habitat suitability for bighorn sheep. Physical features of habitat can be described and vegetation types assigned with present HSI methods, but GIS assessment of habitat permits more precise evaluations and examination of relationships between habitat features. All methods presently used for habitat evaluation are weak for assessing forage species. Better methodology will permit better habitat assessment, which should help reduce transplant failures. Considerable work remains to be done in habitat evaluation and modeling on a landscape basis.

We have learned from several disasters that bighorn can be annihilated by contact with domestic livestock and their diseases (Blaisdell 1982, Sandoval 1988, Weaver 1988). Although these disasters have been costly and emotionally draining for all involved, our understanding of disease transmission to bighorn has improved because of them. Studies of bighorn diseases, disease transmission, and bighorn immunodeficiency will remain important areas of investigation for an indefinite amount of time. Bighorn herds also can be decimated by mountain lion predation. Sixty percent of radiocollared bighorn were killed by lions in a study in southern California (W. Boyce, Univ. California Davis, pers. comm.). In 1997, California's state game commissioners, under pressure from conservation groups, reclassified mountain lions as nongame animals, thereby prohibiting their take and, essentially, their management by the California Department of Fish and Game. Mountain lion populations are increasing in California, which may have serious negative consequences for desert bighorn management.

Recent radiotelemetry studies have demonstrated that small bighorn herds in southern California interact with herds in adjacent mountain ranges as parts of a metapopulation (Schwartz et al. 1986, J. Jaeger 1994). Ramey (1993) found limited genetic differences and relatively little genetic diversity among bighorn nationwide and questioned the current taxonomic status of some subspecies. His studies indicate that there may be clinal variation in genetics and morphology within bighorn across their geographic range of distribution. It would be surprising if this were not true, because clinal variation is present in other mammalian species that have large geographic distributions. Clinal variation is an expression of adaptation to regional environmental and biotic conditions. Ramey's (1993) results and those of Fitzsimmons et al. (1997) lend support to our argument that transplant animals should be selected from the nearest large herd to the transplant site.

Regional, interagency management plans for wildlife are urgently needed to coordinate management actions and to protect large blocks of wildlife habitat from division or degradation (Boyd 1995). With the current national effort to identify areas with high biodiversity within each state (M. J. Scott et al. 1991, 1993), a conceptual framework will be established for regional, interagency management plans. Different legislative mandates for state and federal agencies have regulated how public lands and resources are managed. Management guidelines range from conservation with minimal disturbance to multiple use, resource production, and harvesting. It is understandable that differences of opinion are pronounced between advocates of each system.

The perpetuation of bighorn sheep is a common goal of federal and state land-management agencies, despite differing management philosophies and regulatory mandates. Wildlife is distributed according to habitat availability, not the political boundaries of agencies. Wildlife managers all recognize that their agency shares wildlife with adjacent land owners; this is especially apparent for ungulates.

To safeguard large blocks of wildlife habitat from fragmentation or degradation, federal and state agencies must cooperate to develop regional, interagency management plans. This will not be easily accomplished, but it must be done. Federal and state agencies in California signed an interagency agreement for the preservation of biological diversity in the state (Memorandum of Understanding: "California's Coordinated Regional Strategy to Conserve Biological Diversity," September 1991). This is the first time that a state and the federal government have agreed to work cooperatively in conserving biodiversity on a regional basis across administrative boundaries. This historic agreement helps promote interagency awareness of the need for cooperative management of shared resources. The long-term survival of bighorn sheep dictates that they be safeguarded from additional habitat losses, and this can only be accomplished by multiagency agreements. Bighorn sheep are a national resource whose habitat encompasses a myriad of other species. The protection of wildlife habitat and conservation of biodiversity are intertwined. Bighorn management should have a long-term conservation perspective and should strive to achieve self-regulating populations, while avoiding, whenever possible, actions that will require continuing, long-term intervention.

APPENDIX

Cytogenetics and Genetics

Thomas D. Bunch, Robert S. Hoffmann, and Charles F. Nadler

Chromosomes

Studies of the chromosomes of Holarctic wild sheep *(Ovis)* have established four genetic groupings based on diploid chromosome number (2n), morphology, and other characteristics (Nadler et al. 1973*a,b;* Korobitsyna et al. 1974; Nadler and Bunch 1977; Bunch et al. 1990, 1998*a,b*; Lyapunova et al. 1997). European *(O. musimon)* and Asiatic mouflon *(O. orientalis = gmelinii),* which range from the Mediterranean islands of Corsica, Sardinia, and Cyprus to the Zagros and other mountains of northwestern and southern Iran, have 54 chromosomes. Urial sheep *(O. vignei),* found in northeastern Iran, Turkmenistan, Tadzhikistan, Kazakstan, Afghanistan, Pakistan, and northwestern India (Ladak) have a 2n of 58. The arkar/argali group *(O. ammon)* of Russia, the Central Asian republics, China, and Mongolia maintain 56 chromosomes. The amphiberingian sheep of the subgenus *Pachyceros* include the Siberian snow sheep *(O. nivicola),* which have 52 chromosomes, and the North American sheep *(O. dalli* and *O. canadensis),* which have 54 chromosomes. In North America the ranges of these sheep extend from Alaska to northwestern Mexico (Schmitt and Ulbrich 1968; Wurster and Benirschke 1968; Nadler 1971; Nadler et al. 1971*a*, 1973*a,b;* Korobitsyna et al. 1974; Nadler and Lay 1975; Bunch et al. 1976), and in Siberia from the Putorana Mountains to Chukotka and Kamchatka.

The 2n = 54 karyotype of North American wild sheep contains 3 pairs of biarmed and 23 pairs of acrocentric autosomes. The female has 2 large,

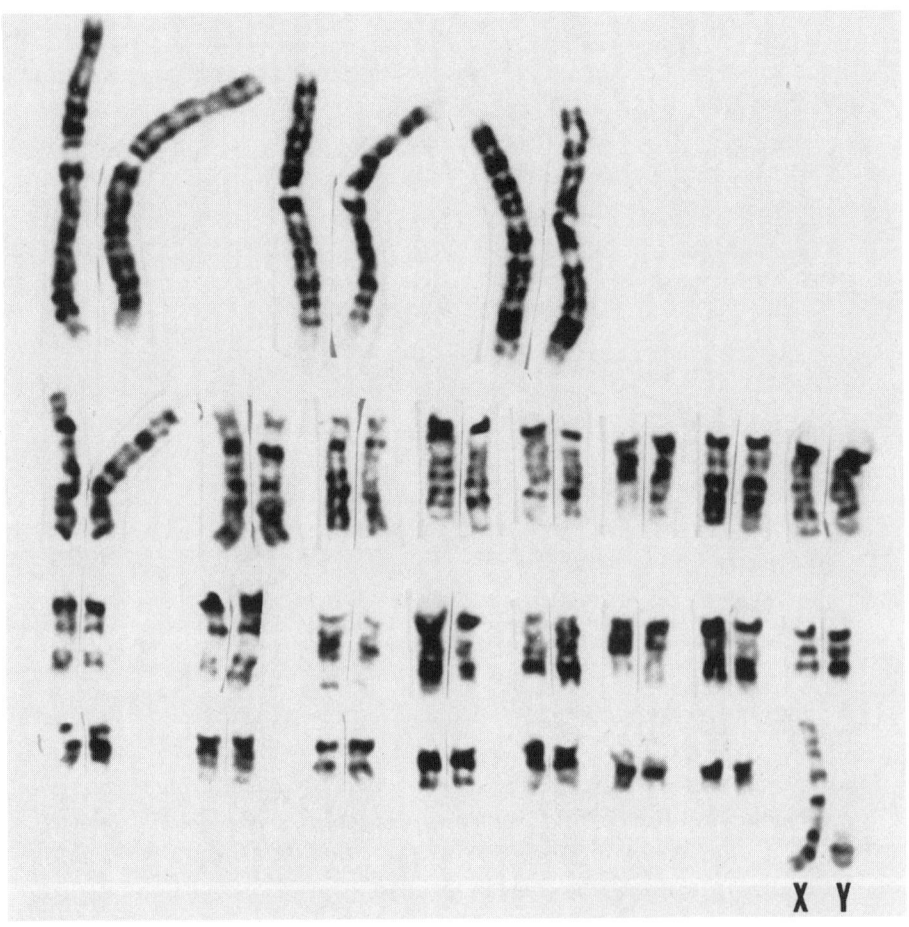

Figure A.1. Typical G-banded karyotype representing all New and Old World wild sheep and domesticated sheep of the genus Ovis *with a diploid (2n) chromosome number of 54. The karyotype consists of 3 pairs of biarmed and 23 pairs of acrocentric autosomes. The sex chromosomes are distinguished as a large, acrocentric X and a small, biarmed Y (male). In a female, sex chromosomes would consist of 2 Xs. (Photograph by Thomas Bunch.)*

acrocentric sex chromosomes and the male has 1 large, acrocentric plus 1 small, biarmed sex chromosome (fig. A.1). Karyotypes of Eurasian sheep with 58, 56, 54, and 52 chromosomes display 1, 2, 3, and 4 pairs of biarmed chromosomes, respectively (fig. A.2).

Cytogenetic Giemsa-banding (G-banding) techniques make it possible to

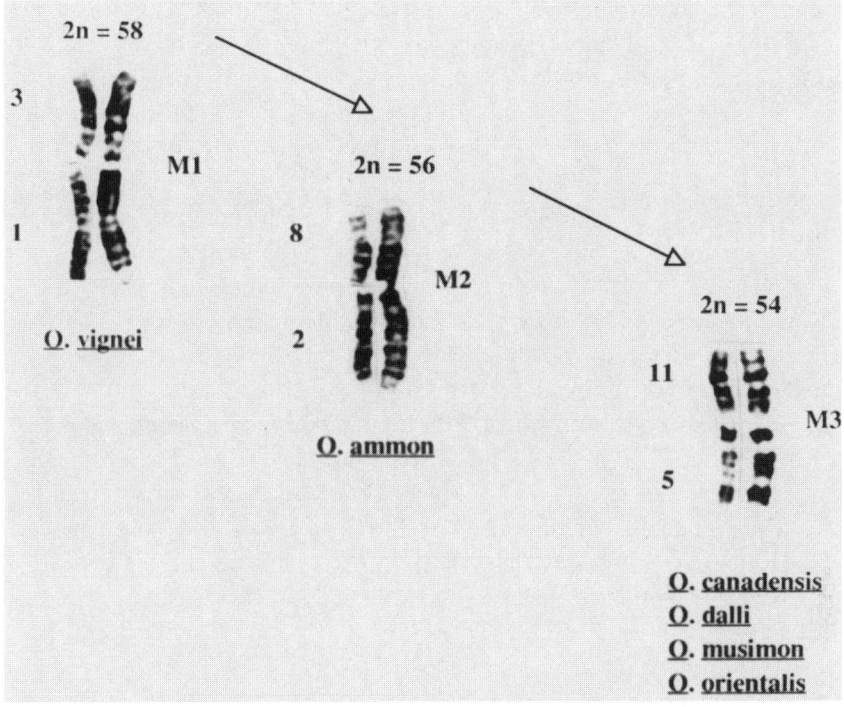

Figure A.2. Specific fusions of acrocentric chromosomes to form biarmed chromosomes brought about the evolution of the 2n = 54 karyotype of North American wild sheep from ancestral stock that once had 60 chromosomes. (Photograph by Thomas Bunch.)

identify discrete morphological segments along a chromosome arm. These techniques have been used to compare the 2n = 54 karyotype of the North American wild sheep to the 2n = 54 karyotype of European and Asiatic mouflons. All G-band karyotypes are similar and the banding segments along the chromosome arms are identical (Nadler et al. 1973a, Bunch et al. 1976). The identical G-band patterns in all members of *Ovis* with 54 chromosomes *(O. aries, O. canadensis, O. dalli, O. musimon,* and *O. orientalis)* suggest that the chromosomes are structurally homologous and therefore these species may have descended from common ancestors (Bunch and Nadler 1980). This homology has been substantiated by the normal bivalent formation during meiosis of Rocky Mountain bighorn–European mouflon hybrids *(O. canadensis* × *O. musimon;* Nadler et al. 1973a).

The 2n = 54 karyotype resulted from autosomal acrocentric fusions that reduced the 60 chromosomes of ancestral stock (Wurster and Benirschke 1968). The ancestral karyotype still persists in close relatives of *Ovis,* the wild goats *Capra,* whose G-band karyotype has served as a reference karyotype to identify acrocentric chromosomes and the order of fusion during the evolution of the 2n = 54 karyotype. Biarmed (metacentric) chromosome 1 (M1) arose from the fusion of acrocentrics 1 and 3; M2 arose from acrocentrics 2 and 8, and M3 arose from acrocentrics 5 and 11 (Menscher et al. 1989, Hayes et al. 1991; fig. A.2). Siberian snow sheep have the most recently evolved karyotype, which formed a pair of biarmed chromosomes from acrocentrics 9 and 19 (Nadler and Bunch 1977, Bunch 1978, Menscher et al. 1989). The M4 translocation occurred after the disruption of the Bering Strait 12,000 years ago (Korobitsyna et al. 1974).

Domestic sheep typically display the 2n = 54 karyotype. Spontaneous centric fusions (T_1, T_2, T_3) during the last 100 years have resulted in 2ns ranging from 53 to 48 (Bruère et al. 1976, Stewart-Scott and Bruère 1987). The acrocentric components that comprise the T translocations differ from the M4 translocation in the Siberian snow sheep.

It has been postulated that karyotypic changes in *Ovis* began in primitive Miocene-Pliocene stock (K. L. Dixon 1979). Two alternative mechanisms have been proposed. In the monophyletic alternative, all 2n = 54 sheep evolved from a single common ancestor via the M1, M2, and M3 centric fusions. In the polyphyletic alternative, the G-band homology of North American and Asiatic wild sheep originated from centric fusions that occurred in different geographic regions, thus engendering identical biarmed chromosomes (Nadler et al. 1973*a*). Although fusions observed in sheep involve certain acrocentrics, diverse, random, chromosomal combinations occur in close relatives of *Ovis,* the bharal *(Pseudois),* and the tahr *(Hemitragus).* The chromosomally homologous lineage in *Ovis* may, therefore, have a monophyletic origin (Bunch and Nadler 1980).

The evolutionary significance of the various karyotypes found in *Ovis* is not known. The chromosomal fusions that resulted in 2ns of 58, 56, 54, and 52 have not prevented interbreeding as in some mammalian taxa (Valdez et al. 1978). However, the monomorphism observed in various taxa of wild sheep suggests a form of selection that has resulted in maintenance of uniform diploid chromosome numbers.

Centric fusions may restrict interchromosomal exchanges during gametogenesis and thereby reduce genetic variability within a gene pool. This may be inconsequential during a generation, but genetic changes associated with centric fusions may have profoundly affected genetic variation among *Ovis* taxa during thousands of years of evolution.

Hemoglobins

Polymorphism of hemoglobin arises regularly in breeds of domestic sheep and has resulted in several electrophoretically separable protein zones (H. Harris and Warren 1955, Evans et al. 1958, Stormont et al. 1968). Hemoglobins HbA and HbB commonly exist in adult sheep. A rare adult type is designated as HbD (Vaskov and Efremov 1967), and the fetal hemoglobin is HbF. Hemoglobin HbC has been associated with severe anemia and is seen in sheep that normally have HbA (van Vliet and Huisman 1964). Hemoglobin HbI appears to be produced under the control of an allele at the HbB locus and is due to a neutral amino acid substitution. This rather common hemoglobin variant was first detected in the Sardinian and Altamurana domestic sheep breeds (Manca et al. 1993). Another polymorphism in an A- and T-rich element at the beta-hemoglobin locus has also been reported in domestic sheep (N. J. Wood et al. 1993). Hemoglobin variants of domestic sheep reportedly differ in oxygen affinity. Sheep may, therefore, differ in reproductive performance and susceptibility to intestinal parasites (Agar et al. 1972, Blunt and Huisman 1975). The distribution of hemoglobin variants has been related to geographic environments (Agar et al. 1972). Hemoglobin types influence physiological function, such as the concentration of reduced glutathione in erythrocytes and whole blood (Rizzi et al. 1988). Hemoglobin-binding haptoglobins have been reported to provide certain selective advantages and disadvantages of the Hb polymorphism, such as biological functions during pathological conditions in the cases of malignant tumors, inflammations, autoimmune diseases, allergic illnesses, affective psychoses, and affective lability favoring addiction (V. Lange 1992). Hemoglobin and other blood proteins have been used as genetic markers for studying relationships between breeds of sheep (Manwell and Baker 1977, Buis and Tucker 1983, Iovenko 1987). Hemoglobin types have been valuable in identifying pedigrees in breeding programs (C.M.A. Baker and Manwell 1983). Paradoxically, HbA in domestic sheep has an adaptive advantage at high altitudes (Evans et al. 1958); in contrast, HbB in wild sheep has been detected at altitudes from below sea level to above 3,000 m.

Hb polymorphism is found in most breeds of sheep, and its distribution has been investigated extensively (Agar et al. 1972, Blunt and Huisman 1975, Bunch and Foote 1976, Reddy et al. 1988, Rizzi et al. 1988). The study of Hb polymorphic distribution in domestic and wild sheep has provided meaningful information about the origin of domestic sheep (Lay et al. 1971, Bunch et al. 1978a, Valdez et al. 1978).

Agar et al. (1972) reviewed the Hb polymorphic distribution of sheep at many locations. In more than 66% of the populations surveyed, HbB was more frequent than HbA. Such percentages were also observed in indigenous domestic breeds (Tan sheep and Mongolian sheep) in China (S. Wang et al. 1989), the major domestic sheep breeds in Iran (Bunch and Foote 1976), the Nellore sheep in India (Reddy et al. 1988), the Askania and Tsigai sheep in the former USSR (Iovenko 1987), and other native sheep breeds in Europe (Rizzi et al. 1988). The HbB type was also prevalent among most breeds of domesticated sheep in the United States (S. Wang et al. 1990).

Except for the European mouflon, which maintains both the HbA and HbB alleles in its gene pool (Bunch et al. 1978a), only HbB has been identified in wild sheep (Lay et al. 1971, Nadler et al. 1971b, Spillett et al. 1975, Bunch and Valdez 1976). Hemoglobin phenotypes have been studied in 100 European mouflons living on the island of Sardinia using isoelectric focusing and high-performance liquid chromatography (HPLC) protein separating techniques (Naitana et al. 1991). The research indicated the presence of two beta-globin alleles, one of which corresponds to beta-B, which was the most common type ($f = 0.94$). None of the mouflon sampled were carriers of the earlier described HbA, but a new polymorphism HbM was observed. The HbB of wild sheep differs from HbB (beta-chain) of domestic sheep at five amino acid sites. The HbA of domesticated sheep and wild sheep differ at only two amino acid sites, which do not overlap (Boyer et al. 1966, Manwell and Baker 1976).

The prevalence of HbB in most breeds of domesticated sheep genotypes and the monomorphism of HbB in wild sheep indicate that the HbA allele may have resulted from a mutation that arose during the early domestication of *Ovis*. This suggests that the European mouflon, which originated on an island, may have arisen from archaic domestic strains (Lay et al. 1971, Bunch et al. 1978a, Valdez et al. 1978). Gregorio et al. (1987) used molecular genetic methods and found that the restriction enzyme patterns of mouflon DNA was identical to that of homologous sheep HbA DNA, which suggests that modern sheep received the HbA haplotype from mouflon (Gregorio et al. 1987). Alternatively, this supports the proposal of Poplin (1979) that the mouflon originated from feral domestic sheep. European mouflons that were severely anaemic were able to synthesize HbC at the expense of HbB, thus suggesting that structural and physiological homologies exist between the mouflon beta-B and domestic sheep beta-A globin genes and between the newly observed beta-M and the beta-B allele of domestic Sardinian sheep (Naitana et al. 1991).

Boyer et al. (1966) hypothesized that the Hb polymorphism resulted from genetic isolation followed by admixture. Manwell and Baker (1976) suggested that the gene pool of domesticated sheep was the result of hybridization, and that the beta-B chain sequence of ancestral sheep was similar to mouflons. According to their hypothesis, HbB resulted from five amino acid substitutions in one genotype, and HbA resulted from two amino acid substitutions in another genotype. Hybridization was responsible for the merging of these two divergent genotypes (Manwell and Baker 1976). Random gene frequency drift then resulted in polymorphisms as observed in domestic sheep. Domestication has facilitated polymorphisms through the hybridization of geographically isolated individuals (Gregorio et al. 1987).

As mentioned above, the HbB of wild sheep differs from HbB (beta-chain) of domestic sheep at five amino acid sites. The HbA of domesticated sheep and wild sheep differ at only two amino acid sites, which do not overlap (Boyer et al. 1966, Manwell and Baker 1976). Sheep HbB and HbA differ by seven amino acids (Manwell and Baker 1976). Because that amino acid substitution occurs at a constant, low rate (Kimura 1983), sheep HbA probably did not evolve from sheep HbB, a process that would require seven amino acid substitutions and the presence of bridging intermediates. It is more likely that wild sheep HbB evolved into domesticated sheep HbA (five amino acid substitutions) or domesticated sheep HbB (five amino acid substitutions). No intermediates between sheep HbA and HbB have been found (Manwell and Baker 1976).

Transferrins

Serum transferrins (Tfs) in *Ovis* are polymorphic, as first reported by Ashton (1958). Since 1958, at least 14 Tf alleles have been described (Nix et al. 1969, Archibald and Webster 1986), each of which controls two electrophoretically separable protein fractions, or a "zone pair," associated with Tf phenotypes. Nine variants in domestic sheep (Tf^I, Tf^A, Tf^G, Tf^B, Tf^C, Tf^D, Tf^M, Tf^E, and Tf^P) are well defined and their genetic control has been established (Tucker 1975, Archibald and Webster 1986). The Tf allelic frequencies have been used to characterize breeds of domestic sheep and to differentiate populations of wild sheep (Ashton and Ferguson 1963, Stormont et al. 1968, Lay et al. 1971, Nadler et al. 1971b, Bunch and Valdez 1976, Jessup and Ramey 1995).

Tf polymorphism is found in all breeds of domesticated sheep and is accompanied by characteristic allelic variation between genetic groups (Tucker 1975). Variation is the result of natural selection, differential selection and management in domestic animals, and genetic drift. The high level of

Table A.1. Transferrin allelic frequencies in North American wild sheep.

GENOTYPE	LOCALITY	NO. OF SPECIMENS	\multicolumn{9}{c}{ALLELES}								
			I	A	G	B	C	D	M	E	P
O. c. canadensis	Montana	14			1 (0.357)			14 (0.500)		13 (0.464)	
O. c. canadensis	Wyoming	9						7 (0.388)		11 (0.611)	
O. c. mexicana	Mexico	10						9 (0.450)		11 (0.611)	
O. c. mexicana	Arizona	20						7 (0.175)		33 (0.825)	
O. c. nelsoni	Arizona	12						1 (0.041)		23 (0.958)	
O. c. nelsoni	Nevada	43								86 (1.000)	
O. c. nelsoni	Utah	25						42 (0.840)		8 (0.160)	
O. c. dalli	Alaska	2			1 (0.250)			3 (0.750)			
O. c. stonei	British Columbia	2					2 (0.500)				
Pooled values					2 (0.007)		2 (0.007)	85 (0.310)		185 (0.675)	

Note: Relative frequencies are shown in parentheses.

polymorphism reflects the worldwide spread of the ovicaprine species (Manwell and Baker 1977).

Many of the Tf variants in wild sheep (e.g., the Dall's sheep, *O. dalli;* bighorn, *O. canadensis;* and mouflon, *O. musimon, O. orientalis,* and *O. vignei*) are the same as those in domesticated sheep (Lay et al. 1971, Nadler et al. 1971b, Tucker and Clarke 1980).

The Tf polymorphisms in North American wild sheep (table A.1) are similar to those in Old World populations of wild sheep (table A.2). The gene pool of North American wild sheep contains the Tf alleles G, C, D, and E, whereas the gene pool of Old World wild sheep contains alleles I and A in addition to G, C, D, and E. Four alleles have been identified in North American wild sheep, of which only Tf^G was found in the Rocky Mountain bighorn *(O. canadensis canadensis)* from Wyoming and the Dall's sheep *(O. dalli)* of Alaska. The closest relative of Dall's sheep, Stone's sheep *(O. d. stonei),* maintained the Tf^C allele. The predominant alleles in the North American wild sheep are Tf^D and Tf^E, whose frequencies are 0.31 and 0.68, respectively.

Genetic isolation has contributed to differences in the frequencies of alleles in Nelson's bighorn *(O. canadensis nelsoni)* in Utah and the same subspecies in Arizona and Nevada. The frequency of the Tf^D allele in Nelson's bighorn in Utah is 0.84, whereas in the Nelson's bighorn of Arizona it is 0.04. The Tf^D allele was not detected in any of the 43 Nelson's bighorn examined from Nevada (Bunch and Valdez 1976).

The higher frequency of the Tf^E allele in the North American bighorn than in the wild sheep of Alaska and British Columbia and the Old World wild sheep may serve as a genetic marker to differentiate these major populations. Interestingly, the Tf^B allele (which is common in most breeds of domestic sheep) has not been observed in wild sheep. The Tf^M and Tf^P alleles are rarely found in domestic sheep and have not been found in wild sheep. If unique Tf allelic frequencies exist in major populations of North American wild sheep, they may provide a method of evaluating phylogenetic relationships and thereby ascertaining the degree of genetic isolation.

The functional significance of Tf variation in mammals has not been definitely established. Ashton (1960) presented evidence that milk production in cattle was related to certain Tf types. Frelinger (1972) suggested that the maintenance of Tf polymorphisms in pigeons was due to the differential inhibition of microbial growth by the different phenotypes. There has also been considerable interest in the microheterogeneity of the Tf molecule (i.e., differences of the binding site for iron and possible donation of iron to the cell, number of carbohydrate antennae, degree of sialyzation as a possible link between microheterogeneity, and physiological and pathophysiological status of the organism; Hristic 1992). It has been reported that

Table A.2. Transferrin allelic frequencies in Old World wild sheep.

GENOTYPE	NO. OF SPECIMENS				ALLELES					
		I	A	G	B	C	D	M	E	P
O. musimon	19		5 (0.131)	2 (0.052)		12 (0.315)	19 (0.500)			
O. orientalis	27	2 (0.037)	11 (0.203)	2 (0.037)		23 (0.425)	13 (0.240)		3 (0.055)	
O. vignei	34	1 (0.014)	2 (0.029)			15 (0.220)	50 (0.735)			
Pooled values		3 (0.019)	18 (0.113)	4 (0.025)		50 (0.313)	82 (0.513)			

Note: Relative frequencies are shown in parentheses.

Tf polymorphism combined with restriction fragment length polymorphisms (RFLPs) of the Tf receptor gene may provide useful markers in population and linkage studies and in studies that have associated body iron stores with susceptibility to genotoxic damage and cancer (Sikstrom and Beckman 1994). Tfs are an indicator of genetic background and have been related to stress endurance in domestic animals (Osta et al. 1994). The Tf phenotype has also been related to systemic-lupus-erythematosis in humans (Jiang and Du 1992). The role of Tf variants in disease susceptibility in wild sheep has not been explored.

The most interesting phenomena associated with Tf polymorphism are its existence in almost all mammalian species, the large number of variants, and the wide variation in distribution (Chen 1992; Juneja and Shibata 1992; Saha et al. 1992; Stratil et al. 1992; Tate et al. 1992a,b; Cizova et al. 1993; Hartl and Ferrand 1993; Nagabuchi et al. 1993; Bell 1994; Dall'Olio et al. 1995; Pepin and N'Guyen 1994; M. Wang and Schreiber 1996). Tf polymorphic proteins may be useful in the investigation of molecular evolution. The study of the molecular structure and function of Tf polymorphic proteins could test the evolutionary theory of neutral mutation–random drift as advanced by Kimura (1983).

Two principles govern evolutionary changes at the molecular level. One principle states that molecules or parts of the molecule that are functionally less important evolve faster than those that are more important. According to the other principle, those mutant substitutions that are less disruptive to the structure and function of a molecule occur more frequently than those that are more disruptive (Kimura 1983).

Tf binds and transports iron from the plasma to the bone marrow and tissue. The existence of many variant alleles in the Tf polymorphic locus is indicative of a high mutational substitution rate in genetic codons. The fact that animals with varied Tf types coexist equally well in the same population under the same ecological conditions indicates that many Tf variants are functionally equivalent.

Considering the variety of Tf proteins, mutations may have involved the non-iron-binding part of the amino acid sequence. Amino acid substitutions may have altered the total electrostatic charges of the molecules to the extent that they can be separated and identified by electrophoresis, without altering the iron-binding efficiency of the molecules. The high substitution rate within the Tf molecule has been the result of many mutant alleles, and random drift in gene frequency has further increased the variation in polymorphic distribution. Differentiation then occurred as alleles were passed from generation to generation, which was enhanced and broadened by geographic isolation. The domestication of species and geographic isolation

Table A.3. Red-cell blood-group factors of Mexican and Nelson's desert bighorn (*O. c. mexicana* and *O. c. nelsoni*) and European mouflon (*O. musimon*).

DOMESTIC SHEEP BLOOD GROUPS	DOMESTIC SHEEP REAGENTS	MEXICAN DESERT BIGHORN	NELSON'S DESERT BIGHORN	EUROPEAN MOUFLON
A	Aa	−	−	+
	Ab	−	−	+
	F16	−	−	−
	F19	−	−	+
C	Ca	−	−	−
	Cb	−	−	+
	F5	−	−	−
	F6	−	−	+
	F32	+	+	+
D	Da	+	−	+
M	Ma	+	+	+
	Mb	−	−	−
	F36	−	−	+
F30[a]	F30	+	+	+
B	Bb	+	+	+
	Bc	−	−	+
	Bd	−	−	+
	Be	−	−	+
	Bf	−	−	+
	Bg	−	−	+
	Bh	−	−	−
	Bi	−	−	+
	F4	−	−	+
	F26	−	−	+
	F35	−	−	+
	F38	−	−	−
R	R	+	+	+
	o	−	−	−
F41	F41	−	−	+

[a] Very faint cross reaction in the desert bighorn.

in wild sheep are responsible for the high level of Tf polymorphism in the genus *Ovis*.

Red-Cell Blood Groups

Five (and possibly six) of the eight blood-group genetic systems recognized in domestic sheep have been observed in desert bighorn (Bunch and N'Guyen 1982; table A.3). The Nelson's bighorn, one of the two subspecies of desert bighorn, lacks the D group. The European mouflon has homologies with all blood-group systems, which indicates that the mouflon is more closely related to domestic sheep than to the bighorn. The desert bighorn and the European mouflon both share the F32, Da, Ma, F30, Bb, and R blood groups. The B, C, F30, M, and R systems have been identified in the domestic goat *(Capra hircus)*. S. Wang et al. (1990) isolated a polymorphic protein character of red blood cells (erythrocyte protein–1, EP-1) in Mexican desert bighorn *(O. c. mexicana)*, mouflon, and domesticated sheep. Only the phenotype EP-1^B was observed in the Mexican bighorn sheep, whereas EP-1^A, EP-1^{AB}, and EP-1^B were observed in all mouflons and domestic sheep.

Further analysis of blood-group systems and blood-protein types may further clarify phylogenetic relationships among wild sheep and their close relatives.

Molecular Genetics

Sequencing of either mitochondrial or nuclear (genomic) DNA has recently been used to analyze relationships among populations and species of mammals to elucidate phylogenetic relationships (Miyamoto et al. 1990, M. F. Smith and Patton 1993). Restriction endonucleases (enzymes) have permitted the detection of intraspecific and interspecific DNA sequence heterogeneity in several species (Lansman et al. 1983, Arnheim et al. 1990). Mitochondrial DNA (mtDNA) sequence analysis has been used to evaluate heterogeneity within and among conspecific populations and particularly to estimate matriarchal phylogenies (Avise et al. 1979, Lansman et al. 1983, Wayne et al. 1990, Cronin 1991). Utilizing morphometric measurements, protein, and mtDNA, Ramey (1991, 1993, 1995) has refuted the recognition of four separate subspecies of mountain sheep in the southwestern United States and Mexico. He recommends that only a single subspecies of desert bighorn sheep *(O. c. nelsoni)* be recognized. These techniques, when further applied to wild sheep, will provide strong evidence of phylogenetic relationships.

Conclusion

At the present time it is not known which criteria (e.g., morphologic [Wehausen and Ramey 1993], chromosomal, serologic, behavioral, palaeontologic, ecological, biochemical, genetic, or a combination of two or more of these) will be the most useful in defining species, subspecies, and other closely related taxa (Huelsenbeck and Rannala 1997). Although the newer DNA technologies have been used to study evolutionary relationships among a number of mammalian species, it is still too early to assess their impact on sheep systematics. A basic, somewhat philosophical question is how much measurable variation constitutes taxonomic discrimination? In nature, individuals of a species generally do not interbreed with another species (Mayr 1963). This is not the case for wild sheep; they will readily interbreed when contact is made between species (Valdez et al. 1978). Reproductive isolation or reduced gene flow between closely related taxa has resulted primarily from geographic separation. The only report of partial reproductive impairment among wild sheep taxa was between desert bighorn sheep and a hybrid wild sheep developed in Texas from Old World wild sheep (*O. ammon nigrimontana* × *O. musimon;* Bunch and Workman 1988). Hybrid lambs were born premature and unless artificially raised, they died between one and eight weeks of age. The authors concluded that it would be doubtful that a similar cross in nature would survive.

The need for strict conservation and management of wild sheep, however, still dictates a conservative approach to lumping taxa. The loss of one of North America's wild sheep taxa *(O. c. auduboni)* stands as a reminder that unique genetic types of wild sheep can be extirpated.

REFERENCES

Adamczewski, J. Z., C. C. Gates, R. J. Hudson, and M. A. Price. 1987. Seasonal changes in body composition of mature female caribou and calves *(Rangifer tarandus groenlandicus)* on an arctic island with limited winter resources. Can. J. Zool. 65:1149–1157.

Adams, L. G., K. L. Risenhoover, and J. A. Bailey. 1982. Ecological relationship of mountain goat and Rocky Mountain bighorn sheep. Proc. Bienn. Symp. North. Wild Sheep and Goat Counc. 3:9–22.

Agar, N. S., J. V. Evans, and J. Roberts. 1972. Red blood cell potassium and haemoglobin polymorphism in sheep. Anim. Breed. Abstr. 40:407–436.

Akeson, J. J., and H. A. Akeson. 1992. Bighorn sheep movements and summer lamb mortality in central Idaho. Proc. Bienn. Symp. North. Wild Sheep and Goat Counc. 8:14–27.

Albon, S. D., F. E. Guiness, and T. H. Clutton-Brock. 1983. The influence of climatic variation on the weights of red deer *(Cervus elaphus)*. J. Zool. 46:295–298.

Alderman, J. A., P. R. Krausman, and B. D. Leopold. 1989. Diel activity of female desert bighorn sheep in western Arizona. J. Wildl. Manage. 53:264–271.

Aldous, M. C., F. C. Craighead Jr., and G. A. Devan. 1958. Some weights and measurements of desert bighorn sheep *(Ovis canadensis nelsoni)*. J. Wildl. Manage. 22:444–445.

Alexander, G. 1968. The environment and neonatal survival in sheep in Australia. Proc. Ecol. Soc. Aust. 3:125–131.

———, and E. E. Shillito. 1977. The importance of odor, appearance and voice in maternal recognition of the young Merino sheep *(Ovis aries)*. Appl. Anim. Behav. 3:127–135.

Alexander, R. D. 1974. The evolution of social behaviour. Annu. Rev. Ecol. Syst. 5:325–383.

Allen, J. A. 1897. Preliminary description of a new mountain sheep from the British Northwest Territory. Bull. Am. Mus. Nat. Hist. 9:111–114.

———. 1902. A new sheep from the Kenai Peninsula. Bull. Am. Mus. Nat. Hist. 16:145–148.

———. 1912. Historical and nomenclature notes on North American sheep. Bull. Am. Mus. Nat. Hist. 31:1–29.

Allen, R. W. 1955. Parasites of mountain sheep in New Mexico with new host records. J. Parasit. 41:583–587.

———. 1960. Diseases and parasites of Barbary and bighorn sheep in the Southeast. Desert Bighorn Counc. Trans. 4:17–22.

———. 1961. Methods of examining bighorn sheep for parasites. Desert Bighorn Counc. Trans. 5:75–79.

———. 1962. Parasitism in bighorn sheep on the Desert Game Range in Nevada. Desert Bighorn Counc. Trans. 6:69–72.

———. 1964. Additional notes on parasites of bighorn sheep on the Desert Game Range, Nevada. Desert Bighorn Counc. Trans. 8:5–9.

Allred, L. G., and W. G. Bradley. 1965. Necrosis and anomalies of the skull in desert bighorn sheep. Desert Bighorn Counc. Trans. 9:75–81.

Alvarez, T. 1976. Status of desert bighorns in Baja California. Desert Bighorn Counc. Trans. 20:18–21.

Archibald, A. L., and J. Webster. 1986. A new transferrin allele in sheep. Anim. Genet. 17:191–194.

Arizona Game and Fish Department. 1980. Big game strategic plans. Arizona Game and Fish Dep., Fed. Aid Wildl. Restor. Proj. FW-11-R-13.

Armentrout, D. J., and W. R. Brigham. 1988. Habitat suitability rating system for desert bighorn sheep in the Basin and Range Province. U.S. Dep. Inter., Bur. Land Manage., Tech. Note 384.

Arnheim, N., T. White, and W. E. Rainey. 1990. Application of PCR. Organismal and population biology. Biol. Sci. 40:174–182.

Arnold, G. W., C.A.P. Boundy, P. D. Morgan, and G. Bartle. 1975. The role of sight and hearing in the lamb in the location and discrimination between ewes. Appl. Anim. Ethol. 1:176–179.

———, and M. L. Dudzinski. 1978. Ethology of free-ranging domestic animals. Elsevier, New York, N.Y.

———, and P. D. Morgan. 1975. Behaviour of the ewe and lamb at lambing and its relationship to lamb mortality. Appl. Anim. Behav. 2:25–46.

Asdell, S. A. 1964. Patterns of mammalian reproduction. Second edition. Comstock Publ. Assoc., Cornell University Press, Ithaca, N.Y.

Ashcroft, G.E.W. 1986a. Sexual segregation and group sizes of California bighorn sheep. M.S. thesis, University of British Columbia, Vancouver.

———. 1986b. Attempted defense of a lamb by a female bighorn sheep. J. Mammal. 67:427–428.

Ashton, G. C. 1958. Polymorphism in the beta-globulins of sheep. Nature 181:849–850.

———. 1960. ß-globulin polymorphism and economic factors in dairy cattle. J. Agric. Sci. 54:321–328.

———, and K. A. Ferguson. 1963. Serum transferrins in Merino sheep. Genet. Res. Camb. 4:240–247.

Augsburger, J. G. 1970. Behavior of Mexican bighorn sheep in the San Andres Mountains, New Mexico. M.S. thesis, New Mexico State University, Las Cruces.

Avise, J. C., G. Giblin-Davidson, J. Laerm, J. C. Patton, and R. A. Lansman. 1979. Mitochondrial DNA clones and matriarchal phylogeny within and among geographic populations of the pocket gopher, *(Geomys pinetis)*. Proc. Nat. Acad. Sci. USA 76:6694–6698.

Ayers, L. A. 1986. The movement patterns and foraging ecology of Dall sheep

(*Ovis dalli dalli*) in the Noatak National Preserve, Alaska. M.S. thesis, University of California, Berkeley.

Bailey, J. A. 1980. Desert bighorn, forage competition and zoogeography. Wildl. Soc. Bull. 8:208–216.

———. 1984. Bighorn zoogeography: response to McCuthchen, Hansen, and Wehausen. Wildl. Soc. Bull. 12:86–89.

———. 1990. Management of Rocky Mountain bighorn sheep herds in Colorado. Colorado Div. Wildl., Spec. Rep. 0(66):1–24.

———, and D. R. Klein. 1997. United States of America. Pages 307–316 *in* D. M. Shackleton, editor. Wild sheep and their relatives, status survey and conservation action plan for Caprinae. IUCN, Gland, Switzerland.

Bailey, V. 1912. A new subspecies of mountain sheep from western Texas and northeastern New Mexico. Proc. Biol. Soc. Wash. 25:109–110.

———. 1936. The mammals and life zones of Oregon. North Am. Fauna 55:1–416.

Baker, C.M.A., and C. Manwell. 1983. Electrophoretic variation of erythrocyte enzymes of domesticated mammals. Pages 367–412 *in* N. S. Agar and P. G. Board, editors. Red blood cells of domestic mammals. Elsevier North-Holland Biochemical Press, Amsterdam.

Baker, L. R. 1967. Variation in the skulls of Nelson bighorn sheep, *Ovis canadensis nelsoni*. M.S. thesis, Nevada Southern University, Las Vegas.

Baldwin, B. A., and E. E. Shillito. 1974. The effects of ablation of the olfactory bulbs on parturition and maternal behaviour in Soay sheep. Anim. Behav. 22:220–223.

Baldwin, R. L. 1969. Mammary growth and lactation. Pages 441–472 *in* H. H. Cole and P. T. Cupps, editors. Reproduction in domestic animals. Academic Press, New York, N.Y.

Barichello, N., and J. Carey. 1988*a*. Snow depth as a likely factor contributing to the decline of a sheep population in the central Yukon. Proc. Bienn. Symp. North. Wild Sheep and Goat Counc. 6:282.

———, and ———. 1988*b*. The effect of wolf reduction on Dall sheep demography in the southwest Yukon. Proc. Bienn. Symp. North. Wild Sheep and Goat Counc. 6:307.

Barker, I. K., A. A. Van Dreumel, and N. Palmer. 1993. The alimentary system. Pages 123–231 *in* K.V.F. Jubb, P. C. Kennedy, and N. Palmer, editors. Pathology of domestic animals. Fourth edition. Academic Press, New York, N.Y.

Barrett, R. H. 1964. Seasonal food habits of the bighorn at the Desert Game Range, Nevada. Desert Bighorn Counc. Trans. 8:85–93.

———. 1967. Some comparisons between the Barbary sheep and the desert bighorn. Desert Bighorn Counc. Trans. 11:16–26.

Baruzzi, R. C., L. F. Marcopito, M.L.C. Serra, F.A.A. Souza, and C. Stabile. 1977. The Kren-Akorore: a recently contacted indigenous tribe. Pages 179–200 *in* P. Hugh-Jones, editor. Health and disease in tribal society. Ciba Foundation Symposia, New Ser. 49. Elsevier North-Holland, New York, N.Y.

Bates, J. W., Jr. 1982. Desert bighorn sheep habitat utilization in Canyonlands National Park. M.S. thesis, Utah State University, Logan.

———, J. C. Pederson, and S. C. Amstrop. 1976. Bighorn sheep range, population trend and movement. Desert Bighorn Counc. Trans. 20:11–12.

———, and G. W. Workman. 1983. Desert bighorn sheep habitat utilization in Canyonlands National Park. Desert Bighorn Counc. Trans. 27:25–28.

Baumen, T. G., and D. R. Stevens. 1978. Winter habitat preferences of bighorn sheep in the Mummy Range, Colorado. Proc. Bienn. Symp. North. Wild Sheep and Goat Counc. 1:320–330.

Bavin, R. L. 1975. Ecology and behavior of the Persian ibex in the Florida Mountains, New Mexico. M.S. thesis, Colorado State University, Fort Collins.

———. 1980. Post-release study of desert bighorn sheep in the Big Hatchet Mountains, New Mexico. Desert Bighorn Counc. Trans. 24:12–14.

———. 1982. Post-release study of desert bighorn sheep in the Big Hatchet Mountains, New Mexico. New Mexico Dep. Game and Fish, Santa Fe.

Bayer, M. B., and N. M. Simmons. 1984. Horn growth in Dall's sheep: a preliminary report. Proc. Bienn. Symp. North. Wild Sheep and Goat Counc. 4:285–294.

Bear, G. D. 1978. Evaluation of fertilizer and herbicide applications on two Colorado bighorn sheep winter ranges. Colorado Div. Wildl., Div. Rep. No. 10, Fed. Aid Wildl. Restor. Proj. W-41-R.

Becker, K., T. Varcalli, E. T. Thorne, and G. B. Butler. 1978. Seasonal distribution patterns of Whiskey Mountain bighorn sheep. Proc. Bienn. Symp. North. Wild Sheep and Goat Counc. 1:1–16.

Becklund, W. W., and C. M. Singer. 1967. Parasites of *Ovis canadensis canadensis* in Montana, with a checklist of the internal and external parasites of Rocky Mountain bighorn sheep in North America. Am. J. Parasitol. 53:157–165.

Bell, K. 1994. Blood protein polymorphisms in the donkey *(Equus asinus)*. Cenim. Genet. 25:109–113.

Bendt, R. H. 1957. Status of bighorn sheep in Grand Canyon National Park and Monument. Desert Bighorn Counc. Trans. 1:16–19.

Berger, J. 1978a. Social development and reproductive strategies in bighorn sheep. Ph.D. dissertation, University of Colorado, Boulder.

———. 1978b. Group size, foraging, and antipredator ploys: an analysis of bighorn sheep decisions. Behav. Ecol. Sociobiol. 4:91–99.

———. 1978c. Maternal defensive behavior in bighorn sheep. J. Mammal. 59:620–621.

———. 1979a. Weaning conflict in desert and mountain sheep *(Ovis canadensis)*: an ecological interpretation. Z. Tierpsychol. 50:188–200.

———. 1979b. Social ontogeny and behavioural diversity: consequences for bighorn sheep *Ovis canadensis* inhabiting desert and mountain environments. J. Zool., Lond. 192:251–266.

———. 1982. Female breeding age and lamb survival in desert bighorn sheep *(Ovis canadensis)*. Mammalia 46:183–190.

———. 1986. Wild horses of the Great Basin: social competition and population size. University of Chicago Press, Chicago, Ill.

———. 1990. Persistence of different sized populations: an empirical assessment of rapid extinctions in bighorn sheep. Conserv. Biol. 4:91–98.

———. 1991. Pregnancy incentives, predation constraints and habitat shifts: experimental and field evidence for wild bighorn sheep. Anim. Behav. 41:61–77.

Berner, L. R., and P. R. Krausman. 1992. Mountain sheep habitat evaluation in Mojave desert scrub. Desert Bighorn Counc. Trans. 36:10–12.

———, ———, and M. C. Wallace. 1992. Habitat selection by mountain sheep in Mojave desert scrub. Desert Bighorn Counc. Trans. 36:13–22.

Berwick, S. H. 1968. Observations on the decline of the Rock Creek, Montana, population of bighorn sheep. M.S. thesis, University of Montana, Missoula.

Binns, W., L. F. James, J. L. Shupe, and G. Everett. 1963. A congenital cyclopian-type malformation in lambs induced by maternal ingestion of a range plant, *Veratrum californicum*. Am. J. Vet. Res. 24:1164–1175.

Black, F. L. 1993. Why did they die? Science 258:1739–1740.

Blaisdell, J. A. 1976. The Lava Beds bighorn—so who worries? Desert Bighorn Counc. Trans. 20:50.

———. 1982. Lava Beds wrap-up, what did we learn? Desert Bighorn Counc. Trans. 26:32–33.

Bleich, V. C. 1993. Sexual segregation in desert-dwelling mountain sheep. Ph.D. dissertation, University of Alaska, Fairbanks.

———, R. T. Bowyer, D. J. Clark, and T. O. Clark. 1992a. An analysis of forage used by mountain sheep in the eastern Mojave Desert, California. Desert Bighorn Counc. Trans. 36:41–47.

———, ———, A. M. Pauli, R. L. Vernoy, and R. W. Anthes. 1990a. Responses of mountain sheep to helicopter surveys. Calif. Fish and Game 76:197–204.

———, ———, and J. D. Wehausen. 1997. Sexual segregation in mountain sheep: resources or predation? Wildl. Monogr. No. 134.

———, M. C. Nicholson, A. T. Lombard, and P. V. August. 1992b. Preliminary tests of a mountain sheep habitat model using a geographic information system. Proc. Bienn. Symp. North. Wild Sheep and Goat Counc. 8:256–263.

———, J. D. Wehausen, and S. A. Holl. 1990b. Desert-dwelling mountain sheep: conservation implications of a naturally fragmented distribution. Conserv. Biol. 4:383–390.

———, ———, J. A. Keay, J. G. Stahmann, and M. W. Berbach. 1990c. Radiotelemetry collars and mountain sheep: a cautionary note. Desert Bighorn Counc. Trans. 34:6–8.

———, ———, R. R. Ramey II, and J. L. Rechel. 1996. Metapopulation theory and mountain sheep: implications for conservation. Pages 353–373 *in* D. R. McCullough, editor. Metapopulations and wildlife conservation. Island Press, Washington, D.C.

Blockey, M. A. deB., R. A. Parr, and B. J. Restall. 1975. Wastage of ova in young merino ewes. Aust. Vet. J. 51:298–302.

Blood, D. A. 1961. An ecological study of California bighorn sheep (*Ovis canadensis californiana* Douglas) in southern British Columbia. M.S. thesis, University of British Columbia, Vancouver.
———. 1963. Some aspects of behavior of a bighorn herd. Can. Field-Nat. 77:77–94.
———. 1967. Food habits of the Ashnola bighorn sheep herd. Can. Field-Nat. 81:23–29.
———, D. R. Flook, and W. D. Wishart. 1970. Weights and growth of Rocky Mountain bighorn sheep in western Alberta. J. Wildl. Manage. 34:451–455.
Blood, D. C., and J. A. Henderson. 1974. Veterinary medicine. Fourth edition. Williams and Wilkins, Baltimore, Md.
Blunt, M. H., H. A. Dawson, and E. T. Thorne. 1972. The birth weights and gestation in captive Rocky Mountain bighorn sheep. J. Mammal. 58:106.
———, and T.H.J. Huisman. 1975. The hemoglobin of sheep. Pages 155–183 *in* M. H. Blunt, editor. The blood of sheep. Springer-Verlag, New York, N.Y.
Boever, W. J., and D. Peters. 1977. Paratuberculosis in two herds of exotic sheep. J. Am. Vet. Med. Assoc. 165:822.
Boone, J. P. 1988. Historical review of exceptional sheep trophies. Pages 121–142 *in* R. Valdez, editor. Wild sheep and wild sheep hunters of the New World. Wild Sheep and Goat International, Mesilla, N. Mex.
Botkin, D. B., P. A. Jordan, A. S. Dominski, H. S. Lwendorf, and G. E. Hutchinson. 1973. Sodium dynamics in a northern ecosystem. Proc. Natl. Acad. Sci. USA 70:2745–2748.
Bottrell, A., B. Gordy, and R. Peterson. 1978. Comparison of chromosomes and blood constituents of Rocky Mountain and California bighorn and Dall and Stone thinhorn sheep. Proc. North. Wild Sheep and Goat Conf. 1:12–14.
Bowyer, R. T., L. Elliot, R. K. Clark, and D. A. Jessup. 1990. Morphometric analysis of *Psoroptes* spp. mites from bighorn sheep, mule deer, rabbits and cattle. J. Parasitol. 76:823–828.
———, and D. M. Leslie Jr. 1992. *Ovis dalli*. Mammalian Species 393:1–7.
Boyce, W. M., and R. A. Brown. 1991. Antigenic characterization of *Psoroptes* spp. (Acari: Psioroptidae) mites from different hosts. J. Parasitol. 77:675–679.
———, L. Elliot, R. K. Clark, and D. A. Jessup. 1990. Morphometric analysis of *Psoroptes* spp. mites from bighorn sheep, mule deer, rabbits and cattle. J. Parasitol. 76:823–828.
———, D. A. Jessup, and R. A. Clark. 1991a. Seriodiagnostic antibody responses to *Psoroptes* infestations in bighorn sheep. J. Wildl. Dis. 27:10–15.
———, J.A.K. Mazet, J. Mellies, R. K. Clark, and D. A. Jessup. 1991b. Kinetic ELISA for detection of anibodies to *Psoroptes* spp. (Acari: Posoroptidae) in bighorn sheep *(Ovis canadensis)*. J. Parasitol. 77:692–696.
———, J. A. Miller, D. A. Jessup, and R. A. Clark. 1992. Use of ivermectin implants for treatment of psoroptic scabies in free-ranging bighorn sheep. J. Zoo Wild. Medicine 23:211–213.

———, and R. L. Zarnke. 1996. Antibody responses to *Psoroptes* spp. mites in Dall's sheep *(Ovis dalli)*. J. Wildl. Dis. 32:711–713.

Boyd, R. J., editor. 1995. Mountain sheep ecosystem management strategies in the 11 western states and Alaska. Bur. Land Manage., Rep. SC/PL-95/001 + 6600.

Boyer, S. H., P. Hathaway, F. Pascasio, C. Orton, J. Bardley, and M. A. Naughton. 1966. Hemoglobins in sheep: multiple differences in amino acid sequences of three beta-chains and possible origins. Science 153:1539–1543.

Bradley, W. G. 1964. The vegetation of the Desert Game Range with special reference to the desert bighorn. Desert Bighorn Counc. Trans. 8:43–67.

———, and L. G. Allred. 1966. A comparative study of dental anomalies in desert bighorn sheep. Desert Bighorn Counc. Trans. 10:78–85.

———, and D. P. Baker. 1967a. Life tables for Nelson bighorn sheep on the Desert Game Range. Desert Bighorn Counc. Trans. 11:142–170.

———, and L. R. Baker. 1967b. Range of variation in Nelson bighorn sheep from the Desert Game Range and its taxonomic significance. Desert Bighorn Counc. Trans. 11:114–140.

———, and J. E. Deacon. 1965. The biotic communities of southern Nevada. Desert Research Institute, Reprint Ser. No. 9. Nevada Southern University, Las Vegas.

Broadbent, R. V. 1969. Nevada's 1968 transplant disappointment. Desert Bighorn Counc. Trans. 13:43–47.

Brody, S. 1945. Bioenergetics and growth. Hafner Publishing, New York, N.Y.

Brown, B. W. 1974. Distribution and population characteristics of bighorn sheep near Thompson Falls in northwestern Montana. M.S. thesis, University of Montana, Missoula.

Brown, K. W., D. D. Smith, D. E. Bernhardt, K. R. Giles, and J. B. Helvie. 1975. Food habits and radionuclide tissue concentrations of Nevada desert bighorn, 1972–1973. Desert Bighorn Counc. Trans. 19:61–68.

———, ———, and R. P. McQuivey. 1977. Food habits of desert bighorn sheep in Nevada 1956–1976. Desert Bighorn Counc. Trans. 21:32–61.

Browning, B. 1960. Preliminary report of the food habits of the wild burro in the Death Valley National Monument. Desert Bighorn Counc. Trans. 4:88–90.

Browning, B. M., and G. Monson. 1980. Food. Pages 80–99 *in* G. Monson and L. Sumner, editors. The desert bighorn: its life history, ecology, and management. University of Arizona Press, Tucson.

Broyles, B. 1995. Desert wildlife water developments: questioning use in the Southwest. Wildl. Soc. Bull. 23:663–675.

Bruère, A. N., H. M. Chapman, P. M. Jaine, and R. M. Morris. 1976. Origin and significance of centric fusions in domestic sheep. J. Hered. 67:149–154.

Brunidge, G. C., L. J. Layne, and T. R. McCabe. 1988. Early pregnancy determination using serum progesterone concentration in bighorn sheep. J. Wildl. Manage. 5:610–612.

Brussard, P. F., and M. E. Gilpin. 1989. Demographic and genetic problems of small populations. Pages 37–48 *in* U. S. Seal, E. T. Thorne, M. A. Bogan, and S. H. Anderson, editors. Conservation biology and the black-footed ferret. Yale University Press, New Haven, Conn.

Buechner, H. K. 1960. The bighorn sheep in the United States, its past, present, and future. Wildl. Monogr. No. 4.

Buis, R. C., and E. M. Tucker. 1983. Relationships between rare breeds of sheep in Netherlands as based on blood typing. Anim. Blood Groups and Biochem. Genet. 14:17–26.

Bunch, T. D. 1978. Fundamental karyotype in domestic and wild species of sheep. J. Hered. 69:77–80.

———. 1979. Skeletal lesions associated with desert bighorn chronic sinusitis. Desert Bighorn Counc. Trans. 23:27–32.

———. 1980. A survey of chronic sinusitis in the bighorn of California. Desert Bighorn Counc. Trans. 24:14–18.

———, and S. D. Allen. 1981. Survey of chronic sinusitis-induced skull anomalies in desert bighorn sheep. J. Am. Vet. Med. Assoc. 179:1150–1152.

———, and W. C. Foote. 1976. Chromosomes, hemoglobins and transferrins of Iranian domestic sheep. J. Hered. 67:167–170.

———, ———, and J. J. Spillett. 1976. Translocations of acrocentric chromosomes and their implications in the evolution of sheep *(Ovis)*. Cytogenet. Cell Genet. 17:122–136.

———, M. Hoefs, and R. L. Glaze. 1984. Upper canines in Dall's sheep *(Ovis dalli dalli)*. J. Wildl. Dis. 20:158–161.

———, J. Karpowitz, J. Connor, and G. W. Workman. 1986. Pregnancy diagnosis in free-ranging Nelson's desert bighorn ewes. Desert Bighorn Counc. Trans. 30:2–4.

———, R. M. Mitchell, and A. Maciulis. 1990. G-banded chromosomes of the Gansu argali *(Ovis ammon jubata)* and their implications in the evolution of the *Ovis* karyotype. J. Hered. 81:227–230.

———, and C. F. Nadler. 1980. Giemsa-band patterns of the tahr *(Hemitragus jemlahicus)* and chromosomal evolution of the tribe Caprini. J. Hered. 71:110–116.

———, and T. C. N'Guyen. 1982. Blood group comparison of European mouflon *(Ovis musimon)* and North American desert bighorn *(Ovis canadensis* spp.). J. Hered. 73:112–114.

———, ———, and J. J. Lauvergne. 1978a. Hemoglobins of the Corsico-Sardinian mouflon *(Ovis musimon)* and their implications for the origin of HbA in domestic sheep *(Ovis aries)*. Ann. Genet. Sel. Anim. 10:503–506.

———, S. R. Paul, and H. E. McCutchen. 1978b. Chronic sinusitis and osteonecrosis in desert bighorn sheep, *(Ovis canadensis nelsoni)*. North Am. Wild Sheep Symp. 261–273.

———, ———, and ———. 1978c. Chronic sinusitis in the desert bighorn *(Ovis canadensis nelsoni)*. Desert Bighorn Counc. Trans. 22:16–20.

———, ———, and ———. 1979. Skeletal lesions associated with desert bighorn chronic sinusitis. Desert Bighorn Counc. Trans. 23:27–32.

———, and R. Valdez. 1976. Comparative morphology, chromosome types, transferrins and hemoglobins of Iranian and North American desert wild sheep. Desert Bighorn Counc. Trans. 20:13–14.

———, N. N. Voronstov, E. A. Lyapunova, and R. S. Hoffmann. 1998a. Confirmation of diploid chromosome number in Seventzov's sheep (*Ovis ammon severtzovi* Nasovon 1914): G-band karotype comparisons within *Ovis*. J. Heredity: 89: 267–269.

———, S. Wang, R. S. Hoffmann, Y. Zhang, A. Lui, and S. Lin. In Press. Diploid chromosome number and karyotype of the Dalai-lamal argali *(Ovis ammon dalai—lamae)*. Encycilia.

———, ———, ———, ———, ———, and ———. 1998b. Diploid chromosome number and karyotype of the Tibetan argali *(Ovis ammon hodgsoni)*. North Am. Colloquium on Gene Mapping and Cytogenetics of Livestock 10:28.

———, and P. M. Webb. 1979. Desert bighorn chronic sinusitis in Arizona. Desert Bighorn Counc. Trans. 23:25–27.

———, and G. W. Workman. 1988. Hybridization of desert bighorn and argali-mouflon wild sheep. Desert Bighorn Counc. Trans. 32:16–18.

———, ———, and R. J. Callan. 1989. Remote body temperature and heart rate monitoring in desert bighorn sheep. Desert Bighorn Counc. Trans. 33:1–5.

———, ———, and T. E. Mock. 1988. Antibody response of desert bighorn rams to Fort Dodge triangle-3 vaccine (BVD, IBR, and PI-3). Desert Bighorn Counc. Trans. 32:11–12.

Bunnell, F. L. 1978. Horn growth and population quality in Dall sheep. J. Wildl. Manage. 42:764–775.

———. 1980a. Factors controlling lambing period of Dall's sheep. Can. J. Zool. 58:1027–1031.

———. 1980b. Weight estimation of Dall's sheep and mountain goats. Wildl. Soc. Bull. 8:291–297.

———. 1982. The lambing period of mountain sheep: synthesis, hypothesis, and tests. Can. J. Zool. 60:1–14.

———, and N. A. Olsen. 1976. Weights and growth of Dall sheep in Kluane Park Reserve, Yukon Territory. Can. Field-Nat. 90:157–162.

———, and ———. 1982. Age-specific natality in Dall's sheep. J. Mammal. 62:379–380.

Burles, D. W., and M. Hoefs. 1984. Winter mortality of Dall sheep, *(Ovis dalli dalli)*, in Kluane National Park, Yukon. Can. Field-Nat. 98:479–484.

———, ———, and N. Barichello. 1984. The influence of winter severity on Dall sheep productivity in southwestern Yukon—a preliminary assessment. Proc. Bienn. Symp. North. Wild Sheep and Goat Counc. 4:67–84.

Burnside, D. M., and B. O. Rowley. 1994. Evaluation of an enzyme-linked immunosorbent assay of diagnosis of paratuberculosis in goats. Am. J. Vet. Res. 55:465–466.

Butts, T. W. 1980. Population characteristics, movements, and distribution patterns of the upper Rock Creek bighorn sheep. Proc. Bienn. Symp. North. Wild Sheep and Goat Counc. 2:115–140.

Callan, R. J., T. D. Bunch, G. W. Workman, and R. E. Mock. 1991. Development of pneumonia in desert bighorn sheep after exposure to a flock of exotic wild and domestic sheep. J. Am. Vet. Med. Assoc. 198:1052–1056.

Campbell, B. H., and R. Remington. 1979. Bighorn use of artificial water sources in the Buckskin Mountains, Arizona. Desert Bighorn Counc. Trans. 23:50–56.

Capelle, K. J. 1966. The occurrence of *(Oestrus ovis)* L. (Diptera: Oestridae) in the bighorn sheep from Wyoming and Montana. J. Parasitol. 52:618–621.

Capp, J. C. 1967. Competition among bighorn sheep, elk and deer in Rocky Mountain National Park, Colorado. M.S. thesis, Colorado State University, Fort Collins.

Castle, W. E. 1940. Genetics of horns in sheep. J. Hered. 31:486–487.

Cater, B. H. 1968. Scabies in desert bighorn sheep. Desert Bighorn Counc. Trans. 12:76–77.

Caughley, G. 1966. Mortality patterns in mammals. Ecology 47:906–919.

———. 1970. Eruption of ungulate populations with emphasis on Himalayan thar in New Zealand. Ecology 51:53–72.

———. 1976. Wildlife management and the dynamics of ungulate populations. Appl. Biol. 1:183–246.

———. 1977. Analysis of vertebrate populations. John Wiley and Sons, New York, N.Y.

Challans, J. A., K. Stevenson, H. W. Reid, and J. M. Sharp. 1994. A rapid method for the extraction and detection of *Mycobacterium avium* subspecies *paratuberculosis* from clinical specimens. Vet. Rec. 134:95–96.

Chalmers, G. A., and M. W. Barrett. 1977. Capture myopathy in pronghorns in Alberta, Canada. J. Am. Vet. Med. Assoc. 171:918–923.

Chappel, R. W. 1978. Bioenergetics of Rocky Mountain bighorn sheep *Ovis canadensis canadensis* Shaw. M.S. thesis, University of Alberta, Edmonton.

———, and R. J. Hudson. 1978. Winter bioenergetics of Rocky Mountain bighorn sheep. Can. J. Zool. 56:2388–2393.

Chen, J. 1992. Transferrin subtypes in 11 South China minority populations. Genet. Acta PRC Genet. Inst. 14:20–26.

Cherniavski, F. B. 1962. On the reproduction and growth of the sheep (*Ovis nivicola* Esch.). Zoologicheskii Zhurnal 12:1556–1566.

———. 1967. Ecology of bighorn sheep and their value as a game animal on the Koryak upland. Problemy Severa 11:128–141.

Child, K. N., K. K. Fujino, and M. W. Warren. 1978. A gray wolf *(Canis lupus columbianus)* and Stone sheep *(Ovis dalli stonei)* fatal predator-prey encounter. Can. Field-Nat. 92:399–401.

Chilelli, M. 1981. Group organization and activity patterns of desert bighorn sheep. M.S. thesis, University of Arizona, Tucson.

———, and P. R. Krausman. 1981. Group organization and activity patterns of desert bighorn sheep. Desert Bighorn Counc. Trans. 25:17–24.

Christian, J. J., and D. E. Davis. 1971. Endocrines, behavior, and population. Pages 69–98 *in* I. A. McLaren, editor. Natural regulation of animal populations. Atherton Press, New York, N.Y.

Cizova, D., F. Karl, A. Stratit, and V. Glasnok. 1993. Intraspecific variation in serum proteins of brown hare (*Lepus europaeus* Pallos, 1778). Comp. Biochem. Physiol. B 106:977–982.

Clark, J. L. 1978. The great arc of the wild sheep. University of Oklahoma Press, Norman.

Clark, R. K., and D. A. Jessup. 1992. The health of mountain sheep in the San Andres Mountains, New Mexico. Desert Bighorn Counc. Trans. 36:30–35.

———, ———, M. D. Kock, and R. A. Weaver. 1985. Survey of desert bighorn sheep in California for exposure-selected infectious disease. J. Am. Vet. Med. Assoc. 187:1175–1179.

———, ———, and R. A. Weaver. 1988. Scabies mite infestation in desert bighorn sheep from California. Desert Bighorn Counc. Trans. 33:13–15.

Clark, T. W., and J. H. Seebeck, editors. 1990. Management and conservation of small populations. Chicago Zoological Society, Brookfield, Ill.

Clutton-Brock, T. H., A. W. Illius, K. Wilson, B. T. Grenfell, A. D. C. MacColl, and S. D. Albon. 1997. Stability and instability in ungulate populations: an empirical analysis. Am. Nat. 149:195–219.

———, M. Major, S. D. Albon, and F. E. Guiness. 1987. Early development and population dynamics in red deer. I. Density-dependent effects on juvenile survival. J. Anim. Ecol. 56:53–67.

———, ———, ———, and ———. 1992. Early development and population fluctuations in Soay sheep. J. Anim. Ecol. 61:381–396.

Cobbett, N. G., and W. C. Mitchell. 1941. Further observations on the life cycle and incidence of the sheep bot *(Oestrus ovis)* in New Mexico and Texas. Am. J. Vet. Res. 2:358–366.

Coggins, V. L., and P. E. Matthews. 1992. Lamb survival and herd status of the Lostine bighorn herd following a *Pasteurella* die-off. Proc. Bienn. Symp. North. Wild Sheep and Goat Counc. 8:147–154.

Cole, F. R. 1969. The flies of western North America. University of California Press, Berkeley.

Collins, M. T., K. B. Kenefick, D. C. Sockett, R. S. Lambrecht, J. McDonald, and J. B. Jorgensen. 1990. Enhanced radiometric detection of *Mycobacterium paratuberculosis* by using filter-concentrated bovine fecal specimens. J. Clin. Microbiol. 28:2514–2519.

———, D. C. Sockett, and S. Ridge. 1991. Evaluation of a commercial enzyme-linked immunosorbent assay of Johne's disease. J. Clin. Microbiol. 29:272–276.

Coney, F. M. 1950. Rocky Mountain bighorn sheep of Montana. Montana Fish and Game Bull. No. 2.

Connell, R. 1954. Contagious ecthyma in Rocky Mountain bighorn sheep. Can. J. Comp. Med. 18:59.

Constan, K. J. 1972. Winter foods and range use of three species of ungulates. J. Wildl. Manage. 36:1068–1075.

Cook, S. F. 1973. The significance of disease in the extinction of the New England Indians. Human Biol. 45:485–508.

Cossio, M. L. 1976. Desert bighorn sheep in Mexico. Pages 138–142 in J. B. Trefethen, editor. The wild sheep in modern North America. The Boone and Crockett Club and Winchester Press, New York, N.Y.

Couey, F. M. 1950. Rocky Mountain bighorn sheep of Montana. Montana Fish and Game Bull. No. 2.

Cousins, D. V., R. J. Evans, and B. R. Francis. 1995. Use of BACTEC radiometric culture method and polymerase chain reaction for the rapid screening of faeces and tissues for *Mycobacterium paratuberculosis*. Aust. Vet. J. 72:458–462.

Cowan, I. M. 1940. Distribution and variation in the native sheep of North America. Am. Midl. Nat. 24:505–580.

———. 1944. Parasites, diseases and injuries of game animals in Banff, Jasper, and Kootenay National Parks. Unpubl. Rep., Natl. Parks Bur., Dep. Mines and Resources, Ottawa.

———. 1947. Range competition between mule deer, bighorn sheep, and elk in Jasper Park, Alberta. Trans. North Am. Wildl. Conf. 12:223–227.

———. 1974. Management implications of behavior in the large herbivorous mammals. Pages 921–934 in V. Geist and F. Walther, editors. The behaviour of ungulates and its relation to management. IUCN Publ., New Ser. No. 24. Morges, Switzerland.

Creeden, P. J., and J. L. Schmidt. 1983. The Colorado desert bighorn introduction project, a status report. Desert Bighorn Counc. Trans. 27:34–36.

Cronin, M. A. 1991. Mitochondrial-DNA phylogeny of deer (Cervidae). J. Mammal. 72:533–566.

Crosbie, P. R., W. L. Goff, D. Stiller, D. A. Jessup, and W. M. Boyce. 1997. The distribution of *Dermacentor hunteri* and *cenaplasma* spp. in bighorn sheep. J. Parasitol. 83:31–37.

Culbertson, J. L., S. M. Howell, and K. W. Rice. 1982. Mountain goat winter range surveys on Kings River, Crow Creek, Portage Valley and Carmen Lake. Unpubl. Rep., U.S. For. Serv., Chugach Natl. For.

———, and R. M. Walker. 1981. Round Mountain fire effects and sheep range survey 1981. Unpubl. Rep., U.S. For. Serv., Chugach Natl. For.

———, ———, and G. Dunne. 1980. Round Mountain fire effects and sheep range survey 1980. Unpubl. Rep., U.S. For. Serv., Chugach Natl. For.

Cunningham, S. C. 1989. Evaluation of bighorn sheep habitat. Pages 135–160 in R. M. Lee, editor. The desert bighorn sheep in Arizona. Arizona Game and Fish Dep., Phoenix.

———, and J. C. deVos. 1992. Mortality of mountain sheep in the Black Canyon area of northwest Arizona. Desert Bighorn Counc. Trans. 36:27–29.

———, L. Hanna, and J. Sacco. 1993. Possible effects of the realignment of U.S. Highway 93 on movements of desert bighorns in the Black Canyon area. Pages 83–100 *in* P. G. Rowlands, C. van Riper, and M. K. Sogge, editors. Proc. First Bienn. Conf. Res. in Colo. Plateau Natl. Parks. Trans. and Proc. Ser. NPS/NRNAU/NRTP-93/10, Natl. Park Serv.

———, and R. D. Ohmart. 1986. Aspects of the ecology of desert bighorn sheep in Carrizo Canyon, California. Desert Bighorn Counc. Trans. 30:14–19.

Curby, C. H. 1981. Habitat selection and activity patterns of Dall sheep ewe groups in the Sheenjek River drainage, Northwest Alaska. M.S. thesis, University of Alaska, Fairbanks.

Dale, A. R. 1987. Ecology and behavior of bighorn sheep, Waterton Canyon, Colorado, 1981–1982. M.S. thesis, Colorado State University, Fort Collins.

Dall'Olio, S., R. Davoli, E. Costosi, P. Zambonelli, and V. Russo. 1995. A B gill polymorphism at the porcine transferrin (Tf) locus. Anim. Genet. 26:211–212.

Dalquest, W. W., and D. F. Hoffmeister. 1948. Mountain sheep from the state of Washington in the collection of the University of Kansas. Trans. Kansas Acad. Sci. 51:224–234.

Dalton, L. B., S. C. Winegardner, and J. W. Bates. 1978. Utah's desert bighorn sheep—a status report, 1978. Desert Bighorn Counc. Trans. 22:6–9.

Dasmann, R. F. 1964. Wildlife biology. First edition. John Wiley and Sons, New York, N.Y.

Davis, J. W., and R. C. Anderson. 1971. Parasitic diseases of wild mammals. Iowa State University Press, Ames.

Davis, W. B. 1938. Summer activity of mountain sheep on Mt. Washburn, Yellowstone National Park. J. Mammal. 19:88–94.

———, and W. P. Taylor. 1939. The bighorn sheep of Texas. J. Mammal. 20:440–455.

Dean, H. C., and J. J. Spillett. 1976. Bighorn in Canyonlands National Park. Desert Bighorn Counc. Trans. 20:15–17.

Decker, J. V. 1970. Scabies in desert bighorn sheep of the Desert National Wildlife Range. Desert Bighorn Counc. Trans. 14:107–108.

Deevey, E. S. 1947. Life tables for natural populations of animals. Q. Rev. Biol. 22:283–314.

DeForge, J. R. 1976. Stress: is it limiting bighorn? Desert Bighorn Counc. Trans. 20:30–31.

———. 1980. Population biology of desert bighorn sheep in the San Gabriel Mountains of California. Desert Bighorn Counc. Trans. 24:29–32.

———. 1984. Population estimate of Peninsular desert bighorn sheep in the Santa Rosa Mountains, California. Desert Bighorn Counc. Trans. 28:41–43.

———, C. W. Jenner, A. J. Plechner, and G. W. Sudmeier. 1979. Decline of bighorn sheep *(Ovis canadensis),* the genetic implications. Desert Bighorn Counc. Trans. 23:63–66.

———, ———, G. W. Sudmeier, R. L. Graham, and S. V. Segreto. 1981. The loss of two populations of desert bighorn sheep in California. Desert Bighorn Counc. Trans. 25:36–38.

——, and J. E. Scott. 1982. Ecological investigations into high lamb mortality. Desert Bighorn Counc. Trans. 26:65–76.

Dehn, M. M. 1986. Vigilance, group size and security in Rocky Mountain elk *(Cervus elaphus nelsoni)*. M. S. thesis, University of Calgary, Alberta.

——. 1990. Vigilance for predators: detection and dilution effects. Behav. Ecol. Sociobiol. 26:337–342.

Delaney, D. E. 1984. Status of desert bighorn sheep in Nevada—1983. Desert Bighorn Counc. Trans. 28:46–47.

Demarchi, D. A. 1970. Effects of grazing on the botanical and chemical composition of range vegetation in the lower Chilcotin River region, British Columbia. M.S. thesis, University of Idaho, Moscow.

——, and H. B. Mitchell. 1973. The Chilcotin River bighorn population. Can. Field-Nat. 87:433–454.

Demarchi, R. A. 1965. An ecological study of the Ashnola bighorn winter ranges. M.S. thesis, University of British Columbia, Vancouver.

——. 1968. Chemical composition of bighorn winter forages. J. Range Manage. 21:385–388.

Deming, O. V. 1952. Tooth development of the Nelson bighorn sheep. Calif. Fish and Game 38:523–529.

——. 1962. Is the desert bighorn a relict species? Desert Bighorn Counc. Trans. 21:21–23.

deVos, J., R. L. Glaze, and T. D. Bunch. 1980. Scabies *(Psoroptic ovis)* in Nelson desert bighorn of northwestern Arizona. Desert Bighorn Counc. Trans. 24:44–46.

Dickinson, T. G., and C. D. Simpson. 1980. Dispersal and establishment of Barbary sheep in southeast New Mexico. Pages 33–45 *in* C. D. Simpson, editor. Proceedings of the symposium on ecology and management of Barbary sheep. Texas Tech University Press, Lubbock.

Divine, D., and C. L. Douglas. 1995. Eagle Mountain bighorn monitoring program: an update. Desert Bighorn Counc. Trans. 39:84–87.

Dixon, J. S. 1938. Birds and mammals of Mount McKinley National Park. U.S. Natl. Park Serv., Fauna Ser. No. 3.

Dixon, K. L. 1979. The influences of Quaternary environments on the evolution and distribution of wild sheep *(Ovis* spp.). Pages 1–13 *in* K. L. Dixon, editor. The domestication of sheep, their ancestors, geography, time period and people involved. International Sheep and Goat Institute, Utah State University Press, Logan.

Dobyns, H. F. 1983. Their numbers become thinned: Native American population dynamics in eastern North America. University of Tennessee Press, Knoxville.

Dodd, N. L. 1983. Ideas and recommendations for maximizing desert bighorn transplant efforts. Desert Bighorn Counc. Trans. 27:12–16.

———, and W. W. Brady. 1986. Cattle grazing influences on vegetation of a sympatric desert bighorn range in Arizona. Desert Bighorn Counc. Trans. 30:8–13.

———, and ———. 1988. Dietary relationships of sympatric desert bighorn sheep and cattle. Desert Bighorn Counc. Trans. 32:1–6.

Douglas, C. L. 1993. Management model for predicting fall lamb:ewe ratios in desert bighorn sheep: Canyonlands National Park, Utah. Pages 64–82 *in* P. G. Rolands, C. van Riper, and M. K. Sogge, editors. Proc. First Bienn. Conf. Res. in Colo. Plateau Natl. Parks. Trans. and Proc. Ser. NPS/NRNAU/NRTP-93/10, Natl. Park Serv.

———, and D. M. Leslie Jr. 1986. Influence of weather and density on lamb survival of desert mountain sheep. J. Wildl. Manage. 50:153–156.

———, and ———. 1996. Feral animals on rangeland. Pages 281–295 *in* P. R. Krausman, editor. Rangeland wildlife. Society for Range Management, Denver, Colo.

———, and C. Norment. 1977. Habitat damage by feral burros in Death Valley. Desert Bighorn Counc. Trans. 21:23–25.

———, and L. D. White. 1978. Radio-telemetric studies of movements in desert bighorn sheep, Joshua Tree National Monument. Desert Bighorn Counc. Trans. 22:21–22.

———, and ———. 1979. Movements of desert bighorn sheep in the Stubbe Spring Area, Joshua Tree National Monument. Desert Bighorn Counc. Trans. 23:71–77.

Douglas, D. 1829. Observations on two undescribed species of North American mammalia *(Cervus leucurus* et *Ovis californianus)*. Zool. J. 4:330–332.

Doyle, T. M. 1956. Johne's disease. Vet. Rec. 68:869–878.

Drewek, J. R. 1970. Population characteristics and behavior of introduced bighorn sheep in Owyhee County, Idaho. M.S. thesis, University of Idaho, Moscow.

Dubash, K., W. P. Shulow, S. Bech-Nielsen, H. F. Still Jr., and R. D. Slemons. 1996. Evaluation of an agar gel immunodiffusion test kit for detection of antibodies to *Mycobacterium paratuberculosis* in sheep. J. Am. Vet. Med. Assoc. 208:401–404.

Dunbar, M. R. 1992. Theoretical concepts of disease versus nutrition as primary factors in population regulation of wild sheep. Proc. Bienn. Symp. North. Wild Sheep and Goat Counc. 8:174–192.

Dunn, W. C. 1991. Evaluation of bighorn sheep habitat in New Mexico. New Mexico Dep. Game and Fish, Final Rep. Fed. Aid Proj. W-1227-R-7, Job 4.

———. 1996. Evaluating bighorn habitat: a landscape approach. U.S. Dep. Inter., Bur. Land Manage., Rep. RS/ST-96/005 + 6600.

Dutt, R. H., and L. F. Bush. 1955. The effect of low environmental temperature on initiation of the breeding season and fertility in sheep. J. Anim. Sci. 14:805–896.

Ebenhard, T. 1991. Colonization in metapopulations: a review of theory and observations. Pages 10–121 *in* M. Gilpin and I. Hanski, editors. Metapopulation dynamics: empirical and theoretical investigations. Academic Press, San Diego, Calif.

Ebert, D. W., and C. L. Douglas. 1993. Desert bighorn movements and habitat use in relation to the proposed Black Canyon Bridge Project, Nevada. Final Rep., U.S. Bur. Reclam., Boulder City, Nev.

Ebert, P. 1978. Bighorn sheep in Oregon. Oregon Wildl. 33:3–6.

Eccles, T. R. 1978. Activity patterns of captive California bighorn sheep *(Ovis canadensis californiana)*. Proc. Bienn. Symp. North. Wild Sheep and Goat Counc. 1:274–292.

——. 1983. Aspects of social organization and diurnal activity patterns of California bighorn sheep *(Ovis canadensis californiana* Douglas 1829). B.C. Min. Envir., Fish and Wildl. Rep. No. R-6.

——, and D. M. Shackleton. 1979. Recent records of twinning in mountain sheep. J. Wildl. Manage. 43:974–976.

——, and ——. 1986. Correlates and consequences of social status in female bighorn sheep. Anim. Behav. 34:1392–1401.

Egorov, O. V. 1967. Wild ungulates of Yakutia. Israel Program for Scientific Translations, Jerusalem.

Elenowitz, A. S. 1983. Habitat use and population dynamics of transplanted desert bighorn sheep in the Peloncillo Mountains, New Mexico. M.S. thesis, New Mexico State University, Las Cruces.

Elliot, D. G. 1904. The land and sea mammals of Middle America and West Indies. Field Columbian Mus. Zool. Ser. 4:83–86.

Elliott, C. L., and J. D. McKendrick. 1984. Food habits of Dall sheep on revegetated coal stripmine soils in Alaska. Proc. Bienn. Symp. North. Wild Sheep and Goat Counc. 4:241–251.

Elliott, J. P. 1978. Range enhancement and trophy production in Stone sheep. Proc. Bienn. Symp. North. Wild Sheep and Goat Counc. 1:113–118.

——. 1985. The status of thinhorn sheep *(Ovis dalli)* in British Columbia. Pages 43–47 *in* M. Hoefs, editor. Wild sheep. Spec. Rep. of the North. Wild Sheep and Goat Counc. Yukon Wildl. Branch, Whitehorse, Canada.

Ellis, F. G. 1941. Idaho mountain sheep survey. Idaho Dep. Fish and Game, Fed. Aid and Wildl. Restor. Rep. W-7-R-C.

Ellstrand, N. C., and D. R. Elam. 1993. Population genetic consequences of small population size: implications for plant conservation. Annu. Rev. Ecol. Syst. 24:217–242.

Erickson, G. L. 1972. The ecology of Rocky Mountain bighorn sheep in the Sun River area of Montana with special reference to summer food habits and range movements. Montana Fish and Game Dep., Fed. Aid and Wildl. Restor. Proj. W-120-R-2 and R-3.

Erickson, J. A. 1970. Sheep report. Alaska Dep. Fish and Game, Fed. Aid Wildl. Restor. Annu. Rep. Vol. 11.

Eshel, I. 1978. On a predator-prey nonzero-sum game and the evolution of gregarious behaviour of evasive prey. Am. Nat. 112:787–795.

Estes, R. D. 1972. The role of the vomeronasal organ in mammalian reproduction. Mammalia 36:315–341.

———. 1974. Social organization in the African Bovidae. Pages 166–205 *in* V. Geist and F. Walther, editors. The behaviour of ungulates and its relation to management. IUCN Publ., New Ser. No. 24. Morges, Switzerland.

Etchberger, R. C., and P. R. Krausman. 1999. Frequency of birth and lambing sites of a small population of mountain sheep. Southwest. Nat. 44: in press.

———, ———, and R. Mazaika. 1989. Mountain sheep habitat characteristics in the Pusch Ridge Wilderness, Arizona. J. Wildl. Manage. 53:902–907.

———, ———, and ———. 1990. Effects of fire on desert bighorn sheep habitat. Pages 53–57 *in* P. R. Krausman and N. S. Smith, editors. Managing wildlife in the Southwest. Arizona Chapter of the Wildlife Society, Phoenix.

Etkin, W. 1964. Cooperation and competition in social behavior. Pages 1–34 *in* W. Etkin, editor. Social behavior and organization among vertebrates. University of Chicago Press, Chicago, Ill.

Eustis, G. P. 1962. Winter lamb surveys on the Kofa Game Range. Desert Bighorn Counc. Trans. 6:83–86.

Evans, J. V., H. Harris, and F. L. Warren. 1958. The distribution of hemoglobin and blood potassium types in British breeds of sheep. Proc. R. Soc. Lond. 149:249–262.

Ewbank, R. 1964. Observations on the suckling habits of twin lambs. Anim. Behav. 12:34–37.

———. 1967. Nursing and suckling behaviour among Clun Forest ewes and lambs. Anim. Behav. 15:251–258.

Fairaizel, S. D. 1980. Population characteristics of transplanted California bighorn sheep in western North Dakota. Proc. Bienn. Symp. North. Wild Sheep and Goat Counc. 2:70–87.

Fairbanks, W. S., J. A. Bailey, and R. S. Cook. 1987. Habitat use by a low-elevation, semicaptive bighorn sheep population. J. Wildl. Manage. 51:912–915.

Farb, P. 1963. Ecology. Life Nature Library. Time, New York, N.Y.

Fedosenko, A. K. 1986. Food habits and habitat use of Putoran snow sheep *(Ovis nivicola borealis)*. Proc. Bienn. Symp. North. Wild Sheep and Goat Counc. 5:381–385.

Festa-Bianchet, M. 1986*a*. Seasonal dispersion of overlapping mountain sheep ewe groups. J. Wildl. Manage. 50:325–330.

———. 1986*b*. Site fidelity and seasonal range use by bighorn rams. Can. J. Zool. 64:2126–2132.

———. 1987. Bighorn sheep, climbing accidents, and implications for mating. Mammalia 51:618–620.

———. 1988*a*. Seasonal range selection in bighorn sheep conflicts between forage quality, forage quantity, and predator avoidance. Oecologia 75:580–586.

———. 1988*b*. Nursing behaviour of bighorn sheep: correlates of ewe age, parasitism, lamb age, birthdate and sex. Anim. Behav. 36:1445–1454.

———. 1988*c*. Birthdate and survival in bighorn lambs *(Ovis canadensis)*. J. Zool., Lond. 214:653–661.

———. 1988*d*. Age-specific reproduction of bighorn ewes in Alberta, Canada. J. Zool. 69:157–160.

———. 1989*a*. Individual differences, parasites, and the costs of reproduction for bighorn ewes *(Ovis canadensis)*. J. Anim. Ecol. 58:785–795.

———. 1989*b*. Survival of male bighorn sheep in southwestern Alberta. J. Wildl. Manage. 53:259–263.

———. 1991*a*. The social system of bighorn sheep: grouping patterns, kinship and female dominance rank. Anim. Behav. 42:71–82.

———. 1991*b*. Numbers of lungworm larvae in feces of bighorn sheep: yearly changes, influence of host sex, and effects on host survival. Can. J. Zool. 69:547–554.

———. 1992. Use of age ratios to predict bighorn sheep population dynamics. Proc. Bienn. Symp. North. Wild Sheep and Goat Counc. 8:227–236.

———, J. T. Jorgenson, W. J. King, K. G. Smith, and W. D. Wishart. 1996. The development of sexual dimorphism: seasonal and lifetime mass changes in bighorn sheep. Can. J. Zool. 74:330–342.

———, ———, M. Lucherini, and D. W. Wishart. 1995. Life history consequences of variation in age of primiparity in bighorn ewes. Ecology 76:871–881.

Feuerstein, R. L., R. L. Schmidt, C. D. Hibler, and W. H. Rutherford. 1980. Bighorn sheep mortality in the Taylor River–Almont Triangle area, 1978–1979: a case study. Colorado Div. Wildl., Spec. Rep. No. 48.

Fisher, A. S. 1992. Status of desert bighorn sheep in New Mexico, 1991. Desert Bighorn Counc. Trans. 36:80.

Fitzsimmons, N. N., and S. W. Buskirk. 1992. Effective population sizes for bighorn sheep. Proc. Bienn. Symp. North. Wild Sheep and Goat Counc. 8:1–7.

———, ———, and M. H. Smith. 1997. Genetic changes in reintroduced Rocky Mountain bighorn sheep populations. J. Wildl. Manage. 61(3):863–872.

Fletcher, I. C. 1971. Relationships between frequency of suckling, lamb growth and postpartum oestrous behaviour in ewes. Anim. Behav. 19:101–111.

Foreyt, W. J. 1989. Fatal *Pasteurella haemolytica* pneumonia in bighorn sheep after direct contact with clinically normal domestic sheep. Am. J. Vet. Pres. 50:341–344.

———, V. Coggins, and T. Parker. 1985. *Psoroptes ovis* (Acari: Psoroptidae) in a Rocky Mountain bighorn sheep *(Ovis canadensis canadensis)* in Idaho. J. Wildl. Dis. 21:456–457.

———, and D. A. Jessup. 1982. Fatal pneumonia of bighorn sheep following association with domestic sheep. J. Wildl. Dis. 18:163–168.

———, and R. M. Silflow. 1996. Attempted protection of bighorn sheep *(Ovis canadensis*) from pneumonia using a nonlethal cyotoxic strain of *Pasteurella haemolytica,* biotype A, serotype. J. Wildl. Dis. 32:315–321.

———, K. P. Snipes, and R. W. Kasten. 1994. Fatal pneumonia following inoculation of healthy bighorn sheep with *Pasteurella haemolytica* from healthy domestic sheep. J. Wildl. Dis. 30:137–145.

Forrester, D. J. 1971. Bighorn sheep lungworm-pneumonia complex. Pages 158–173 *in* J. W. Davis and R. C. Anderson, editors. Parasitic diseases of wild mammals. Iowa State University Press, Ames.

———, and R. S. Hoffmann. 1963. Growth and behavior of a captive bighorn lamb. J. Mammal. 14:116–118.

Free, J. C., R. M. Hansen, and P. L. Sims. 1970. Estimating the dry-weights of food plants in feces of herbivores. J. Range Manage. 23:300–302.

Frelinger, J. A. 1972. The maintenance of transferrin polymorphism in pigeons *(Saccharnyces cerevisiae)* egg white. Proc. Natl. Acad. Sci. USA 69:326–329.

French, C. E., L. C. McEwan, N. D. Magruder, R. H. Ingram, and R. W. Swift. 1955. Nutritional requirements of white-tailed deer for growth and antler development. Pennsylvania State University Agricultural Experiment Station, Bull. No. 600.

Frisch, R. E. 1980. Fatness, puberty, and fertility. Nat. Hist. 89:16–27.

Frisina, M. R. 1974. Ecology of bighorn sheep in the Sun River area of Montana during fall and spring. M.S. thesis, Montana State University, Bozeman.

Gallizioli, S. 1977. Overgrazing on desert bighorn ranges. Desert Bighorn Counc. Trans. 21:21–23.

Gambaryan, P. P. 1974. How animals run. John Wiley and Sons, New York, N.Y.

Gasaway, W. C., and W. E. Heimer. n.d. A comparison of two Dall sheep populations: forage quality, rumen function, and bioenergetics. Unpubl. Rep., Alaska Dep. Fish and Game.

———, R. O. Stephenson, J. L. Davis, P.E.K. Shepherd, and O. E. Burris. 1983. Interrelationships of wolves, prey, and man in interior Alaska. Wildl. Monogr. No. 84.

Geist, V. 1966*a.* Validity of horn segment counts in ageing bighorn sheep. J. Wildl. Manage. 30:634–645.

———. 1966*b.* The evolutionary significance of mountain sheep horns. Evolution 20:558–566.

———. 1966*c.* The evolution of horn-like organs. Behavior 27:175–214.

———. 1967. A consequence of togetherness. Nat. Hist. 76:24–31.

———. 1968. On the interrelation of external appearance, social behaviour and social structure of mountain sheep. Z. Tierpsychol. 25:199–215.

———. 1971. Mountain sheep: a study in behavior and evolution. University of Chicago Press, Chicago, Ill.

———. 1974. On the relationship of ecology and behaviour in the evolution of ungulates: theoretical considerations. Pages 235–246 *in* V. Geist and F. Walther, editors. The behaviour of ungulates and its relation to management. IUCN Publ., New Ser. No. 24. Morges, Switzerland.

———. 1975. On the management of mountain sheep: theoretical considerations. Pages 77–105 *in* J. B. Trefethen, editor. The wild sheep in modern North America. The Boone and Crockett Club and Winchester Press, New York, N.Y.

———. 1978. Life strategies, human evolution, environmental design. Springer-Verlag, New York, N.Y.

———. 1981. On reproductive strategies in ungulates and some problems of adaptation. Pages 111–132 *in* G.G.E. Scudder and J. L. Reveal, editors. Evolution today. Proceedings of the Second International Congress of Systemists and Evolutionary Biologists. Carnegie-Mellon University, Pittsburgh, Pa.

———. 1982. Adaptive behavioral strategies. Pages 219–277 *in* J. W. Thomas and D. E. Toweill, editors. Elk of North America. Stackpole Books, Harrisburg, Pa.

———. 1985*a*. On evolutionary patterns in the Caprinae with comments on the punctuated mode of evolution, gradualism and a general model of mammalian evolution. Pages 15–30 *in* S. Lovari, editor. The biology and management of mountain ungulates. Croom Helm, London, U.K.

———. 1985*b*. On Pleistocene bighorn sheep: some problems of adaptation and relevance to today's American megafauna. Wildl. Soc. Bull. 13:351–359.

———. 1985*c*. Game ranching: threat to wildlife conservation in North America. Wildl. Soc. Bull. 13:594–598.

———. 1986. The paradox of the great Irish stags. Nat. Hist. 95:54–64.

———. 1987*a*. Bergmann's rule is invalid. Can. J. Zool. 65:1035–1038.

———. 1987*b*. On speciation in Ice Age mammals, with special reference to cervids and caprids. Can. J. Zool. 65:1067–1084.

———. 1987*c*. On the evolution of the Caprinae. Pages 3–40 *in* H. Soma, editor. The biology and management of *Capricornis* and related mountain antelopes. Croom Helm, London, U.K.

———. 1988. How markets in wildlife meat and parts, and the sale of hunting privileges jeopardize wildlife conservation. Conserv. Biol. 2:1–12.

———. 1989. Environmentally guided phenotype plasticity in mammals and some of its consequences to theoretical and applied biology. Pages 153–176 *in* N. M. Bruton, editor. Alternative styles of animals. Kluwer Academic Publisher, Dordrecht, Netherlands.

———. 1991*a*. On the taxonomy of giant sheep (*Ovis ammon* Linnaeus, 1766). Can. J. Zool. 69:706–723.

———. 1991*b*. Bones of contention revisited: did antlers evolve with sexual selection as a consequence of neonatal security strategies? Appl. Anim. Behav. Sci. 29:453–469.

———. 1993. How bison live. Alberta 3:1–19.

———. 1996. Bison nation. Voyageur Press, Stillwater, Minn.

———. 1998. Deer of the world. Stackpole Books, Mechanicsburg, Penn.

———, and M. Bayer. 1988. Sexual dimorphism in the Cervidae and its relation to habitat. J. Zool., London 214:45–53.

———, and R. G. Petocz. 1977. Bighorn sheep in winter: do rams maximize reproductive fitness by spatial separation and habitat segregation from ewes? Can. J. Zool. 55:1802–1810.

———, R. E. Stemp, and R. H. Johnson. 1985. Heart rate telemetry on bighorn sheep *(Ovis canadensis)* as a means to investigate disturbance. Pages 91–99 *in* N. G. Bayfield and G. C. Barrow, editors. The ecological impact of outdoor recreation on mountain areas in Europe and North America. Recreation Ecol. Res. Group Rep. No. 9. Wye College, U.K.

George, J. M. 1969. Variation in the time of parturition of Merino and Dorset horn ewes. J. Agric. Sci. 73:295–299.

Georgi, J. R., and M. E. Georgi. 1990. Parasitology for veterinarians. W. B. Saunders, Harcourt Brace Javanovich, Indianapolis, Ind.

Gilchrist, D. 1990. Montana—land of giant rams. Outdoor Expeditions and Books, Hamilton, Mont.

———. 1992. Why is Montana the land of giant rams? Proc. Bienn. Symp. North. Wild Sheep and Goat Counc. 8:8–13.

Gilpin, M. E., and I. Hanski, editors. 1989. Metapopulation dynamics: empirical and theoretical investigations. Academic Press, San Diego, Calif.

———, and M. E. Soule. 1986. Minimum viable populations: processes of species extinction. Pages 19–34 *in* M. E. Soule, editor. Conservation biology: the science of scarcity and diversity. Sinauer Associates, Sunderland, Mass.

Ginnett, T. F., and C. L. Douglas. 1982. Food habits of feral burros and desert bighorn sheep in Death Valley National Monument. Desert Bighorn Counc. Trans. 26:81–87.

Gionfriddo, J. P., and P. R. Krausman. 1986. Summer habitat use by mountain sheep. J. Wildl. Manage. 50:331–336.

Glaze, R. L., T. D. Bunch, and P. Webb. 1981. Aberrations of the tooth arcade and mandible in desert bighorn sheep. Desert Bighorn Counc. Trans. 25:33–35.

———, M. Hoefs, and T. D. Bunch. 1982. Aberrations of the tooth arcade and mandible in Dall's sheep from southwestern Yukon. J. Wildl. Dis. 18:305–309.

Godley, W. C., R. L. Wildon, and V. Hurst. 1966. Effect of controlled environment on the reproductive performance of ewes. J. Anim. Sci. 25:212–216.

Godman, J. D. 1826. American natural history. Vol. 2. Part 1. Mastology. H. C. Carey and I. Lea, Philadelphia, Pa.

Goff, W. L., D. A. Jessup, K. A. Waldrup, J. W. Thomford, P. A. Conrad, W. M. Boyce, J. R. Gorham, and G. A. Wagner. 1993. The isolation and partial characterization of a *Babesia* spp. from desert bighorn sheep *(Ovis canadensis nelsoni)*. J. Eucaryotic Microbiol. 40:237–243.

Goldman, E. A. 1937. A new mountain sheep from Lower California. Proc. Biol. Soc. Wash. 50:29–32.

Good, J. R. 1974. Recorded body weight of two California bighorn rams. J. Wildl. Manage. 38:575–576.

Goodson, N. J. 1978. Status of bighorn sheep in Rocky Mountain National Park. M.S. thesis, Colorado State University, Fort Collins.

———. 1982. Effects of domestic sheep grazing on bighorn sheep populations: review. Proc. Bienn. Symp. North. Wild Sheep and Goat Counc. 3:287–313.

———. 1994. Persistence and population size in mountain sheep: why different interpretations? Conserv. Biol. 8:617–618.

———, and D. R. Stevens. 1996. Management implications of an intensive study of winter foraging ecology of bighorn sheep. Proc. Bienn. Symp. North. Wild Sheep and Goat Counc. 8:58–67.

———, ———, and J. A. Bailey. 1991. Winter-spring foraging ecology and nutrition of bighorn sheep on Montana ranges. J. Wildl. Manage. 55:422–433.

Graf, W. 1980. Habitat protection and improvement. Pages 310–319 in G. Monson and L. Sumner, editors. The desert bighorn: its life history, ecology, and management. University of Arizona Press, Tucson.

Grant, C. 1980. The desert bighorn and aboriginal man. Pages 7–39 in G. Monson and L. Sumner, editors. The desert bighorn: its life history, ecology, and management. University of Arizona Press, Tucson.

Green, H. U. 1949. The bighorn sheep of Banff National Park. Natl. Parks Serv., Ottawa.

Gregorio, P. D., A. Rando, and P. Masina. 1987. Differences in the DNA restriction patterns between sheep with HbA and HbB. Anim. Genet. 18:241–247.

Gross, J. E. 1960. History, present, and future status of the desert bighorn sheep (*Ovis canadensis mexicana*) in the Guadalupe Mountains of southeastern New Mexico and northwestern Texas. Desert Bighorn Counc. Trans. 4:66–71.

Groves, C. P. 1989. Feral mammals of the Mediterranean islands: documents of early domestication. Pages 46–58 in J. Clutton-Brock, editor. The walking larder. Unwin Hyman, London, U.K.

Grubb, P. 1974. Mating activity and the social significance of rams in a feral sheep community. Pages 457–476 in V. Geist and F. Walther, editors. The behaviour of ungulates and its relation to management. IUCN Publ., New Ser. No. 24. Morges, Switzerland.

———, and P. A. Jewell. 1973. The rut and the occurrence of oestrus in the Soay sheep on St. Kilda. J. Reprod. Fertil. 19(Suppl.):491–502.

Gunn, R. G. 1968. A note on difficult birth in Scottish hill flocks. Anim. Prod. 10:213–215.

Guthrie, R. D. 1968. Paleoecology of the large-mammal community in interior Alaska during the Late Pleistocene. Am. Midl. Nat. 79:346–363.

———. 1984. Mosaics, allochemics and nutrients: an ecological theory of Late Pleistocene megafaunal extinctions. Pages 259–298 in P. S. Martin and R. G. Klein, editors. Quarternary extinctions. University of Arizona Press, Tucson.

———. 1989. Frozen fauna of the mammoth steppe. University of Chicago Press, Chicago, Ill.

Guymon, J. G., and J. W. Bates. 1984. Utah's desert bighorn sheep status report, 1984. Desert Bighorn Counc. Trans. 28:49–50.

Guzman, G., Jr. 1961. Vegetation zones of the territory of Baja California in relation to wildlife. Desert Bighorn Counc. Trans. 5:68–74.

Hafez, E.S.E., editor. 1952. The behaviour of domestic animals. Williams and Wilkins, Baltimore, Md.

Haigh, J. C., R. R. Stewart, G. Wobeser, and P. S. MacWilliams. 1977. Capture myopathy in a moose. J. Am. Vet. Med. Assoc. 171:924–926.

Hailey, T. L. 1964. Status of transplanted bighorns in Texas. Desert Bighorn Counc. Trans. 8:113–116.

———, R. G. Marburger, R. M. Robinson, and K. A. Clark. 1972. Disease losses in desert bighorn sheep, Black Gap area. Desert Bighorn Counc. Trans. 16:79–83.

Hall, E. R. 1981. The mammals of North America. Vol. 2. Second edition. Ronald Press, New York, N.Y.

Hamilton, K. S., S. A. Holl, and C. L. Douglas. 1982. An evaluation of the effects of recreational activity on bighorn sheep in the San Gabriel Mountains, California. Desert Bighorn Counc. Trans. 26:50–55.

Hamilton, W. D. 1971. Geometry of the selfish herd. J. Theor. Biol. 31:295–311.

Hanks, J. 1972. Growth of the African elephant *(Loxodonta africana)*. E. Afr. Wildl. J. 10:251–252.

Hanley, T. A. 1982. The nutritional basis for food selection by ungulates. J. Range Manage. 35:146–151.

Hansen, C. G. 1967. Bighorn sheep populations of the Desert Game Range. J. Wildl. Manage. 31:693–706.

———. 1980*a*. Physical characteristics. Pages 52–63 *in* G. Monson and L. Sumner, editors. The desert bighorn: its life history, ecology, and management. University of Arizona Press, Tucson.

———. 1980*b*. Habitat. Pages 64–79 *in* G. Monson and L. Sumner, editors. The desert bighorn: its life history, ecology, and management. University of Arizona Press, Tucson.

———. 1980*c*. Population dynamics. Pages 217–235 *in* G. Monson and L. Sumner, editors. The desert bighorn: its life history, ecology, and management. University of Arizona Press, Tucson.

———. 1980*d*. Habitat evaluation. Pages 320–335 *in* G. Monson and L. Sumner, editors. The desert bighorn: its life history, ecology, and management. University of Arizona Press, Tucson.

———, and O. V. Deming. 1980. Growth and development. Pages 152–171 *in* G. Monson and L. Sumner, editors. The desert bighorn: its life history, ecology, and management. University of Arizona Press, Tucson.

Hansen, M. C. 1982*a*. Status and habitat preference of California bighorn sheep on Sheldon National Wildlife Refuge, Nevada. M.S. thesis, Oregon State University, Corvallis.

———. 1982*b*. Desert bighorn sheep: another view. Wildl. Soc. Bull. 10:133–140.

Hansen, R. M. 1971. Estimating plant composition of wild sheep diets. Trans. North Am. Wild Sheep Conf. 1:108–115.

———. 1975. Foods of the hoary marmot on the Kenai Peninsula, Alaska. Am. Midl. Nat. 94:348–353.

———, and P. S. Martin. 1973. Ungulate diets in the lower Grand Canyon. J. Range Manage. 26:380–381.

Hardin, G. 1960. The competitive exclusion principle. Science 131:1291–1297.

Harper, W. L. 1984. Pregnancy rate and early lamb survival of California bighorn sheep (*Ovis canadensis californiana,* Douglas 1871) in the Ashnola watershed, British Columbia. M.S. thesis, University of British Columbia, Vancouver.

———, and R.D.H. Cohen. 1985. Accuracy of Doppler ultrasound in diagnosing pregnancy in bighorn sheep. J. Wildl. Manage. 49:793–796.

Harris, A. H. 1977. Wisconsin age environments in the northern Chihuahuan Desert: evidence from the higher vertebrates. Pages 23–52 *in* R. H. Wauer and D. H. Riskind, editors. Symp. Biol. Resourc. Chihuahuan Desert region. Sul Ross State University, Alpine, Texas.

———, and P. Mundel. 1974. Size reduction in bighorn sheep *(Ovis canadensis)* at the close of the Pleistocene. J. Mammal. 55:678–680.

Harris, H., and F. L. Warren. 1955. Occurrence of electrophoretically distinct haemoglobins in ruminants. Biochem. F. 60:29.

Harris, L. K. 1992. Recreation in mountain sheep habitat. Ph.D. dissertation, University of Arizona, Tucson.

———, R. H. Gimblett, and W. W. Shaw. 1995. Multiple use management: using a geographic information system (GIS) model to conserve wildlife habitat in a high recreation use area. J. Soc. Nat. Resour. 8:559–572.

Harrison, S. 1990. Cougar predation on bighorn sheep in the Junction Wildlife Management Area, British Columbia. M.S. thesis, University of British Columbia, Vancouver.

———, and D. A. Hebert. 1988. Selective cougar predation on the Junction Wildlife Management Area. Proc. Bienn. Symp. North. Wild Sheep and Goat Counc. 5:292–306.

Hartl, G. B. In Press. Molecular genetics of the Caprinae—implications for phylogeny, conservation and management. *In* N. Franco, editor. CIC-Caprinae atlas. International Council for Game and Wildlife Conservation, Paris, France.

———, H. Burger, R. Willing, and F. Suchentrunk. 1990. On the biochemical systematics of the Caprini and the Rupicaprini. Biochemical Sys. Ecol. 18:175–182.

———, and N. Ferrand. 1993. Genetic polymorphism of transferrin (Tf) and the haemoglobin alpha chain (HbA) in the brown hare *(Lupus europaeus)*. Anim. Genet. 24:439–440.

Hass, C. C. 1984. "Cooperative" nursing by bighorn ewes on the National Bison Range. Proc. Bienn. Symp. North. Wild Sheep and Goat Counc. 4:252–269.

———. 1989. Bighorn lamb mortality: predation, inbreeding, and population effects. Can. J. Zool. 67:699–705.

———. 1990. Alternative maternal-care patterns in two herds of bighorn sheep. J. Mammal. 71:24–35.

———. 1991. Social status in female bighorn sheep *(Ovis canadensis):* expression, development and reproductive correlates. J. Zool., Lond. 225:509–523.

———, and D. A. Jenni. 1991. Structure and ontogeny of dominance relationships among bighorn rams. Can. J. Zool. 69:471–476.

Hayes, H., E. Petit, and B. Dutrillaux. 1991. Comparison of RGB-bonded karyotypes of cattle, sheep, and goats. Cytogenet. Cell Genet. 57:51–55.

Hebert, D. M. 1973. Altitudinal migration as a factor in the nutrition of bighorn sheep. Ph.D. dissertation, University of British Columbia, Vancouver.

———, and I. M. Cowan. 1971. Natural salt licks as a part of the ecology of the mountain goat. Can. J. Zool. 49:605–610.

———, and S. Harrison. 1988. The impact of coyote predation on lamb mortality patterns at the Junction Wildlife Management Area. Proc. Bienn. Symp. North. Wild Sheep and Goat Counc. 5:283–291.

Heimer, W. E. 1972. Sheep report. Pages 1–17 (plus 5 appendixes) *in* L. Nichols and W. Heimer, editors. Alaska Dep. Fish and Game, Fed. Aid Wildl. Restor. Annu. Proj. Prog. Rep. Vol. 13.

———. 1973. Dall sheep movements and mineral lick use. Alaska Dep. Fish and Game, Fed. Aid. Wildl. Restor. Final Rep.

———. 1974. The importance of mineral licks to Dall sheep in interior Alaska and its significance to sheep management. Trans. North. Wild Sheep Counc. 3:49–63.

———. 1978. Alternate year reproduction in a low quality, declining Dall sheep population: management considerations. Proc. Bienn. Symp. North. Wild Sheep and Goat Counc. 1:30–41.

———. 1980*a*. Can population quality be related to population density through nutrition? Proc. Bienn. Symp. North. Wild Sheep and Goat Counc. 2:288–309.

———. 1980*b*. Interior sheep studies. Alaska Dep. Fish and Game, Fed. Aid Wildl. Restor. Annu. Proj. Prog. Rep. Vol. 5.

———. 1988. A magnesium-driven hypothesis of Dall sheep mineral lick use: preliminary tests and management relevance. Proc. Bienn. Symp. North. Wild Sheep and Goat Counc. 6:135–140.

———, F. J. Mauer, and S. W. Keller. 1994. The effects of physical geography on Dall sheep habitat quality and home range size. Trans. North. Wild Sheep and Goat Counc. 9:144–148.

———, and A. C. Smith. 1972. Appendix V. Pages 1–3 *in* L. Nichols and W. Heimer, editors. Sheep report. Alaska Dep. Fish and Game, Fed. Aid Wildl. Restor. Annu. Proj. Prog. Rep. Vol. 13.

———, and ———. 1974. A preliminary look at Dall ram horn growth in Alaska and its implications. Trans. North. Wild Sheep Counc. 3:128–139.

———, and ———. 1975. Dall ram horn growth and population quality and their significance to Dall sheep management in Alaska. Alaska Dep. Fish and Game, Tech. Bull. No. 4.

———, and R. O. Stephenson. 1982. Responses of Dall sheep populations to wolf control in interior Alaska. Proc. Bienn. Symp. North. Wild Sheep and Goat Counc. 3:320–329.

———, and S. M. Watson. 1982. Differing reproductive patterns in Dall sheep: population strategy or management artefact? Bienn. Symp. North. Wild Sheep and Goat Counc. 3:330–338.

———, ———, and T. C. Smith. 1984. Excess ram mortality in a heavily hunted Dall sheep population. Proc. Bienn. Symp. North. Wild Sheep and Goat Counc. 4:425–432.

Helvie, J. B. 1971. Bighorns and fences. Desert Bighorn Counc. Trans. 16:3–8.

Hemming, J. E. 1961. Mandibular dentition and horn development as criteria of age in the Dall sheep, *Ovis dalli dalli* Nelson. M.S. thesis, University of Montana, Missoula.

———. 1969. Cemental deposition, tooth succession and horn development as criteria of age in Dall sheep. J. Wildl. Manage. 33:552–558.

Hengel, D. A., S. H. Anderson, and W. G. Hepworth. 1992. Population dynamics, seasonal distribution and movement patterns of the Laramie Peak bighorn sheep herd. Proc. Bienn. Symp. North. Wild Sheep and Goat Counc. 8:83–96.

Henshaw, J. 1970. Conflict between Dall sheep and caribou. Can. Field-Nat. 84:388–390.

Hibler, C. P., R. E. Lange, and C. J. Metzger. 1972. Transplacental transmission of *Protostrongylus* spp. in bighorn sheep. J. Wildl. Dis. 8:389.

———, C. J. Metzger, T. R. Spracker, and R. E. Lange. 1974. Further observations on *Protostrongylus* spp. infection by transplacental transmission in bighorn sheep. J. Wildl. Dis. 10:39–41.

Hickey, F. 1960. Death and reproductive rates of sheep in relation to flock culling selection. N.Z. J. Agric. Res. 3:332–344.

Hickey, W. O. 1975. Bighorn sheep ecology. Idaho Dep. Fish and Game, Fed. Aid Wildl. Restor. Proj. W-160-R-2.

———. 1978. Bighorn sheep ecology. Idaho Dep. Fish and Game, Fed. Aid Wildl. Restor. Proj. W-160-R-5.

Hjeljord, O. G. 1971. Feeding ecology and habitat preference of the mountain goat in Alaska. M.S. thesis, University of Alaska, Fairbanks.

Hoban, P. A. 1990. A review of desert bighorn sheep in the San Andres Mountains, New Mexico. Desert Bighorn Counc. Trans. 34:14–22.

Hobbs, N. T., and R. A. Spowart. 1984. Effects of prescribed fire on nutrition of mountain sheep and mule deer during winter and spring. J. Wildl. Manage. 48:551–560.

———, and D. M. Swift. 1985. Estimates of habitat carrying capacity incorporating explicit nutritional constraints. J. Wildl. Manage. 49:814–822.

Hodge, R. W. 1966. The relative pasture intake of grazing lambs at two levels of milk intake. Aust. J. Exp. Agric. Anim. Husb. 6:314–316.

Hoefs, M. 1971. Food selection by Dall's sheep (*Ovis dalli dalli* Nelson). Pages 759–786 *in* V. Geist and F. Walther, editors. The behaviour of ungulates and its relation to management. IUCN Publ., New Ser. No. 24. Morges, Switzerland.

———. 1974. Abnormal dentition in Dall sheep (*Ovis dalli dalli* Nelson). Can. Field-Nat. 88:227–229.

———. 1978. Twinning in Dall sheep. Can. Field-Nat. 92:292–293.

———. 1984a. Population dynamics and horn growth characteristics of Dall sheep *(Ovis dalli)* and their relevance to management. Pages 99–115 *in* R. Olson, R. Hastings, and F. Geddes, editors. Northern ecology and resource management. University of Alberta Press, Edmonton.

———. 1984b. Productivity and carrying capacity of a subarctic sheep winter range. Arctic 37:141–147.

———, and N. Barichello. 1984. Comparison between a hunted and an unhunted Dall sheep population—a preliminary assessment of the impact of hunting. Proc. Bienn. Symp. North. Wild Sheep and Goat Counc. 4:433–466.

———, and M. Bayer. 1983. Demographic characteristics of an unhunted Dall sheep *(Ovis dalli dalli)* population in southwest Yukon, Canada. Can. J. Zool. 61:1346–1357.

———, and V. C. Brink. 1978. Forage production and utilization of a Dall sheep winter range, southwest Yukon Territory. Proc. Bienn. Symp. North. Wild Sheep and Goat Counc. 1:87–105.

———, and I. M. Cowan. 1979. Ecological investigation of a population of Dall sheep (*Ovis dalli dalli* Nelson). Syesis 12, Suppl. 1.

———, H. Hoefs, and D. Burles. 1986. Observations on Dall sheep, *Ovis dalli dalli,* grey wolf, *Canis lupus pambasilens,* in Kluane Lake area, Yukon. Can. Field-Nat. 100:78–84.

———, and R. Konig. 1984. Reliability of aging old Dall sheep ewes by the horn annulus technique. J. Wildl. Manage. 48:980–982.

———, and T. Nette. 1982. Horn growth and horn wear in Dall rams and their relevance to management. Proc. Bienn. Symp. North. Wild Sheep and Goat Counc. 3:143–156.

Hoffecker, J. F., W. R. Powers, and T. Goebel. 1993. The colonization of Beringia and the peopling of the New World. Science 259:46–53.

Hogg, J. T. 1984. Mating in bighorn sheep: multiple creative male strategies. Science 225:526–529.

———. 1987. Intrasexual competition and mate choice in Rocky Mountain bighorn sheep. Ethology 75:119–144.

———. 1988. Copulatory tactics in relation to sperm competition in Rocky Mountain bighorn sheep. Behav. Ecol. Sociobiol. 22:49–59.

Holechek, J. L., R. D. Pieper, and C. H. Herbel. 1995. Range management. Second edition. Prentice Hall, Englewood Cliffs, N.J.

Holl, S. A. 1982. Evaluation of bighorn sheep habitat. Desert Bighorn Counc. Trans. 26:47–49.

———, and V. C. Bleich. 1983. San Gabriel Mountain sheep: biological and management considerations. U.S. For. Serv., San Bernardino Natl. For., Admin. Rep.

———, ———, and B. Rios. 1980. San Gabriel bighorn sheep. U.S. For. Serv., San Bernardino Natl. For., Annu. Rep.

Holt, B. S., W. H. Miller, and B. F. Wakeling. 1992. Composition and quality of mountain sheep diets in the Superstition Mountains, Arizona. Desert Bighorn Counc. Trans. 36:36–40.

Honess, R. F., and N. M. Frost. 1942. A Wyoming bighorn sheep study. Wyoming Game and Fish Dep. Bull. No. 1.

———, and K. Winter. 1956a. Diseases of wildlife in Wyoming. Wyoming Game and Fish Comm. Bull. No. 9.

———, and ———. 1956b. What about the bighorn? Wyo. Wildl. 20:20–23.

Hopkins, D. M. 1967. The Cenozoic history of Beringia—a synthesis. Pages 451–484 *in* D. M. Hopkins, editor. The Bering land bridge. Stanford University Press, Stanford, Calif.

———, J. V. Matthews Jr., C. S. Schweger, and S. B. Young, editors. 1982. Paleoecology of Beringia. Academic Press, New York, N.Y.

Horak, I. G., and M. J. Butt. 1977. Parasites of domestic and wild animals in South Africa. *Oestrus* spp. and *Gedoelstia hassleri* in the Blesbok. Onderstepoort J. Vet. Res. 44:113–118.

Horejsi, B. L. 1972. Behavioral differences in bighorn lambs (*Ovis canadensis canadensis* Shaw) during years of high and low survival. Trans. North. Wild Sheep Counc. 2:51–73.

———. 1976. Suckling and feeding behaviour in relation to lamb survival in bighorn sheep *(Ovis canadensis).* Ph.D. dissertation, University of Calgary, Alberta.

Horn, H. S. 1971. Regulation of animal numbers: a model counter-example. Pages 51–57 *in* I. A. McLaren, editor. Natural regulation of animal populations. Atherton Press, New York, N.Y.

Hornaday, W. T. 1901. Notes on the mountain sheep of North America, with a description of a new species. N.Y. Zool. Soc. Annu. Rep. 5:77–122.

Hornocker, M. G. 1969. Defensive behaviour in female bighorn sheep. J. Mammal. 50:128.

Hristic, V. 1992. Transferrins. Glas. Spr. Akad. Nauka. Med. 1992:37–55.

Hudson, R. J., D. M. Hebert, and V. C. Brink. 1976. Occupational patterns of wildlife on a major East Kootenay winter-spring range. J. Range Manage. 29:38–43.

Hudson, W. E., editor. 1991. Landscape linkages and biodiversity. Island Press, Washington, D.C.

Huelsenbeck, J. P., and B. Rannala. 1997. Phylogenetic methods come of age: testing hypotheses in an evolutionary context. Science 276:227–232.

Huggard, D. J. 1993. Effect of snow depth on predation and scavenging by grey wolves. J. Wildl. Manage. 57:382–388.

Iovenko, V. N. 1987. Comparison of the genetic structures of sheep populations using biochemical polymorphisms. Anim. Breed. Abstr. 57:1684.

Irvine, C. A. 1969. Factors affecting the desert bighorn in southeastern Utah. Desert Bighorn Counc. Trans. 13:6–13.

Irwin, L. L., J. G. Cook, D. E. McWhirter, S. G. Smith, and E. B. Arnett. 1993. Assessing winter dietary quality in bighorn sheep via fecal nitrogen. J. Wildl. Manage. 57:413–421.

Jaeger, E. C. 1957. The North American deserts. Stanford University Press, Stanford, Calif.

Jaeger, J. 1994. Distribution, movements and demography of mountain sheep in the Kingston and Clark Mountain ranges of California. M.S. thesis, University of Nevada, Las Vegas.

Jakimchuk, R. D., D. J. Vernam, and L. G. Sopuck. 1984. The relationship between Dall sheep and the trans-Alaska Pipeline in the northern Brooks Range. Renewable Resources Consulting Service, Sidney, B.C.

James, F. C. 1983. Environmental component of morphological differentiation in birds. Science 211:184–186.

James, M. T. 1947. The flies that cause myiasis in man. U.S.D.A. Misc. Publ. No. 631.

Jarman, P. J. 1974. The social organization of antelope in relation to their ecology. Behaviour 48:215–267.

———. 1982. Prospects for interspecific comparison in sociobiology. Pages 323–342 *in* King's College Sociobiology Group, editors. Current problems in sociobiology. Cambridge University Press, Cambridge, U.K.

———, and M. V. Jarman. 1979. The dynamics of ungulate social organisation. Pages 185–220 *in* A.R.E. Sinclair and M. Norton-Griffiths, editors. Serengeti: dynamics of an ecosystem. University of Chicago Press, Chicago, Ill.

Jense, G. K., J. W. Bates, and J. A. Robertson. 1979. Utah bighorn sheep status report. Desert Bighorn Counc. Trans. 23:89–91.

Jessup, D. A. 1985. Diseases of domestic livestock which threaten bighorn sheep populations. Desert Bighorn Counc. Trans. 29:29–33.

———, R. K. Clark, and N. D. Kock. 1991. Contagious echtyma in free-ranging desert bighorn sheep: Iatrongenic and zoonotic cases. Pages 215–217 *in* Proceedings of the American Association of Zoo Veterinarians, Calgary, Canada.

———, B. T. Osburn, and W. P. Heuschele. 1984. Blue tongue in California's wild ruminants: distribution and pathology. Annu. Meet. U.S. Anim. Health Assoc. 88:616–630.

———, and R. R. Ramey II. 1995. Genetic variation of bighorn sheep as measured by blood protein electropheresis. Desert Bighorn Counc. Trans. 39:17–25.

Jewell, P. A., and P. Grubb. 1974. The breeding cycle, the onset of oestrus, and conception in Soay sheep. Pages 224–241 *in* P. A. Jewell, C. Milner, and J. M.

Boyd, editors. Island survivors: the ecology of Soay sheep of St. Kilda. Athlone Press, London, U.K.

Jiang, Z., and C. Du. 1992. Studies of the relationship between transferrin genetic polymorphism and diseases. PCR Genet. Acta 19:198–202.

John, R. T. 1968. Results of Utah's 1967 desert bighorn sheep hunt. Desert Bighorn Counc. Trans. 12:1–4.

Johnson, B. K., and D. L. Smith. 1980. Food habits and forage preferences of bighorn sheep in alpine and subalpine communities. Proc. Bienn. Symp. North. Wild Sheep and Goat Counc. 6:1–17.

Johnson, E. L. 1957. Disease and mechanical injury to desert bighorn sheep. Desert Bighorn Counc. Trans. 1:38–42.

Johnson, J. D. 1975. An evaluation of the summer range of bighorn sheep (*Ovis canadensis canadensis* Shaw) on Ram Mountain, Alberta. M.S. thesis, University of Calgary, Alberta.

Johnston, A., L. M. Bezeau, and S. Smoliak. 1968. Chemical composition of alpine tundra plants. J. Wildl. Manage. 32:773–777.

Jones, F. F. 1963. Sheep investigations. Alaska Dep. Fish and Game, Proj. Rep. W-6-R-4.

Jones, F. L. 1950. A survey of the Sierra Nevada bighorn. Sierra Club Bull. 35:29–76.

———. 1980. Competition. Pages 197–216 *in* G. Monson and L. Sumner, editors. The desert bighorn: its life history, ecology, and management. University of Arizona Press, Tucson.

Jones, R. L., and H. C. Hanson. 1985. Mineral licks: geography and biochemistry of North American ungulates. Iowa State University Press, Ames.

Joralemon, D. 1982. New World depopulation and the case of disease. J. Anthropol. Res. 38:108–127.

Jorgensen, M. C., and R. E. Turner. 1975. Desert bighorn of the Anza-Borrego Desert State Park. Desert Bighorn Counc. Trans. 19:51–53.

Jorgensen, P. 1974. Vehicle use at a desert bighorn watering area. Desert Bighorn Counc. Trans. 18:18–24.

Jorgenson, J. T. 1992. Seasonal changes in lamb:ewe ratios. Proc. Bienn. Symp. North. Wild Sheep and Goat Counc. 8:219–226.

———, M. Festa-Bianchet, J. M. Gaillad, and W. D. Wishart. 1997. Effects of age, sex, disease, and density on survival of bighorn sheep. Ecology 78:1019–1032.

———, and W. D. Wishart. 1984. Growth rates of Rocky Mountain bighorn sheep on Ram Mountain, Alberta. Proc. Bienn. Symp. North. Wild Sheep and Goat Counc. 4:270–284.

Joyce, J. P., and P. V. Rattray. 1970. The intake and utilization of milk and grass by lambs. Proc. N.Z. Soc. Anim. Prod. 30:94–105.

Juneja, R. K., and T. Shibata. 1992. Genetic polymorphism and close linkage of two plasma protein loci in dogs. Anim. Genet. 23:143–150.

Keeler, R. F., and W. Binns. 1971. Teratogenic compounds of *Veratrum califor-*

nicum as a function of plant part, stage, and site of growth. Phytochemistry 10:1765–1769.

Kelly, W. E. 1979. A comparison of three bighorn areas on the Humboldt National Forest. Desert Bighorn Counc. Trans. 23:37–39.

———. 1980. Predator relationships. Pages 186–196 *in* G. Monson and L. Sumner, editors. The desert bighorn: its life history, ecology, and management. University of Arizona Press, Tucson.

Kennedy, C. A. 1948. Golden eagle kills bighorn lamb. J. Mammal. 29:68–69.

Kilpatric, J. 1982. Texas desert bighorn sheep status report—1982. Desert Bighorn Counc. Trans. 26:102–104.

Kimura, M. 1983. The neutral theory of molecular evolution. Cambridge University Press, Cambridge, U.K.

King, M. M., and G. W. Workman. 1982. Desert bighorn on BLM lands in southeastern Utah. Desert Bighorn Counc. Trans. 26:104–106.

Kinzer, H. G., W. P. Meleney, R. E. Lange Jr., and W. E. Houghton. 1983. Preliminary evaluation of ivermectin for control of *Psoroptes ovis* in desert bighorn sheep. J. Wildl. Dis. 19:52–54.

Kistner, T. P., S. M. Matlock, D. Wyse, and G. E. Mason. 1977. Helminth parasites of bighorn sheep in Oregon. J. Wildl. Dis. 13:125–130.

———, G. E. Reynolds, L. D. Koller, D. O. Trainer, and D. L. Eastman. 1975. Clinical and serological findings on the distribution of blue tongue and epizootic hemorrhagic disease virus in Oregon. Am. Assoc. Vet. Lab. Diag. 19:135–148.

Kjemtrup, A., J. Thomford, I. Gardner, P. A. Conrad, D. A. Jessup, and W. M. Boyce. 1995. Seroprevalence of two *Babesia* spp. isolates in selected bighorn sheep *(Ovis canadensis)* and mule deer *(Odocoileus hemionus)* populations in California. J. Wildl. Dis. 31:467–471.

Klein, D. R. 1953. A reconnaissance study of the mountain goat in Alaska. M.S. thesis, University of Alaska, Fairbanks.

———. 1964. Range-related differences in growth of deer reflected in skeletal ratios. J. Mammal. 45:226–235.

———. 1965. Ecology of deer range in Alaska. Ecol. Monogr. 35:259–284.

———. 1969. Food selection by North American deer and their response to overutilization of preferred plant species. Pages 25–46 *in* A. Watson, editor. Animal populations in relation to their food resources. Blackwell, Oxford, U.K.

Klingel, J., M. Lenarz, and R. Quimby. 1974. Dall sheep. *In* R. D. Jakimchuk, editor. Mammal studies in northeastern Alaska with emphasis within the Canning River drainage. Vol. 24. Renewable Resources Consulting Service, Sidney, B.C.

Konig, R., and M. Hoefs. 1984. Volume and density of horns of Dall rams. Proc. Bienn. Symp. North. Wild Sheep and Goat Counc. 4:295–309.

Kontrimavichus, V. I., editor. 1986. Beringia in the Cenozoic era. A. A. Balkema, Rotterdam, Holland.

Kornet, C. A. 1978. Status and habitat use of California bighorn sheep on Hart Mountain, Oregon. M.S. thesis, Oregon State University, Corvallis.

Korobitsyna, K. V., C. F. Nadler, N. N. Vorontsov, and R. S. Hoffmann. 1974. Chromosomes of the Siberian snow sheep, *Ovis nivicola,* and implications concerning the origin of amphiberingian wild sheep (subgenus *Pachyceros*). Quaternary Res. 4:235–245.

Kovach, S. D. 1979. An ecological survey of the White Mountain Peak bighorn. Desert Bighorn Counc. Trans. 23:57–61.

Krausman, P. R. 1985. Impacts of the Central Arizona Project on desert mule deer and desert bighorn sheep. Final Rep. 9-730-X069, U.S. Bur. Reclam., Phoenix, Ariz.

———. 1993a. The exit of the last wild mountain sheep. Pages 242–250 *in* G. P. Nabhan, editor. Counting sheep. University of Arizona Press, Tucson.

———. 1993b. Field research methods. Field Research Methods Workshop. Wildlife Institute of India, Dehra Dun.

———. 1997. The influence of landscape scale on the management of desert bighorn sheep. Pages 349-367 *in* J. A. Bissonette, editor. Wildlife and landscape ecology. Springer-Verlag, New York, N.Y.

———, J. Bailey, V. Bleich, D. Armentrout, and R. Ramey. 1992. What is a minimum viable population? Desert Bighorn Counc. Trans. 36:68–74.

———, B. Bobek, F. Whiting, and W. Brown. 1988. Dry matter and energy intake in relation to digestibility in desert bighorn sheep. Acta Theriol. 33:121–130.

———, and R. C. Etchberger. 1993. Effectiveness of mitigation features for desert ungulates along the Central Arizona Project. Final Rep. 9-CS-32-00350, U.S. Bur. Reclam., Phoenix, Ariz.

———, and ———. 1995. Response of desert ungulates to a water project in Arizona. J. Wildl. Manage. 59:292–300.

———, and ———. 1996. Desert bighorn sheep and water: a bibliography. U.S. Geol. Surv. Coop. Park Studies Unit, Univ. Ariz., Spec. Rep. No. 13.

———, ———, and R. M. Lee. 1993a. Mountain sheep population persistence in Arizona. Conserv. Biol. 7:219.

———, and J. J. Hervert. 1983. Mountain sheep responses to aerial surveys. Wildl. Soc. Bull. 11:372–375.

———, and B. D. Leopold. 1986. The importance of small populations of desert bighorn sheep. Trans. North Am. Wild. Nat. Resour. Conf. 51:52–61.

———, G. Long, R. F. Seegmiller, and S. G. Torres. 1989. Relationships between desert bighorn sheep and habitat in western Arizona. Wildl. Monogr. No. 102.

———, ———, and L. Tarango. 1996. Desert bighorn sheep and fire, Santa Catalina Mountains, Arizona. Pages 162–168 *in* P. F. Ffolliott, L. F. DeBano, M. B. Baker Jr., G. J. Gottfried, G. Sols-Garza, C. B. Edminster, D. G. Neary, L. S. Allen, and R. H. Hamre, tech. coords. Effects of fire on the Madrean Province ecosystems. U.S. For. Serv., Rep. RM-GTR-289.

———, J. R. Morgart, and M. Chilelli. 1984. Annotated bibliography of desert bighorn sheep literature, 1897–1983. Southwest Natural History Association, Phoenix, Ariz.

——, W. W. Shaw, R. C. Etchberger, and L. K. Harris. 1995. The decline of bighorn sheep in the Santa Catalina Mountains, Arizona. Pages 245–250 *in* L. F. DeBano, P. F. Ffolliott, A. Ortego-Rubio, G. J. Gottfried, R. H. Hamre, and C. B. Edminster, tech. coords. Biodiversity and management of the Madrean Archipelago: the Sky Islands of the southwestern United States and northeastern Mexico. U.S. For. Serv., Rep. RM-GTR-264.

——, ——, and J. L. Stair. 1979. Bighorn sheep in the Pusch Ridge Wilderness Area, Arizona. Desert Bighorn Counc. Trans. 23:40–46.

——, S. Torres, L. L. Ordway, J. J. Hervert, and M. Brown. 1985. Diel activity of ewes in the Little Harquahala Mountains, Arizona. Desert Bighorn Counc. Trans. 29:24–26.

——, M. C. Wallace, C. L. Hayes, and D. W. DeYoung. 1998. Effects of jet aircraft on mountain sheep. J. Wildl. Manage. 62:in press.

——, ——, M. E. Weisenberger, D. W. DeYoung, and O. E. Maughan. 1993*b*. Effects of simulated aircraft noise on heart-rate and behavior of desert ungulates. U.S. Air Force AL/OE-TR-1993-0185, U.S. Gov. Print. Off., 1994-550-057/81063.

——, ——, M. J. Zine, L. R. Berner, C. L. Hayes, and D. W. DeYoung. 1993*c*. The effects of low altitude aircraft on mountain sheep heart-rate and behavior. U.S. Air Force AL/OE-TR-1993-0184, U.S. Gov. Print. Off., 1994-550-057/81065.

Krebs, J. R., and N. B. Davies. 1987. An introduction to behavioural ecology. Second edition. Sinauer Associates, Sunderland, Mass.

Krull, W. H. 1969. Notes in veterinary parasitology. University Press of Kansas, Lawrence.

Kurten, B., and E. Anderson. 1980. Pleistocene mammals of North America. Columbia University Press, New York, N.Y.

Lacy, R. C. 1987. Loss of genetic diversity from managed populations: interacting effects of drift, mutation, immigration, selection, and population subdivision. Conserv. Biol. 1:143–157.

Ladewig, J., and B. L. Hart. 1980. Flehmen and vomeronasal organ function in male goats. Physiol. Behav. 24:1067–1071.

Lande, R. 1995. Mutation and conservation. Conserv. Biol. 9:782–791.

Lange, R. E. 1973. Epidemiology of lungworms (*Protostrongylus stition* and *P. nuohir*) in Rocky Mountain bighorn sheep *(Ovis canadensis canadensis)*. M.S. thesis, Colorado State University, Fort Collins.

——, A. V. Sandoval, and W. P. Meleney. 1980. Psoroptic scabies in bighorn sheep *(Ovis canadensis mexicana)* in New Mexico. J. Wildl. Dis. 16:77–82.

Lange, V. 1992. Haptoglobin polymorphism—not only a genetic marker. Anthropol. Ang. 50:281–302.

Lansman, R. A., J. C. Avise, C. F. Aquadro, J. F. Shapira, and S. W. Daniel. 1983.

Extensive variation in mitochondrial DNAs among geographic populations of the deer mouse, *Peromyscus maniculatus*. Evolution 37:1–16.

Lathrop, E. W., and P. G. Rowlands. 1983. Plant ecology in deserts: an overview. Pages 113–152 *in* R. H. Webb and H. G. Wilshire, editors. Environmental effects of off-road vehicles. Springer-Verlag, New York, N.Y.

Latson, F. E. 1977. The distribution and ecology of intermediate host snents of *Protostrongylus* spp., lungworms of bighorn sheep on Pike's Peak, Colorado. M.S. thesis, Colorado State University, Fort Collins.

Lawrence, W. E. 1956. Congenital infection with *Mycobacterium johnei* in cattle. Vet. Rec. 68:312–314.

Lay, D. M., C. F. Nadler, and J. D. Hassinger. 1971. The transferrins and hemoglobins of wild Iranian sheep (*Ovis* Linnaeus). Comp. Biochem. Physiol. B 40:521–529.

Lees, J. L. 1966. Variations in the time of onset of the breeding season in ewes. J. Agric. Sci. 67:173–179.

Lenarz, M. S. 1979. Social structure and reproductive strategy in desert bighorn sheep *(Ovis canadensis mexicana)*. J. Mammal. 60:671–678.

———, and W. Conley. 1980. Demographic considerations in reintroduction programs of bighorn sheep. Acta Theriol. 25:71–80.

———, and ———. 1982. Reproductive gambling in bighorn sheep *(Ovis):* a simulation. J. Theor. Biol. 98:1–7.

Lent, P. C. 1974. Mother-infant relationships in ungulates. Pages 14–55 *in* V. Geist and F. Walther, editors. The behaviour of ungulates and its relation to management. IUCN Publ., New Ser. No. 24. Morges, Switzerland.

Leslie, D. M., Jr. 1977. Home range, group size, and group integrity of the desert bighorn sheep in the River Mountains, Nevada. Desert Bighorn Counc. Trans. 21:25–28.

———. 1978. Differential utilization of water sources by desert bighorn sheep in the River Mountains, Nevada. Desert Bighorn Counc. Trans. 22:23–26.

———. 1980. Remnant populations of desert bighorn sheep as a source for transplantation. Desert Bighorn Counc. Trans. 24:36–44.

———, and C. L. Douglas. 1979. Desert bighorn of the River Mountains, Nevada. Wildl. Monogr. No. 66.

———, and ———. 1980. Human disturbance at water sources of desert bighorn sheep. Wildl. Soc. Bull. 8:284–290.

———, and ———. 1982. Simulated demography of the River Mountain herd. Desert Bighorn Counc. Trans. 26:97–99.

———, and ———. 1986. Modeling demographics of bighorn sheep: current abilities and missing links. Trans. North Amer. Wildl. Nat. Resourc. Conf. 51:62–73.

Lewis, R. J., G. A. Chalmers, M. W. Barrett, and R. Bhatnagar. 1977. Captive myopathy in elk in Alberta, Canada: a report of three cases. J. Am. Vet. Med. Asoc. 171:927–932.

L'Heureux, N., M. Festa-Bianchet, and T. Jorgenson. 1996. Effects of visible signs of contagious ecthyma on mass and survival of bighorn lambs. J. Wildl. Manage. 32:286–292.

Light, J. T., Jr., and R. Weaver. 1973. Report on bighorn sheep habitat study in the area for which an application was made to expand the Mt. Baldy winter sports facility. U.S. For. Serv., San Bernardino Natl. For.

Lightman, S. 1979. The responsibility of intervention in isolated societies. Pages 303–314 in P. Hugh-Jones, editor. Health and disease in tribal society. Ciba Foundation Symposia, New Ser. 49. Elsevier North-Holland, New York, N.Y.

Lima, S. L. 1987. Vigilance while feeding and its relation to the risk of predation. J. Theor. Biol. 124:303–316.

Linderman, S. 1972. Appendix III. Pages 1–13 in L. Nichols and W. Heimer, editors. Sheep report. Alaska Dep. Fish and Game, Fed. Aid Wildl. Restor. Annu. Proj. Prog. Rep. Vol. 13.

Lindsay, D. R. 1965. The importance of olfactory stimuli in the mating behaviour of the ram. Anim. Behav. 13:75–78.

Livingston, C. W., and W. T. Hardy. 1960. The longevity of contagious ecthyma virus. J. Am. Vet. Med. Assoc. 137:651.

Long, T. A., R. L. Cowan, C. W. Wolfe, T. Rader, and R. W. Swift. 1959. Effect of seasonal feed restriction on antler development in white-tailed deer. Pennsylvania State University Agricultural Experiment Station, Prog. Rep. No. 209:1–11.

Lopez, F.M.C., and V. M. Orihuela G. 1976. Behavior of the desert bighorn *(Ovis canadensis weemsi)* in Baja California. Desert Bighorn Counc. Trans. 20:24–25.

Louden, A.S.I., A. S. McNeilly, and J. A. Milne. 1983. Nutrition and control of fertility in red deer. Nature 302:145–147.

Luckhurst, A. J. 1973. Stone sheep and their habitat in the northern Rocky Mountain foothills of British Columbia. M.S. thesis, University of British Columbia, Vancouver.

Luedke, A. J., R. H. Jones, and M. M. Jochim. 1967. Transmission of bluetongue between sheep and cattle by *Culicoides variipennis*. Am. J. Vet. Res. 28:457–460.

Luikart, G., and F. W. Allendorf. 1996. Mitochondrial-DNA variation and genetic-population structure in Rocky Mountain bighorn sheep *(Ovis canadensis canadensis)*. J. Mammal. 77:109–123.

———, and J. M. Cornuet. 1997. Usefulness of microsatellite DNA markers for detecting population bottlenecks in mountain sheep. World Conf. Mt. Ungulates 2:33.

Lyapunova, E. A., T. D. Bunch, N. N. Voronsov, and R. S. Hoffmann. 1997. Chromosome sets and the taxonomy of Severtsov wild sheep *(O. ammon severtzovi)*. Russian J. Zool. 1:387–396.

Lynch, J. J., and G. Alexander. 1976. The effect of gramineous wind breaks on behaviour and lamb mortality among shorn and unshorn Merino sheep during lambing. Appl. Anim. Ethol. 2:305–325.

MacArthur, R. A., V. Geist, and R. H. Johnston. 1982a. Cardiac and behavioral responses of mountain sheep to human disturbance. J. Wildl. Manage. 46:351–358.

——, ——, and ——. 1982b. Physiological correlates of social behaviour in bighorn sheep: a field study using electrocardiogram telemetry. J. Zool., Lond. 196:401–415.

——, R. H. Johnston, and V. Geist. 1979. Factors influencing heart rate in free-ranging bighorn sheep: a physiological approach to the study of wildlife harassment. Can. J. Zool. 57:2010–2021.

MacArthur, R. H., and E. O. Wilson. 1967. The theory of island biogeography. Princeton University Press, Princeton, N.J.

MacCallum, B. 1988. Seasonal and spatial distribution of bighorn sheep at an open pit mining site in the Alberta foothills. Proc. Bienn. Symp. North. Wild Sheep and Goat Counc. 6:106–120.

——, and V. Geist. 1992. Mountain restoration: soil and surface wildlife habitat. GeoJournal 27:23–46.

Macnab, J. 1983. Wildlife management as scientific experimentation. Wildl. Soc. Bull. 11:397–401.

——. 1985. Carrying capacity and related slippery shibboleths. Wildl. Soc. Bull. 13:403–410.

Main, M. B., and B. E. Coblentz. 1990. Sexual segregation among ungulates: a critique. Wildl. Soc. Bull. 18:204–210.

Manca, L., A. Di-Luccia, E. Pieragostini, S. Naitona, and B. Mosala. 1993. Haemoglobin I: a new beta-globin chain variant found in sheep of Italian breeds. Anim. Genet. 24:203–204.

Manwell, C., and C.M.A. Baker. 1976. Protein polymorphisms in domesticated species: evidence for hybrid origin? Pages 105–130 *in* S. Karlin and E. Nevo, editors. Population genetics and ecology. Academic Press, New York, N.Y.

——, and ——. 1977. Genetic distance between the Australian Merino and the Poll Dorset sheep. Genet. Res. 29:239–253.

Marsh, H. 1938. Pneumonia in Rocky Mountain bighorn sheep. J. Mammal. 19:214–219.

Martin, L. D., and B. M. Gilbert. 1978. Excavations at Natural Trap Cave. Trans. Nebraska Acad. Sci. 6:107–116.

Maruyama, M., and K. Kimura. 1980. Genetic variability and effective population size when local extinction and recolonization of subpopulations are frequent. Proc. Natl. Acad. Sci. USA 77:6710–6714.

Mayr, E. 1948. The bearing of the new systematics on genetical problems: the nature of species. Adv. Genet. 2:205.

——. 1963. Animal species and evolution. Belknap Press of Harvard University Press, Cambridge, Mass.

Mazaika, R., P. R. Krausman, and R. C. Etchberger. 1992. Nutritional carrying capacity of desert bighorn sheep habitat in the Pusch Ridge Wilderness, Arizona. Southwest. Nat. 37:372–378.

Mazet, J.A.K., W. M. Boyce, J. Mellies, I. Gardner, R. K. Clark, and D. A. Jessup. 1992. Exposure to *Psoroptes* spp. mites is common among bighorn *(Ovis canadensis)* populations in California. J. Wildl. Dis. 28:542–547.

McCann, J. L. 1956. Ecology of mountain sheep. Am. Midl. Nat. 56:297–324.

McCarty, C. W., and J. A. Bailey. 1992. Measuring visual obstruction caused by discrete objects in bighorn sheep habitat. Desert Bighorn Counc. Trans. 36:1–3.

———, and ———. 1994. Habitat requirements of desert bighorn sheep. Colorado Div. of Wildl., Spec. Rep. 69:1–36.

McClymont, C. L., K. N. Wynne, P. K. Briggs, and M. C. Franklin. 1957. Sodium chloride supplementation of high-grain diets for fattening Merino sheep. Aust. J. Agric. Res. 8:83.

McCullough, D. R., and E. R. Schneegas. 1966. Winter observations on the Sierra Nevada bighorn sheep. Calif. Fish and Game 52:68–84.

McCutchen, H. E. 1976. Status of Zion National Park desert bighorn restoration project, 1975. Desert Bighorn Counc. Trans. 20:52–54.

———. 1981. Desert bighorn zoogeography and adaptation in relation to historic land use. Wildl. Soc. Bull. 9:171–179.

———. 1982. Behavioral ecology of reintroduced desert bighorns, Zion National Park, Utah. Ph.D. dissertation, Colorado State University, Fort Collins.

———. 1987. Diurnal summer activity patterns of captive desert bighorn sheep. Desert Bighorn Counc. Trans. 31:15–17.

———. 1988. Desert bighorn sheep nursing behavior in relation to population quality and lamb survival. Desert Bighorn Counc. Trans. 32:7–10.

McDonald, J. N. 1978. The North American bison: a revised classification and interpretation of their evolution. Ph.D. dissertation, University of California, Los Angeles.

———. 1984. An extinct muskox mummy from near Fairbanks, Alaska: a progress report. Biol. Pap. Univ. Alaska Spec. Rep. 4:148–152.

———, C. E. Ray, and C. R. Harrington. 1991. Taxonomy and zoogeography of the musk ox genus *Praeovibos* Staudinger, 1908. Pages 285–314 *in* J. R. Purdue, W. E. Klippel, and B. W. Styles, editors. Beamers, bobwhites and blue-points: tribute to the career of Paul W. Parmalee. Illinois State Mus. Sci. Pap. Vol. 23. Illinois State Museum, Springfield.

McEwan, E. H. 1975. The adaptive significance of the growth patterns in cervids compared with other ungulate species. Zool. Zh. 54:1221–1232.

McKee, N. R., J. F. Karpowitz, and J. Cresto. 1995. Status of desert bighorn sheep in Utah—1994. Desert Bighorn Counc. Trans. 39:80–81.

McLaren, I. A., editor. 1971. Natural regulation of animal populations. Atherton Press, New York, N.Y.

McMichael, T. J. 1964. Relationships between desert bighorn and feral burros in the Black Mountains of Mohave County. Desert Bighorn Counc. Trans. 8:29–35.

McQueen, D. S., and E. G. Russell. 1979. Culture of *Mycobacterium paratuberculosis* from bovine foetuses. Censt. Vet. J. 55:203–204.

McQuivey, R. P. 1978. The desert bighorn sheep of Nevada. Nevada Dep. Wildl., Biol. Bull. No. 6.

McWhirter, D., A. Smith, E. Merrill, and L. Irwin. 1992. Foraging behavior and vegetation responses to prescribed burning on bighorn sheep winter range. Proc. Bienn. Symp. North. Wild Sheep and Goat Counc. 8:264–278.

Mendoza, V. J. 1976. The bighorn sheep of the state of Sonora. Desert Bighorn Counc. Trans. 20:25–26.

Mensch, J. L. 1969. Desert bighorn *(O. canadensis nelsoni)* losses in a natural trap tank. Calif. Fish and Game 55:237–238.

Mensher, S. H., T. D. Bunch, and A. Maciulis. 1989. High-resolution of G-banded karyotype and idiogram of the goat: a sheep-goat G-banded comparison. J. Hered. 80:150–155.

Merriam, C. H. 1901. Two new bighorns and a new antelope from Mexico and the United States. Proc. Biol. Soc. Wash. 15:29–32.

———. 1916. *Ovis sheldoni,* a new mountain sheep from Sierra del Rosario, Sonora, Mexico. Proc. Biol. Soc. Wash. 15:260–263.

Merritt, M. F. 1974. Measurement of utilization of bighorn sheep habitat in the Santa Rosa Mountains. Desert Bighorn Counc. Trans. 18:4–17.

Metcalf, H. E. 1977. Bluetongue and related diseases. Mimeo. U.S.D.A., APHIS, Vet. Serv., Arthropod-Borne Animal Research Laboratory, Denver, Colo.

Miller, G. D., and W. S. Gaud. 1989. Composition and variability of desert bighorn sheep diets. J. Wildl. Manage. 53:597–606.

Miller, R. 1976. Models, metaphysics and long-lived species. Bull. Ecol. Soc. Am. 57:2–6.

Miura, S., I. Kita, and M. Sugimura. 1987. Horn growth and reproductive history in female Japanese serow. J. Mammal. 68:826–836.

Miyamoto, M. M., F. Kraus, and O. A. Ryder. 1990. Phylogeny and evolution of antlered deer determined from mitochondrial DNA sequences. Proc. Natl. Acad. Sci. USA 85:6127–6132.

Moehlman, P. D. 1974. Behavior and ecology of feral asses *(Equus asinus)*. Ph.D. dissertation, University of Wisconsin, Madison.

Monson, G. 1964. Long-distance and nighttime movements of desert bighorn sheep. Desert Bighorn Counc. Trans. 8:11–17.

———. 1980. Distribution and abundance. Pages 40–51 *in* G. Monson and L. Sumner, editors. The desert bighorn: its life history, ecology, and management. University of Arizona Press, Tucson.

———, and L. Sumner, editors. 1980. The desert bighorn: its life history, ecology, and management. University of Arizona Press, Tucson.

Monson, P. A., and G. Post. 1972. Experimental transmission of *Protostrangylus stilesi* to bighorn sheep hybrids. J. Parasitol. 58:29–33.

Moore, T. D. 1958. Transplanting and observations of transplanted bighorn sheep. Desert Bighorn Counc. Trans. 2:43–46.

———. 1961. The Texas desert bighorn sheep transplant. Desert Bighorn Counc. Trans. 5:53–55.

Morgan, J. K. 1970. Ecology of the Morgan Creek and East Fork of the Salmon River bighorn sheep herds and management of bighorn sheep in Idaho. M.S. thesis, Utah State University, Logan.

Morgan, P. D., C.A.P. Boundy, G. W. Arnold, and D. R. Lindsay. 1975. The roles played by the senses of the ewe in location and recognition of lambs. Appl. Anim. Ethol. 1:143–150.

Morgantini, L. E., and E. Bruns. 1988. Attraction of bighorn sheep to wellsites and other man-made mineral licks along the eastern slopes of Alberta: a management concern. Proc. Bienn. Symp. North. Wild Sheep and Goat Counc. 6:135–140.

———, and R. J. Hudson. 1981. Sex differential in use of the physical environment by bighorn sheep *(Ovis canadensis)*. Can. Field-Nat. 95:69–74.

Morgart, J. R., and P. R. Krausman. 1983. Early breeding in bighorn sheep. Southwest. Nat. 28:460–461.

———, ———, W. H. Brown, and F. M. Whiting. 1986. Chemical analysis of mountain sheep forage in the Virgin Mountains, Arizona. Univ. Ariz., College Agric. Tech. Bull. No. 257.

Morrison, D. C. 1972. Habitat utilization by mule deer in relation to cattle and California bighorn sheep in the Ashnola River Valley, British Columbia. M.S. thesis, University of British Columbia, Vancouver.

Moser, C. A. 1962. The bighorn sheep of Colorado. Colorado Game and Fish Dep., Tech. Bull. No. 10.

Mouton, R. J., R. M. Lee, and R. J. Olding. 1991. A desert bighorn sheep decline in Aravaipa Canyon, Arizona. Desert Bighorn Counc. Trans. 35:18–19.

Munoz, J. R. 1981. Movements and mortalities of desert bighorn sheep in the San Andres Mountains, New Mexico. Desert Bighorn Counc. Trans. 25:64–65.

———. 1982. Movements and mortalities of desert bighorn sheep in the San Andres Mountains, New Mexico. Desert Bighorn Counc. Trans. 26:107–108.

Murie, A. 1944. The wolves of Mount McKinley. U.S. Natl. Park Serv., Fauna Ser. No. 5.

Murphy, E. C. 1974. An age structure and re-evaluation of the population dynamics of Dall sheep *(Ovis dalli dalli)*. M.S. thesis, University of Alaska, Fairbanks.

———, F. J. Singer, and L. Nichols. 1990. Effects of hunting on survival and productivity of Dall sheep. J. Wildl. Manage. 54:284–290.

———, and K. R. Whitten. 1976. Dall sheep demography in McKinley Park and a re-evaluation of Murie's data. J. Wildl. Manage. 40:597–609.

Muschenheim, A. L., E. T. Thorne, E. S. Williams, S. H. Anderson, and F. C. Wright. 1990. Psoroptic scabies in Rocky Mountain bighorn sheep *Ovis canadensis* from Wyoming. J. Wild. Dis. 26:554–557.

Nadler, C. F. 1971. Chromosomes of the Dall sheep, *Ovis dalli dalli* (Nelson). J. Mammal. 51:461–463.

———, and T. D. Bunch. 1977. G-band pattern of the Siberian snow sheep *(Ovis nivicola)* and their relationship to chromosomal evolution in sheep. Cytogenet. Cell Genet. 19:108–117.

———, R. S. Hoffmann, and A. Woolf. 1973a. G-band patterns as chromosomal markers and the interpretation of chromosomal evolution in wild sheep *(Ovis)*. Experientia 29:117–119.

———, K. V. Korobitsyna, R. S. Hoffmann, and N. N. Vorontsov. 1973b. Cytogenetic differentiation, geographic distribution, and domestication in Palearctic sheep *(Ovis)*. Z. Saugetierkund. 38:109–125.

———, and D. M. Lay. 1975. Chromosomes of some Asian wild sheep *(Ovis)* and goats *(Capra)*. Chrom. Info. Serv. 18:28–31.

———, ———, and J. D. Hassinger. 1971a. Cytogenetic analysis of wild sheep populations in northern Iran. Cytogenetics 10:137–152.

———, A. Woolf, and K. E. Harris. 1971b. The transferrins and hemoglobins of bighorn sheep *(Ovis canadensis),* Dall sheep *(Ovis dalli),* and mouflon *(Ovis musimon)*. Comp. Biochem. Physiol. B 40:567–570.

Nagabuchi, M., Y. Kawamoto, T. Nishikawa, and M. Nishimura. 1993. Polymorphism of transferrin found in the laboratory rat and wild rats in Japan. Biochem. Genet. 31:147–154.

Naitana, S., S. Ledda, E. Cocco, L. Monca, and B. Masala. 1991. Haemoglobin phenotypes of the wild European mouflon sheep living on the island of Sardinia. Anim. Genet. 22:67–75.

National Research Council. 1964. Nutrient requirements of domestic animals. Natl. Acad. Sci., Natl. Res. Counc., Publ. No. 1193.

———. 1968. Nutrient requirements of sheep. Natl. Acad. Sci., Publ. No. 1.

Needham, A. E. 1964. The growth process in animals. Sir Isaac Pitman and Sons, London, U.K.

Neel, J. V. 1979. Health and disease in unaccultured Amerindian populations. Pages 155–168 *in* P. Hugh-Jones, editor. Ciba Foundation Symposia, New Ser. 49. Elsevier North-Holland, New York, N.Y.

Neiland, K. A. 1978. Contagious ecthyma: a potential wildlife disease problem in Alaska. Unpubl. Rep., Alaska Dep. Fish and Game.

Nelson, E. W. 1884. A new geographical race of mountain sheep *(Ovis montana dalli* var. n.) from Alaska. Proc. U.S. Natl. Mus. 7:12–13.

Nette, T., D. Burles, and M. Hoefs. 1984. Observations of golden eagle, *Aquila chrysaetos,* predation on Dall sheep, *Ovis dalli dalli,* lambs. Can. Field-Nat. 98:252–254.

Nichols, L. 1971. Sheep report. Pages 1–25 *in* L. Nichols and A. Smith, editors. Sheep report. Alaska Dep. Fish and Game, Fed. Aid Wildl. Restor. Annu. Proj. Prog. Rep. Vol. 12.

———. 1972. Sheep reports. Pages 1–51 *in* L. Nichols and W. Heimer, editors. Sheep report. Alaska Dep. Fish and Game, Fed. Aid Wildl. Restor. Annu. Proj. Prog. Rep. Vol. 13.

———. 1973. Sheep report. Alaska Dep. Fish and Game, Fed. Aid Wildl. Restor. Annu. Proj. Prog. Rep. Vol. 14.

———. 1975. Report from Alaska. Pages 8–13 *in* J. B. Trefethan, editor. The wild sheep in modern North America. The Boone and Crockett Club and Winchester Press, New York, N.Y.

———. 1976. An experiment in Dall sheep management: progress report. Trans. Second North Am. Wild Sheep Conf. 2:16–34.

———. 1978a. Dall sheep reproduction. J. Wildl. Manage. 42:570–580.

———. 1978b. Dall's sheep. Pages 173–198 in J. L. Schmidt and D. L. Gilbert, editors. Big game of North America. Stackpole Books, Harrisburg, Pa.

———. 1984. Some effects of a full-curl law on Dall's sheep. Proc. Bienn. Symp. North. Wild Sheep and Goat Counc. 4:389–394.

———. 1988. Simple methods of comparing winter snow conditions on alpine and subalpine ranges of Dall's sheep and mountain goats in Alaska. Proc. Bienn. Symp. North. Wild Sheep and Goat Counc. 6:330–335.

Nievergelt, B. 1966. Der Alpensteinbock (*Capra ibex* L.) in seinem Lebensraum. Mammalia Depicta. Verlag Paul Parey, Berlin.

Nix, C. E., R. Bogart, and D. A. Price. 1969. Genetics of plasma transferrin in five breeds of sheep. J. Hered. 60:97–100.

Noss, R. F. 1987. Corridors in real landscapes: a reply to Simberloff and Cox. Conserv. Biol. 1:159–164.

Oftedal, O. T. 1985. Pregnancy and lactation. Pages 215–238 in R. J. Hudson, and R. G. White, editors. Bioenergetics of wild herbivores. CRC Press, Boca Raton, Fla.

Ogren, H. A. 1965. Barbary sheep. New Mexico Dep. Game and Fish, Bull. No. 13.

Oldemeyer, J. L., W. J. Barmore, and D. L. Gilbert. 1971. Winter ecology of bighorn sheep in Yellowstone National Park. J. Wildl. Manage. 35:257–269.

Olding, R. J. 1984. Arizona bighorn sheep status report—1984. Desert Bighorn Counc. Trans. 28:51–53.

Olech, L. A. 1979. Summer activity rhythms of peninsular bighorn sheep in Anza-Borrego Desert State Park, San Diego County, California. Desert Bighorn Counc. Trans. 23:33–36.

Onderka, D. K., S. A. Rawluk, and W. D. Wishart. 1988. Susceptibility of Rocky Mountain bighorn sheep and domestic sheep to pneumonia induced by bighorn and domestic livestock strains of *Pasteurella haemolytica*. Can. J. Vet. Res. 52:439–444.

———, and W. D. Wishart. 1988. Experimental contact transmission of *Pasteurella haemolytica* from clinically normal domestic sheep causing pneumonia in Rocky Mountain bighorn sheep. J. Wildl. Dis. 24:663–667.

Osgood, W. H. 1909. Biological investigations in Alaska and Yukon Territory. North Am. Fauna 30:1–96.

———. 1913. The name of the Rocky Mountain sheep. Proc. Biol. Soc. Wash. 26:57–62.

———. 1914. Dates for *Ovis canadensis, Ovis cervina,* and *Ovis montana.* Proc. Biol. Soc. Wash. 27:1–4.

Osta, R., C. Rodellar, S. Garcia-Belenguer, M. Gascon, and P. Zaragoza. 1994. Effect of genetic factors on the criteria of stress in fighting bulls. Vet. Res. 25:442–449.

Packard, F. M. 1946. An ecological study of the bighorn sheep in Rocky Mountain National Park, Colorado. J. Mammal. 27:3–28.

Pallister, G. L. 1974. The seasonal distribution and range use of bighorn sheep in the Beartooth Mountains, with special reference to the West Rosebud and Stillwater herds. Montana Fish and Game Dep., Fed. Aid Wildl. Restor. Proj. W-120-R-5.

Palmer, L. J. 1941. Dall sheep in the Mount Hayes region. Unpubl. Rep., U.S. Fish and Wildl. Serv.

Palmer, N. 1993. Bones and joints. Pages 1–181 *in* K.V.P. Jubb, P. C. Kennedy, and N. Palmer, editors. Pathology of domestic animals. Vol. 1. Fourth edition. Academic Press, New York, N.Y.

Paul, S. R., and T. D. Bunch. 1978. Chronic frontal sinusitis and osteolysis in Nelson's desert bighorn sheep *(Ovis canadensis nelsoni)*. J. Am. Vet. Med. Assoc. 173:1178–1180.

Pees, W., and H. Hemmer. 1980. Hirngroesse und Activitaet bei Wildschafen und Hausschafen (Gattung *Ovis*). Saugetierkund. Mitt. 28:39–45.

Pepelko, W. E., and M. T. Clegg. 1965. Studies of mating behaviour and some factors influencing sexual response in the male sheep, *Ovis aries*. Anim. Behav. 13:249–258.

Pepin, L., and T. C. N'Guyen. 1994. Blood groups and protein polymorphisms in five goat breeds *(Capra hircus)*. Anim. Genet. 25:333–336.

Petocz, R. G. 1973. The effect of snow cover on the social behaviour of bighorn rams and mountain goats. Can. J. Zool. 51:987–993.

Phillips, J. C. 1928. A sportsman's scrapbook. Houghton Mifflin, Boston, Mass.

Pianka, E. 1978. Evolutionary ecology. Harper and Row, New York, N.Y.

Picton, H. D. 1984. Climate and the prediction of reproduction of three ungulates. J. Appl. Ecol. 21:869–879.

Pielou, E. C. 1991. After the Ice Age. University of Chicago Press, Chicago, Ill.

Pillmore, R. E. 1961. Investigations of diseases and parasites affecting game animals. Study of the lung nematodes of bighorn sheep. Comparative studies of infection intensities in wild and laboratory populations. Colorado Div. Wildl., Fed. Aid Wildl. Restor. Rep. W-95-R-4.

Pitzman, M. S. 1970. Birth behavior and lamb survival in mountain sheep in Alaska. M.S. thesis, University of Alaska, Fairbanks.

Plummer, A. P. 1972. Selection. Pages 121–126 *in* C. M. McKell, J. P. Blaisdell, and J. R. Goodin, editors. Wildland shrubs—their biology and utilization. U.S.D.A. For. Range Exp. Stn. Tech. Rep. INT-1.

Poplin, F. 1979. Origene du mouflon de Corse dans une nouvelle perspective paleontologique; Par. Marronnage. Ann. Genet. Et Select. Anim. 11:133–144.

Post, G. 1962. Pasteurellosis of Rocky Mountain bighorn sheep *(Ovis canadensis canadensis)*. J. Wildl. Dis. 23:1.

———. 1971. The pneumonia complex in bighorn sheep. Trans. North Am. Wild Sheep Conf. 1:98–106.

Potts, M. K. 1937. Hemorrhagic septicemia in the bighorn of Rocky Mountain National Park. J. Mammal. 18:105–106.

Price, R. 1972. Effect of human disturbance on Dall sheep. Alaska Coop. Wildl. Res. Unit., Univ. Alaska, Fairbanks, Final Rep. No. 23.

Priede, I. G., and S. M. Swift, editors. 1992. Wildlife telemetry: remote monitoring and tracking of animals. Ellis Horwood, New York, N.Y.

Pulliam, H. R., and T. Caraco. 1984. Living in groups: is there an optimal group size? Pages 122–147 in J. R. Krebs and N. B. Davies, editors. Behavioural ecology: an evolutionary approach. Second edition. Blackwell, Oxford, U.K.

Pulling, A. von S. 1945. Non-breeding in bighorn sheep. J. Wildl. Manage. 9:155–156.

Purdy, K. G., and W. W. Shaw. 1981. An analysis of recreational use patterns in desert bighorn habitat: the Pusch Ridge Wilderness case. Desert Bighorn Counc. Trans. 25:1–5.

Quinn, J. F., and A. Hastings. 1987. Extinction in subdivided habitats. Conserv. Biol. 1:198–208.

Rachlow, J. L., and R. T. Bowyer. 1991. Interannual variation in timing and synchrony of parturition in Dall's sheep. J. Mammal. 72:487–492.

Ramey, R. R. 1991. New evidence for the reclassification of bighorn sheep in the southwestern United States and Mexico. Unpubl. abstract presented at the Thirty-fifth Annual Desert Bighorn Council Meeting, Las Cruces, N. Mex.

———. 1993. Evolutionary genetics and systematics of North American mountain sheep: implications for conservation. Ph.D. dissertation, Cornell University, Ithaca, N.Y.

———. 1995. Mitochondrial DNA variation, population structure, and the evaluation of mountain sheep in the southwestern United States and Mexico. Mol. Ecol. 4:429–439.

Ramsay, M. A., and R.M.F.S. Sadleir. 1979. Detection of pregnancy in living bighorn sheep by progestin determination. J. Wildl. Manage. 43:970–973.

Rausch, R. L. 1961. A review of the distribution of Holarctic recent mammals. Pacific Sci. Congr. 10:29–43.

Reddy, V. P., V.R.C. Reddy, and K. K. Reddy. 1988. Genetics of haemoglobin types and their influence on growth traits in Nellore and Nellore × Dorset halfbreed sheep. Indian J. Anim. Sci. 58:1234–1236.

Reid, R. L., and S. C. Miles. 1962. Studies on the carbohydrate metabolism of sheep: the adrenal response to psychological stress. Aust. J. Agr. Res. 13:282–295.

Remington, R. R. 1981. Arizona bighorn sheep status report. Desert Bighorn Counc. Trans. 25:44–46.

———. 1982. Age and weight relationships of desert bighorn sheep captured in Arizona during 1981–82. Desert Bighorn Counc. Trans. 26:38–42.

———. 1983. Arizona bighorn sheep status report, 1983. Desert Bighorn Counc. Trans. 27:39–41.

Reneau, J., and S. C. Reneau, editors. 1993. Records of North American big game. The Boone and Crockett Club, Missoula, Mont.

Riemann, H., M. R. Zaman, R. Ruppaner, O. Aalund, J. B. Jorgensen, H. Worsaal, and D. Behymer. 1979. Paratuberculosis in cattle and free-living deer. J. Am. Vet. Med. Assoc. 174:841–843.

Riggs, R. A. 1977. Winter habitat use patterns and populations of bighorn sheep in Glacier National Park. M.S. thesis, University of Idaho, Moscow.

Risenhoover, K. L., and J. A. Bailey. 1985. Foraging ecology of mountain sheep: implications for habitat management. J. Wildl. Manage. 49:797–804.

Rizzi, R., A. Caroli, P. Bolla, A. Acciaioli, and G. Pagnacco. 1988. Variability of reduced glutathione levels in Masseses ewes and its effect on daily milk production. J. Dairy Res. 55:345–353.

Robbins, C. T., and A. N. Moen. 1975. Milk consumption and weight gain of white-tailed deer. J. Wildl. Manage. 39:355–360.

Robinette, W. L. 1966. Mule deer home range and dispersal in Utah. J. Wildl. Manage. 30:335–349.

Robinson, R. M., T. L. Hailey, C. W. Livingston, and J. W. Thomas. 1967. Blue tongue in the desert bighorn sheep. J. Wildl. Manage. 31:165–168.

Rolstad, J. 1991. Consequences of forest fragmentation for the dynamics of bird populations: conceptual issues and the evidence. Pages 149–163 *in* M. Gilpin and I. Hanski, editors. Metapopulation dynamics: empirical and theoretical investigations. Academic Press, San Diego, Calif.

Rominger, E. M., A. R. Dale, and J. A. Bailey. 1988. Shrubs in the summer diet of Rocky Mountain bighorn sheep. J. Wildl. Manage. 52:47–50.

Roseneau, D. G., and P. M. Stern. 1974. Distribution of moose, muskox, and sheep in northeastern Alaska, 1972. *In* R. D. Jakimchuk, editor. Distribution of moose, sheep, muskox, and furbearing mammals in northeastern Alaska. Vol. 6. Renewable Resources Consulting Service, Sidney, B.C.

Rowland, M. M., and J. L. Schmidt. 1981. Transplanting desert bighorn sheep—a review. Desert Bighorn Counc. Trans. 25:25–28.

Russo, J. P. 1956. The desert bighorn in Arizona. Arizona Game and Fish Dep., Bull. No. 1.

Rutter, N. W., V. Geist, and D. M. Shackleton. 1972. A bighorn sheep skull 9280 years old from British Columbia. J. Mammal. 53:641–644.

Ryder, T. J., E. S. Williams, K. W. Mills, K. H. Bowles, and E. T. Thorne. 1992. Effect of pneumonia on population size and lamb recruitment in Whiskey Mountain bighorn sheep. Proc. Bienn. Symp. North. Wild Sheep and Goat Counc. 8:136–146.

Sadleir, R.M.F.S. 1969. The ecology of reproduction in wild and domestic mammals. Methuen, London, U.K.

———. 1987. Reproduction in female cervids. Pages 123–144 *in* C. M. Wemmer, editor. Biology and management of the Cervidae. Smithsonian Institution Press, Washington, D.C.

Saha, N., J. S. Tay, A. C. Roy, M. K. Das, M. Roy, B. Dey, S. Banerjee, and B. N. Mukherjee. 1992. Genetic study of five populations of Bihar, India. Hum. Biol. 64:175–186.

Samuel, W. M., G. A. Chalmers, J. G. Stelfox, A. Loewen, and J. J. Thomsen. 1975. Contagious ecthyma in bighorn sheep and mountain goats in western Canada. J. Wildl. Dis. 11:26–31.

Sanchez, D. R. 1976. Analysis of stomach contents of bighorn sheep in Baja California. Desert Bighorn Counc. Trans. 20:21–22.

Sandoval, A. V. 1979a. Preferred habitat of desert bighorn sheep in the San Andres Mountains, New Mexico. M.S. thesis, Colorado State University, Fort Collins.

———. 1979b. Evaluation of historic desert bighorn sheep ranges. New Mexico Dep. Game and Fish, Santa Fe.

———. 1979c. Bighorn sheep status report from New Mexico. Desert Bighorn Counc. Trans. 23:82–87.

———. 1980. Management of a psoroptic scabies epizootic in bighorn sheep *(Ovis canadensis mexicana)* in New Mexico. Desert Bighorn Counc. Trans. 24:21–28.

———. 1981. New Mexico bighorn sheep status report. Desert Bighorn Counc. Trans. 25:66–68.

———. 1982. Evaluation of the Peloncillo and Alamo Hueco Mountains, New Mexico, as potential desert bighorn habitat. New Mexico Dep. Game and Fish, Santa Fe.

———. 1988. Bighorn sheep die-off following association with domestic sheep: case history. Desert Bighorn Counc. Trans. 32:36–38.

Sausmann, K. 1982. Survival of captive born *Ovis canadensis* in North American zoos. Desert Bighorn Counc. Trans. 26:26–31.

Schallenberger, A. D. 1966. Food habits, range use and interspecific relationships of bighorn sheep in the Sun River area, west-central Montana. M.S. thesis, Montana State University, Bozeman.

———. 1972. Management and research on bighorn sheep, Sun River area, Montana. Trans. North. Wild Sheep Counc. 2:2–13.

Schaller, G. B. 1977. Mountain monarchs: wild sheep and goats of the Himalaya. University of Chicago Press, Chicago, Ill.

Schinkel, P. G. 1954. The effect of the ram on the incidence and occurrence of oestrus in ewes. Aust. Vet. J. 30:189–195.

Schmitt, J., and F. Ulbrich. 1968. Die chromosomen verschiedener Caprini, Simpson, 1945. Z. Saugetierkund. 33:180–186.

Scholander, P. F., V. Walters, R. Hock, and L. Irving. 1950. Body insulation of some arctic and tropical mammals and birds. Biol. Bull. 99:225–236.

Schwantje, H. M. 1988. Causes of bighorn sheep mortality and dieoff—literature review. Wildl. Wildl. Branch, Minist. Environ., Gov. B.C., Victoria, Working Rep. No. WR-35.

Schwartz, O. A., V. C. Bleich, and S. A. Holl. 1986. Genetics and the conservation of mountain sheep. Conserv. Biol. 37:179–190.

Scott, J. E., R. R. Remington, and J. C. deVos Jr. 1990. Numbers, movements, and disease status of bighorn in southwestern Arizona. Desert Bighorn Counc. Trans. 34:9–13.

Scott, M. J., B. Csuti, and S. Caicco. 1991. Gap analysis: assessing protection needs. Pages 15–26 *in* W. E. Hudson, editor. Landscape linkages and biodiversity. Island Press, Washington, D.C.

——, F. Davis, B. Csuti, R. Noss, B. Butterfield, C. Groves, H. Anderson, S. Caicco, F. D'Erchia, T. C. Edwards Jr., J. Ulliman, and R. G. Wright. 1993. Gap analysis: a geographic approach to protection of biological diversity. Wildl. Monogr. No. 123.

Scott, R. V., E. F. Chatelain, and W. A. Elkins. 1950. The status of Dall sheep and caribou in Alaska. Trans. North Am. Wildl. Conf. 15:612–625.

Scotton, B. D. 1996. Estimation of neonatal mortality rate and determination of neonatal mortality causes in Dall sheep in the Central Alaska Range Unit 20A. Alaska Dep. Fish and Game, Fed. Aid Wildl. Restor. Annu. Proj. Rep. W-24-4.

Seegmiller, R. F., P. R. Krausman, W. H. Brown, and F. M. Whiting. 1990. Nutritional composition of desert bighorn sheep forage in the Harquahala Mountains, Arizona. Desert Plants 10:87–90.

——, and R. D. Ohmart. 1981. Ecological relationships of feral burros and desert bighorn sheep. Wildl. Monogr. No. 78.

——, and C. D. Simpson. 1979. The Barbary sheep: some conceptual implications of competition with desert bighorn. Desert Bighorn Counc. Trans. 23:47–49.

Seip, D. R. 1983. Foraging ecology and nutrition of Stone's sheep. Ph.D. dissertation, University of British Columbia, Vancouver.

——, and F. L. Bunnell. 1984. Body weights and measurements of Stone's sheep. J. Mammal. 65:513–514.

——, and ——. 1985*a*. Nutrition of Stone's sheep on burned and unburned ranges. J. Wildl. Manage. 49:397–405.

——, and ——. 1985*b*. Foraging behaviour and food habits of Stone's sheep. Can. J. Zool. 63:1638–1646.

Seton, E. T. 1929. The bighorn. Pages 519–573 *in* E. T. Seton, editor. Lives of the game animals. Vol. 3. Part 2. Doubleday, Doran Co., Garden City, N.Y.

Shackleton, D. M. 1973. Population quality and bighorn sheep (*Ovis canadensis canadensis* Shaw). Ph.D. dissertation, University of Calgary, Alberta.

——. 1976. Variability in physical and social maturation between bighorn sheep populations. Trans. North. Wild Sheep Counc. 4:1–8.

——. 1985. *Ovis canadensis*. Mammalian Species No. 230. American Society of Mammalogists.

——. 1991. Social maturation and productivity in bighorn sheep: are young males incompetent? Appl. Anim. Behav. Sci. 29:173–184.

——, N. Barichello, A. Gunn, D. H. Hebert, and F. Harper. 1997. Pages 296–303 *in* D. M. Shackleton, editor. Conservation of wild sheep and goats and their relatives: status survey and conservation action plan for Caprinae. IUCN, Gland, Switzerland.

———, and J. Haywood. 1985. Early mother-young interactions in California bighorn sheep, *Ovis canadensis californiana*. Can. J. Zool. 63:868–875.

———, and D. A. Hutton. 1971. An analysis of the mechanisms of brooming in mountain sheep horns. Z. Säugetierk. 36:342–350.

———, R. G. Peterson, J. Haywood, and A. Bottrell. 1984. Gestation period in *Ovis canadensis*. J. Mammal. 65:337–338.

———, and C. C. Shank. 1984. A review of the social behavior of feral and wild sheep and goats. J. Anim. Sci. 58:500–509.

Shank, C. C. 1977. Cooperative defense in bighorn sheep. J. Mammal. 58:243–244.

———. 1979. Sexual dimorphism and the ecological niche of wintering Rocky Mountain bighorn sheep. Ph.D. dissertation, University of Calgary, Alberta.

———. 1982. Age-sex differences in the diets of wintering Rocky Mountain bighorn sheep. Ecology 63:627–633.

Shannon, N.H.R., R. J. Hudson, V. C. Brink, and W. D. Kitts. 1975. Determinants of spatial distribution of Rocky Mountain bighorn sheep. J. Wildl. Manage. 39:387–401.

Shaw, G. 1804. Naturalist's miscellany. Vol. 15. Plate 6. Nodder and Co., London, U.K.

Sheldon, C. 1919. The wilderness of the Upper Yukon. Second edition. Charles Scribner's Sons, New York, N.Y.

———. 1930. The wilderness of Denali. Charles Scribner's Sons, New York, N.Y.

Shillito, E. E. 1975. A comparison of the role of vision and hearing in lambs finding their own dams. Appl. Anim. Ethol. 1:369–377.

Sigurdsson, B. 1954. Paratuberculosis (Johne's disease) of sheep in Iceland: immunological studies and observations on its mode of spread. Br. Bet. J. 110:307–322.

Sikstrom, C., and L. Beckman. 1994. RsaI and BcII polymorphism of the transferrin receptor gene. Hum. Hered. 44(6):312–315.

Simberloff, D., and J. Cox. 1987. Consequences and costs of conservation corridors. Conserv. Biol. 1:63–71.

Simmons, N. M. 1963. A desert bighorn study. Part 1. Desert Bighorn Counc. Trans. 7:72–83.

———. 1969. Heat stress and bighorn behavior in the Cabeza Prieta Game Range, Arizona. Desert Bighorn Counc. Trans. 13:56–63.

———. 1980. Behavior. Pages 124–144 *in* G. Monson and L. Sumner, editors. The desert bighorn: its life history, ecology, and management. University of Arizona Press, Tucson.

———. 1982. Seasonal ranges of Dall's sheep, Mackenzie Mountains, Northwest Territories. Arctic 35:512–518.

Simmons, N. M., M. B. Bayer, and L. O. Sinkey. 1981. Dall's sheep demography in the Mackenzie Mountains. Unpubl. Rep., N.W.T. Wildlife Serv.

———, ———, and ———. 1984. Demography of Dall's sheep in the Mackenzie Mountains, Northwest Territories. J. Wildl. Manage. 48:156–162.

Simpson, C. D., editor. 1980. Proceedings of the symposium on ecology and management of Barbary sheep. Texas Tech University Press, Lubbock.

———, and L. J. Krysl. 1981. Status and distribution of Barbary sheep in the southwest United States. Desert Bighorn Counc. Trans. 25:9–15.

———, ———, D. B. Hampy, and G. G. Gray. 1978. The Barbary sheep: a threat to desert bighorn survival. Desert Bighorn Counc. Trans. 22:26–31.

Sinclair, A.R.E. 1989. Population regulation in animals. Pages 197–241 *in* J. M. Cherrett, editor. Ecological concepts: the contribution of ecology to an understanding of the natural world. Blackwell, Oxford, U.K.

Singer, F. J. 1984. Some population characteristics of Dall sheep in six Alaska national parks and preserves. Proc. Bienn. Symp. North. Wild Sheep and Goat Counc. 4:1–10.

———, D. Johnson, and R. Quimby. 1983. Dall sheep numbers and distribution in the Noatak National Preserve, 1983. U.S. Natl. Park Serv., Nat. Resourc. Surv. and Inventory Rep. AR-83.

———, and L. Nichols. 1992. Trophy hunting of Dall sheep in Alaska: an evaluation of the biological implications. Proc. Bienn. Symp. North. Wild Sheep and Goat Counc. 8:28–48.

Singer, R. S., D. A. Jessup, I. A. Gardner, and W. M. Boyce. 1997. Pathogen exposure patterns among sympatric populations of bighorn sheep, mule deer, and cattle. J. Wildl. Dis. 33:377–382.

Skögland, T. 1984. The effects of food and maternal conditions on fetal growth and size in wild reindeer. Rangifer 4:39–46.

———. 1991. What are the effects of predators on large ungulate populations? Oikos 61:401–411.

Smiley, F. E. 1978. Changes in the cursorial ability of Wyoming Holocene bison. Wyo. Contrib. Anthropol. 42:105–126.

Smith, D. R. 1954. The bighorn sheep in Idaho: its status, life history and management. Idaho Game and Fish Dep. Wildl. Bull. No. 1.

———, and P. Krausman. 1987. Diets of desert bighorn in the Virgin Mountains, Arizona. Desert Bighorn Counc. Trans. 31:11–14.

Smith, F. V., C. Van-Toller, and T. Boyes. 1966. The critical period in the attachment of ewes and lambs. Anim. Behav. 14:120–125.

Smith, K. G., and W. D. Wishart. 1978. Further observations of bighorn sheep non-trophy seasons in Alberta and their management implications. Proc. Bienn. Symp. North. Wild Sheep and Goat Counc. 1:52–74.

Smith, M. F., and J. L. Patton. 1993. The diversification of South American murid rodents: evidence from mitochrondrial DNA sequence for the Akodontine tribe. Biol. H. Kimm. Soc. 50:149–177.

Smith, T. S. 1991. The bighorn sheep of Bear Mountain: ecological investigations and management recommendations. Ph.D. dissertation, Brigham Young University, Provo, Utah.

———, and J. T. Flinders. 1992. Evaluation of mountain sheep habitat in Zion National Park, Utah. Desert Bighorn Counc. Trans. 36:4–9.

Soulsby, E.J.L. 1982. Helminths, arthropods, and protozoa of domesticated animals. Seventh edition. Lea and Febiger, Philadelphia, Pa.

Spalding, D. J. 1966. Twinning in bighorn sheep. J. Wildl. Manage. 30:207.

———, and J. N. Bone. 1970. The California bighorn sheep of the south Okanagan Valley, British Columbia. Fish and Wildl. Branch, Victoria, B.C., Wildl. Manage. Publ. No. 3.

Spencer, C. C. 1943. Notes on the life history of Rocky Mountain bighorn sheep in the Tarryall Mountains of Colorado. J. Mammal. 24:1–11.

Spillett, J. J., W. C. Foote, and T. D. Bunch. 1975. Chromosome and blood analysis of wild and domestic sheep. Desert Bighorn Counc. Trans. 19:46–50.

Spraker, T. R. 1974. Lamb mortality. Trans. North. Wild Sheep Counc. 3:102–103.

———. 1975. Capture myopathy of bighorn sheep. Trans. North. Wild Sheep Counc. 3:113–116.

———. 1977. Fibrinous pneumonia of bighorn sheep. Desert Bighorn Counc. Trans. 21:17–18.

———. 1979. The pathogenesis of pulmonary protostrongylosis in bighorn lambs. Ph.D. dissertation, Colorado State University, Fort Collins.

———, C. P. Hibler, G. G. Schoonveld, and W. S. Adney. 1984. Pathological changes and microorganisms found in bighorn sheep during stress-related die-off. J. Wildl. Dis. 20:319–327.

Stabel, J. R. 1996. Production of gamma-interferon by peripheral blood mononuclear cells: an important diagnostic tool for detection of subclinical paratuberculosis. J. Vet. Diag. Invest. 8:345–350.

Stanger, M. C., J. Cresto, G. W. Workman, and T. D. Bunch. 1986. Desert bighorn sheep—riverboat interactions in Cataract Canyon, Utah. Desert Bighorn Counc. Trans. 30:5–7.

Steel, T. L., and G. W. Workman. 1990. Influence of forage quantity on microhabitat use by desert bighorn sheep. Desert Bighorn Counc. Trans. 34:1–5.

Steinkamp, M. J. 1990. The effect of seasonal cattle grazing on California bighorn sheep habitat use. M.S. thesis, Utah State University, Logan.

Stelfox, J. G. 1971. Bighorn sheep in the Canadian Rockies: a history, 1800–1970. Can. Field-Nat. 85:101–122.

———. 1975. Range ecology of Rocky Mountain bighorn sheep in Canadian National Parks. Ph.D. dissertation, University of Montana, Missoula.

———. 1976. Range ecology of Rocky Mountain bighorn sheep. Can. Wildl. Serv., Rep. Ser. No. 39.

———, and J. McGillis. 1970. Seasonal growth patterns of bighorns correlated with range condition and endoparasite loads. Trans. North. Wild Sheep Counc. 1:35–38.

———, and R. D. Taber. 1969. Big game in the northern Rocky Mountain coniferous forests. Pages 197–222 in R. D. Taber, editor. Coniferous forest of the northern Rocky Mountains. Foundation Center for Natural Resources, University of Montana, Missoula.

Stemp, R. E. 1983. Responses of bighorn sheep to environmental factors and harassment. M.S. thesis, University of Calgary, Alberta.

Stephenson, R. O., D. V. Grangaard, and J. Burch. 1991. Lynx, *Felis lynx,* predation on red foxes, *Vulpes vulpes,* caribou, *Rangifer tarandus,* and Dall sheep, *Ovis dalli,* in Alaska. Can. Field-Nat. 2:255–262.

Stevens, D. R., and N. J. Goodson. 1993. Assessing effects of removals for transplanting on a high-elevation bighorn sheep population. Conserv. Biol. 7:908–915.

Stewart, S. T. 1975. Ecology of the West Rosebud and Stillwater bighorn sheep herds, Beartooth Mountains, Montana. Montana Fish and Game Dep., Fed. Aid Wildl. Restor. Proj. W-120-R-6 and R-7.

——. 1980. Mortality patterns in a bighorn sheep population. Proc. Bienn. Symp. North. Wild Sheep and Goat Counc. 2:313–330.

——. 1982. Late parturition in bighorn sheep. J. Mammal. 63:154–155.

Stewart-Scott, I., and A. N. Bruère. 1987. Distribution of heterozygous translocations and aneuploid spermatocyte frequency in domestic sheep. J. Hered. 78:37–40.

Stock, A. D., and W. L. Stokes. 1969. A re-evaluation of Pleistocene bighorn sheep from the Great Basin and their relationship to living members of the genus *Ovis.* J. Mammal. 50:805–807.

Stockstad, D. S., M. S. Morris, and E. C. Lory. 1953. Chemical characteristics of natural licks used by big game animals in western Montana. Trans. North Am. Wildl. and Nat. Resourc. Conf. 18:247–258.

Stokes, W. L., and K. C. Condie. 1961. Pleistocene bighorn sheep from the Great Basin. J. Paleontol. 35:598–609.

Stormont, C., Y. Suzuki, G. E. Bradford, and P. King. 1968. A survey of hemoglobins, transferrins and certain red cell antigens in nine breeds of sheep. Genetics 60:363–371.

Stratil, A., D. Cizova, E. Gabrisova, and R. Pakorny. 1992. Inter- and intraspecific differences in serum proteins of different species and sub-species of zebras. Comp. Biochem. Physiol. B 103:589–599.

Streeter, R. G. 1969. A literature review on bighorn sheep population dynamics. Colorado Div. Game, Fish and Parks, Spec. Rep. No. 24.

Sugden, L. G. 1961. The California bighorn in British Columbia with particular reference to the Churn Creek herd. B.C. Dep. Rec. and Conserv., Victoria.

Summerfield, B. L. 1974. Population dynamics and seasonal movement patterns of Dall sheep in the Atigun Canyon area, Brooks Range, Alaska. M.S. thesis, University of Alaska, Fairbanks.

Sushkin, P. O. 1925. The wild sheep of the world and their distribution. J. Mammal. 6:145–157.

Sweensy, R. W., R. H. Whitlock, and C. L. Buckley. 1995. Evaluation of a

commercial enzyme-linked immunosorbent assay for the diagnosis of paratuberculosis in dairy cattle. J. Vet. Diag. Invest. 7:488–493.
Swihart, R. K., N. A. Slade, and B. J. Bergstrom. 1988. Relating body size to the rate of home range use in mammals. Ecology 69:393–399.
Taber, R. D. 1969. Criteria of sex and age. Pages 325–401 *in* R. G. Giles Jr., editor. Wildlife management techniques. Third edition. The Wildlife Society, Washington, D.C.
Tankersley, N. G. 1984. Mineral lick use by Dall sheep in the Watana Creek hills, Alaska. Proc. Bienn. Symp. North. Wild Sheep and Goat Counc. 4:211–230.
Tate, M. L., K. G. Dodds, K. J. Thomas, and K. M. McEwan. 1992*a*. Genetic polymorphism of plasminogen and vitamin D binding protein in red deer (*Cervus elaphus* L.). Anim. Genet. 23:303–313.
———, H. C. Manly, and A. Schmack. 1992*b*. Genetic polymorphism in sheep plasma detected using antibodies to human plasminogen. Anim. Genet. 23: 385–389.
Taylor, R. A. 1962. Characteristics of horn growth in bighorn rams. M.S. thesis, University of Montana, Missoula.
Taylor, R. E. 1976. Mortality of Nevada bighorn sheep from pneumonia. Desert Bighorn Counc. Trans. 20:51–52.
Technical Staff Desert Bighorn Council. 1990. Guidelines for management of domestic sheep in the vicinity of desert bighorn habitat. Desert Bighorn Counc. Trans. 34:33–35.
Thibault, C., M. Courot, L. Martinet, P. Mauleon, S. du Mesnil du Buisson, R. Ortavant, J. Pelletier, and J. P. Signoret. 1966. Regulation of breeding season and oestrous cycles by light and external stimuli in some mammals. J. Anim. Sci. 25(Suppl.):119–142.
Thompson, R. W., and J. C. Turner. 1982. Temporal geographic variation in the lambing season of bighorn sheep. Can. J. Zool. 60:1781–1793.
Thorne, E. T., and T. J. Walthall. 1983. Diagnosis of diseases in wildlife. Wyoming Game and Fish Dep., Fed. Aid Proj. FW-3-R-28.
Todd, J. W. 1972*a*. Foods of Rocky Mountain bighorn sheep in southern Colorado. M.S. thesis, Colorado State University, Fort Collins.
———. 1972*b*. A literature review of bighorn sheep food habits. Colorado Dep. Game, Fish and Parks Coop. Wildl. Res. Unit, Spec. Rep. No. 27.
———, and R. M. Hansen. 1973. Plant fragments in the feces of bighorns as indicators of food habits. J. Wildl. Manage. 37:363–366.
Tompkins, T., and M. J. Bryant. 1972. Mating behaviour in a small flock of lowland sheep. Anim. Prod. 5:203–210.
Torres, S. G., V. C. Bleich, and J. D. Wehausen. 1994. Status of bighorn sheep in California, 1993. Desert Bighorn Counc. Trans. 38:17–28.
Trefethen, J. B., editor. 1975. The wild sheep in modern North America. The Boone and Crockett Club and Winchester Press, New York, N.Y.

Triesmann, M. 1975. Predation and the evolution of gregariousness. Anim. Behav. 23:779–825.
Tucker, E. M. 1975. Genetic markers in the plasma and red blood cells. Pages 123–153 in M. H. Blunt, editor. The blood of sheep. Spring-Verlag, New York, N.Y.
———, and S. W. Clarke. 1980. Comparative aspects of biochemical polymorphisms in the blood of Caprinae species and their hybrids. Anim. Blood Groups and Biochem. Genet. 11:163–183.
Turner, J. C. 1970. Water consumption of desert bighorn sheep. Desert Bighorn Counc. Trans. 14:189–197.
———. 1973. Water, energy and electrolyte balance in the bighorn sheep, *Ovis canadensis*. Ph.D. dissertation, University of California, Riverside.
———. 1976. Initial investigations into the reproductive biology of the desert bighorn ram, *Ovis canadensis nelsoni, O. c. cremnobates*. Proc. Bienn. Symp. North. Wild Sheep and Goat Counc. 4:22–25.
———. 1979a. Adaptive strategies of selective fatty acid deposition in the bone marrow of desert bighorn sheep. Comp. Biochem. Physiol. A 62:599–604.
———. 1979b. Osmotic fragility of desert bighorn sheep red blood cells. Comp. Biochem. Physiol. A 64:167–175.
———. 1982. Chronic cranal osteitis in a Rocky Mountain bighorn sheep. J. Wildl. Dis. 18:373–379.
———, and C. G. Hansen. 1980. Reproduction. Pages 145–151 in G. Monson and L. Sumner, editors. The desert bighorn: its life history, ecology, and management. University of Arizona Press, Tucson.
———, and R. A. Weaver. 1980. Water. Pages 100–112 in G. Monson and L. Sumner, editors. The desert bighorn: its life history, ecology, and management. University of Arizona Press, Tucson.
Uhazy, L. S., and J. C. Holmes. 1971. Helminths of the Rocky Mountain sheep in western Canada. Can. J. Zool. 49:507–512.
———, ———, and J. G. Stelfox. 1971. Lungworms in the Rocky Mountain bighorn sheep of western Canada. Can. J. Zool. 51:817–824.
Valdez, R. 1982. The wild sheep of the world. Wild Sheep and Goat International, Mesilla, N. Mex.
———. 1988. Wild sheep and wild sheep hunters of the New World. Wild Sheep and Goat International, Mesilla, N. Mex.
———. 1997. Mexico. Pages 303–307 in D. M. Shackleton, editor. Wild sheep and their relatives, status survey and conservation action plan for Caprinal. IUCN, Gland, Switzerland.
———, and L. V. Alamia. 1977. Fecund mouflons. Nat. Hist. 88:72–77.
———, C. F. Nadler, and T. D. Bunch. 1978. Evolution of wild sheep in Iran. Evolution 32:56–72.
Vandenberge, R. J., C. S. Caldes, and M. K. Haderlie. 1984. A management update, Kofa National Wildlife Refuge. Desert Bighorn Counc. Trans. 28:26–29.
Van Dyke, W. A. 1978. Population characteristics and habitat utilization of

bighorn sheep, Steens Mountain, Oregon. M.S. thesis, Oregon State University, Corvallis.

van Dyne, G. M., and H. F. Heady. 1965. Botanical composition of sheep and cattle diets on a mature annual range. Hilgardia 36:465–492.

Van Valdenburgh, B., and F. Hertel. 1993. Tough times at La Brea: tooth breakage in large carnivores of the late Pleistocene. Science 261:456–459.

van Vliet, G., and T.H.J. Huisman. 1964. Changes in the haemoglobin types of sheep as a response to anaemia. Biochem. 93:401–409.

Vaskov, B., and G. Efremov. 1967. Fourth haemoglobin type in sheep. Nature 216:593–594.

Venegas, M. 1757. Noticia de la California. Vol. 1. Imprenta Viuda de Manuel Fernandez, Madrid.

Viereck, L. A. 1963. Range survey. Pages 8–27 *in* F. F. Jones, R. F. Batchelor, H. R. Merriam, and L. A. Viereck, editors. Sheep and goat investigations. Alaska Dep. Fish and Game, Proj. W-6-R-3, Work Plan E, Vol. 3.

Vincent, J. 1974. The management of large mammals in Natal, with special reference to utilization for stocking or restocking purposes. Pages 900–908 *in* V. Geist and F. Walthers, editors. The behavior of ungulates and its relation to management. IUCN Publ., New Ser. No. 24. Morges, Switzerland.

Wakeling, B. F., and W. H. Miller. 1989. Bedsite characteristics of desert bighorn sheep in the Superstition Mountains, Arizona. Desert Bighorn Counc. Trans. 33:6–8.

———, and ———. 1990. A modified habitat suitability index for desert bighorn sheep. Pages 58–66 *in* P. R. Krausman and N. S. Smith, editors. Managing wildlife in the Southwest. Arizona Chapter of the Wildlife Society, Phoenix.

Wakelyn, L. A. 1987. Changing habitat conditions on bighorn sheep ranges in Colorado. J. Wildl. Manage. 51:904–912.

Walters, J. E. 1979. Bighorn sheep population estimate for the South Tonto Plateau—Grand Canyon. Desert Bighorn Counc. Trans. 23:96–106.

———, and R. M. Hansen. 1978. Evidence of feral burro competition with desert bighorn sheep in Grand Canyon National Park. Desert Bighorn Counc. Trans. 22:10–16.

Walther, F. R. 1984. Communication and expression in hoofed mammals. Indiana University Press, Bloomington.

Walton, J. S., J. R. McNeilly, A. S. McNeilly, and F. J. Cunningham. 1977. Changes in concentrations of follicle stimulating hormone, prolactin and progesterone ion the plasma of ewes during the transition from anestrus to breeding activity. J. Endocrinol. 75:127–136.

Wang, M., and A. Schreiber. 1996. Low blood protein heterozygosity in zoo-bred hamadryas baboons. Biochem. Mol. Biol. 113:403–409.

Wang, S., W. C. Foote, and T. D. Bunch. 1990. A polymorphic protein character of red blood cells (EP-1) in domesticated and wild sheep. J. Hered. 81:206–208.

———, M. Zhenzhong, W. Suifang, and J. Qingwen. 1989. Study on geographical differences of the blood biochemical characteristics of Tan sheep and other sheep. Pages 39–58 *in* Studies on ecological geography of Tan sheep. Science Press, Beijing, China.

Wang, X. 1984. Late Pleistocene bighorn sheep *(Ovis canadensis)* of natural Trap Cave, Wyoming. M.A. thesis, University of Kansas, Lawrence.

Warrick, G. D., and P. R. Krausman. 1987. Foraging behavior of female mountain sheep in western Arizona. J. Wildl. Manage. 51:99–104.

———, and ———. 1989. Barrel cacti consumption by desert bighorn sheep. Southwest. Nat. 34:483–486.

Watson, R. H., and H. M. Redford. 1960. Influence of rams on the onset of oestrus in Merino ewes in the spring. Aust. J. Agric. Res. 11:65–71.

Watson, S. M., and W. E. Heimer. 1984. An age-specific winter die-off in Dall sheep in Alaska. Proc. Bienn. Symp. North. Wild Sheep and Goat Counc. 4:61–66.

Watts, T. J. 1979. Detrimental movement patterns in a remnant population of bighorn sheep *(Ovis canadensis mexicana)*. M.S. thesis, New Mexico State University, Las Cruces.

Wayne, R. K., A. Meyer, N. Lehman, B. Van Valkenburgh, P. W. Kat, T. K. Fuller, D. Girman, and S. J. O'Brian. 1990. Large sequence divergence among mitochondrial DNA genotypes within populations of East African black-backed jackals. Proc. Natl. Acad. Sci. USA 87:1772–1776.

Weaver, R. A. 1972. Conclusion of the bighorn investigation in California. Desert Bighorn Counc. Trans. 16:56–65.

———. 1973. California's bighorn management plan. Desert Bighorn Counc. Trans. 17:22–42.

———. 1982. Bighorn in California—a plan to determine current status and trends. California Dep. Fish and Game, Sacramento.

———. 1988. Status of bighorn sheep in California, 1987. Desert Bighorn Counc. Trans. 31:20.

———, and J. Hall. 1971. Bighorn sheep in Joshua Tree National Monument and adjacent areas. California Dep. Fish and Game, Fed. Aid Proj. W-51-R.

———, and J. L. Mensch. 1970. Desert bighorn sheep in northern Inyo and southern Mono Counties. California Dep. Fish and Game, Fed. Aid Proj. W-51-R.

Weeks, H. P., and C. M. Kirkpatrick. 1976. Adaptations of white-tailed deer to naturally occurring sodium deficiencies. J. Wildl. Manage. 40:610–625.

Wehausen, J. D. 1984. Comment on desert bighorn as relicts: further considerations. Wildl. Soc. Bull. 12:82–85.

———. 1987. Some probabilities associated with sampling for diseases in bighorn sheep. Desert Bighorn Counc. Trans. 31:8–10.

———. 1996. Effects of mountain lion predation on bighorn sheep in the Sierra Nevada and Granite Mountains of California. Wildl. Soc. Bull. 24:471–479.

———, V. C. Bleich, B. Blong, and T. L. Russi. 1987. Recruitment dynamics in a southern California mountain sheep population. J. Wildl. Manage. 51:86–98.

———, and R. R. Ramey III. 1993. A morphometric reevaluation of the peninsular bighorn subspecies. Desert Bighorn Counc. Trans. 37:1–10.

Weisenberger, M. E., P. R. Krausman, M. C. Wallace, D. W. DeYoung, and O. E. Maughan. 1996. Effects of simulated jet aircraft noise on heart rate and behavior of desert ungulates. J. Wildl. Manage. 60:52–61.

Welch, R. D. 1969. Behavioral patterns of desert bighorn sheep in south-central New Mexico. Desert Bighorn Counc. Trans. 13:114–129.

Welles, R. E., and F. B. Welles. 1961. The bighorn of Death Valley. U.S. Natl. Park Serv., Fauna Ser. No. 6.

Welsh, G. W. 1971. What's happening to our sheep. Desert Bighorn Counc. Trans. 15:63–73.

——, and T. D. Bunch. 1982. Three-year observation of psoroptic scabies in desert bighorn sheep from northwestern Arizona. Desert Bighorn Counc. Trans. 26:42–44.

——, and ——. 1983. Psoroptic scabies in desert bighorn sheep *(Ovis canadensis nelsoni)* from northwestern Arizona. J. Wildl. Dis. 19:342–344.

Werren, J. H., and P. D. Taylor. 1984. The effects of population recruitment on sex ratio selection. Am. Nat. 124:143–148.

Whitehead, P. E., and E. H. McEwen. 1980. Progesterone levels in peripheral plasma of Rocky Mountain bighorn ewes *(Ovis canadensis)* during the oestrous cycle and pregnancy. Can. J. Zool. 58:1105–1108.

Whitten, K. R. 1975. Habitat relationships and population dynamics of Dall sheep *(Ovis dalli dalli)* in Mt. McKinley National Park, Alaska. M.S. thesis, University of Alaska, Fairbanks.

Wiener, J. G. 1975. Nutrient cycles, nutrient limitation and vertebrate populations. The Biologist 75:104–124.

Wikeem, B. M. 1984. Forage selection by California bighorn sheep and the effects of grazing on an *Artemisia-Agropyron* community in southern British Columbia. Ph.D. dissertation, University of British Columbia, Vancouver.

——, and M. D. Pitt. 1979. Interpreting diet preference of California bighorn sheep on native rangeland in south-central British Columbia. Rangelands 1:200–202.

——, and ——. 1987. Evaluation of methods to determine use of browse by California mountain sheep. Wildl. Soc. Bull. 15:430–433.

——, and ——. 1992. Diet of California bighorn sheep: assessing optimal foraging habitat. Can. Field-Nat. 106:327–335.

Wilcove, D. S., C. H. McLellan, and A. P. Dobson. 1986. Habitat fragmentation in the temperate zone. Pages 237–256 *in* M. E. Soule, editor. Conservation biology: the science of scarcity and diversity. Sinauer Associates, Sunderland, Mass.

Williams, E. S. 1981. Spontaneous and experimental infection of wild ruminants with *Mycobacterium paratuberculosis*. Ph.D. dissertation, Colorado State University, Fort Collins.

——. 1982. Paratuberculosis (Johne's Disease). Pages 91–94 *in* E. T. Thorne, N. Kingston, W. R. Jolley, and R. C. Bergstom, editors. Diseases of wildlife in Wyoming. Second edition. Wyoming Game and Fish Dep., Cheyenne.

———, J. C. De Martini, and S. P. Snyder. 1985. Lymphocyte blastogenesis, complement fixation, and fecal culture as diagnostic tests for paratuberculosis in North American wild ruminants and domestic sheep. Am. J. Vet. Res. 46: 2317–2321.

———, S. P. Snyder, and K. L. Martin. 1983. Pathology of spontaneous and experimental infection of North American wild ruminants with *Mycobacterium paratuberculosis*. Vet. Pathol. 20:274–291.

———, T. R. Spraker, and G. G. Schoonveld. 1979. Paratuberculosis (Johne's disease) in bighorn sheep and mountain goats in Colorado. J. Wildl. Dis. 15:221–227.

Williamson, J. A., and B. Charlesworth. 1976. The effect of age of founder on the probability of survival of a colony. J. Theor. Biol. 56:175–190.

Wilson, E. O. 1975. Sociobiology: the new synthesis. Harvard University Press, Cambridge, Mass.

Wilson, L. O. 1968. Distribution and ecology of desert bighorn sheep in southeastern Utah. Utah Dep. Nat. Resour. and Div. Fish and Game, Publ. No. 68-5.

———. 1975. Discussion following management of mountain sheep: theoretical considerations. Pages 99–105 *in* J. B. Trefethen, editor. The wild sheep in modern North America. The Boone and Crockett Club and Winchester Press, New York, N.Y.

———. 1976. Biases in bighorn research relating to food preferences and determining competition between bighorn and other herbivores. Desert Bighorn Counc. Trans. 20:46–48.

———, and C. L. Douglas. 1982. Revised procedures for capturing and re-establishing desert bighorn. Desert Bighorn Counc. Trans. 26:1–7.

Wilson, M. 1980. Morphological dating of late Quarternary bison on the Northern Plains. Can. J. Anthropol. 1:81–85.

Winfield, C. J., W. Brown, and I.A.M. Lucas. 1969. Sheltering behaviour at lambing by Welsh Mountain Ewes. Anim. Prod. 11:101–105.

Winkler, C. K. 1977. Status of the Texas desert bighorn program. Desert Bighorn Counc. Trans. 21:4.

Winters, J. F. 1980. Summer habitat and food utilization by Dall's sheep and their relation to body and horn size. M.S. thesis, University of Alaska, Fairbanks.

Wishart, W. D. 1958. The bighorn sheep of the Sheep River Valley. M.S. thesis, University of Alberta, Edmonton.

———, and D. Brochu. 1982. An evaluation of horn and skull characteristics as a measure of population quality in Alberta bighorns. Proc. Bienn. Symp. North. Wild Sheep and Goat Counc. 3:127–142.

Witham, J. H. 1983. Desert bighorn sheep in southwestern Arizona. Ph.D. dissertation, Colorado State University, Fort Collins.

———, R. R. Remington, and E. L. Smith. 1982. Desert bighorn summer mortality in southwestern Arizona, 1979. Desert Bighorn Counc. Trans. 26:44–46.

———, and E. L. Smith. 1979. Desert bighorn movements in a southwestern Arizona mountain complex. Desert Bighorn Counc. Trans. 23:20–24.

Wobeser, G., J.E.C. Bellamy, B. G. Boysen, P. S. MacWilliams, and W. Runge. 1976. Myopathy and myoglobinuria in a wild white-tailed deer. J. Am. Vet. Med. Assoc. 169:971–974.

Wood, A. J., I. McTaggart-Cowan, and H. C. Nordan. 1962. Periodicity of growth in ungulates as shown by deer of the genus *Odocoileus*. Can. J. Zool. 40:594–603.

Wood, N. J., S. H. Phua, and B. P. Davidson. 1993. Polymorphism in an A- and T-rich element in the beta-haemoglobin locus in sheep. Anim. Genet. 24:71.

Woodard, T. N., C. Hibler, and W. Rutherford. 1972. Bighorn lamb mortality investigations in Colorado. Trans. North. Wild Sheep Counc. 2:44–47.

Woodgerd, W. 1964. Population dynamics of bighorn sheep on Wildhorse Island. J. Wildl. Manage. 28:381–391.

Woodward, S. L., and R. D. Ohmart. 1976. Habitat use and fecal analysis of feral burros *(Equus asinus)*, Chemehuevi Mountains, California, 1974. J. Range Manage. 29:482–485.

Woolf, A., T. O'Shea, and D. L. Gilbert. 1970. Movements and behavior of bighorn sheep on summer ranges in Yellowstone National Park. J. Wildl. Manage. 34:446–450.

Wright, G. M., J. S. Dixon, and B. H. Thompson. 1933. A preliminary survey of faunal relations in National Parks of the U.S. U.S. Natl. Park Serv., Fauna Ser. No. 1.

Wurster, D. H., and K. Benirschke. 1968. Chromosome studies in the superfamily Bovoidea. Chromosoma 25:152–171.

Wynne-Edwards, V. C. 1963. Intergroup selection in the evolution of social systems. Nature 200:623–628.

Yoakum, J. 1964. Bighorn food habit-range relationships in the Silver Peak Range, Nevada. Desert Bighorn Counc. Trans. 8:95–102.

Young, S. P., and R. H. Manville. 1960. Records of bighorn hybrids. J. Mammal. 41:523–525.

Zine, M. J., P. R. Krausman, M. C. Wallace, and L. R. Berner. 1992. Resource use by mountain sheep in a large enclosure. Desert Bighorn Counc. Trans. 36:23–26.

Zipf, G. K. 1949. Human behavior and the principle of least effort. Addison-Wesley, Cambridge, Mass.

ABOUT THE CONTRIBUTORS

WALTER M. BOYCE received a D.V.M. from Auburn University in 1981 and a Ph.D. from Purdue University in 1988. He is an associate professor in the Department of Pathology, Microbiology, and Immunology in the School of Veterinary Medicine at the University of California, Davis. His research interests include population health and ecology of free-ranging wildlife, especially bighorn sheep.

THOMAS D. BUNCH received both his B.S. and M.S. from Brigham Young University in zoology (1968) and ecology and parasitology (1969), respectively. In 1979 he received a Ph.D. from Utah State University in wildlife biology, cytogenetics, and toxicology. He completed his postdoctoral training in reproductive physiology and technologies of embryo biology. Dr. Bunch is currently a professor in the Department of Animal, Dairy and Veterinary Sciences, Utah State University. His research interests are cytogenetics and molecular genetics in domestic and wild sheep and the development of new technologies for animal breeding. His current research is on the cytogenetics and molecular genetics of the argaliform wild sheep of Asia and the development of cloning strategies to rescue rare or endangered species of wild sheep.

FRED L. BUNNELL holds the forest renewal of British Columbia chair in applied conservation biology and is director of the Centre for Applied Conservation at the University of British Columbia. Dr. Bunnell studied at the University of British Columbia (B.S.F. with Honors, 1965), Eidengenossiche Technische Hochschule, Zurich (Fachhörer, 1966), and the University of California, Berkeley (Ph.D., 1973). In 1971, he joined the faculty at the University of British Columbia, where he teaches and conducts research in the Department of Forest Sciences. His primary research interests are applying the theories of conservation biology to forested systems, developing approaches to maintain biological diversity in managed forests, determining habitat relations of ungulates, and evaluating population dynamics of large carnivores. For his contributions to science he has received four national and/or international awards, including the Canadian Institute of Forestry's Gold Medal for Scientific Achievement in Forestry. Most recently, he was acknowledged as Outstanding Scientist of 1994 for internationally recognized scientific accomplishments of outstanding importance in the Pacific Northwest, and in 1997 he received the O. C. Wallmo Award for his work with deer. He is a happily married parent and serves on the Ministry and Counsel of the Religious Society of Friends (Quakers).

CHARLES L. DOUGLAS is the unit leader of the United States Geological Survey/Biological Resources Division Cooperative Park Studies Unit and an adjunct professor of Biology at the University of Nevada, Las Vegas. He received a B.A. in biology from Antioch College in 1959, an M.S. from Dartmouth College in 1961, and a Ph.D. in zoology from the University of Kansas in 1967. Since 1973, he and his graduate students have studied desert bighorn sheep in Death Valley, Joshua Tree, and Canyonland National Parks and Lake Mead National Recreation Area. He has served as curator of zoology at the University of Texas, professor and chairman of the Center for Man and Environment at Prescott College, and is a founding partner of Southwestern Environmental Consultants. He served for 10 years as editor of the *Desert Bighorn Council Transactions,* and is currently serving as the council's treasurer.

RICHARD C. ETCHBERGER received his B.S. in ecology from Unity College, Maine, in 1986, and earned his M.S. and Ph.D. at the University of Arizona by 1993. He is currently an assistant professor in the Department of Fisheries and Wildlife, Uintah Basin Branch Campus, Utah State University. His research interests include ungulate ecology, natural resource education, ecosystem restoration, and wildlife habitat use.

VALERIUS GEIST began his studies in 1961 by spending almost two years alone with Stone's sheep in the wilderness of the Spazisi in Northern British Columbia. This was followed by a two-year study of bighorn in Banff National Park and one fall studying Dall's sheep in what is now Kluanae National Park in the Yukon. He received his Ph.D. in 1966, and after postdoctoral study with Konrad Lorenz in Germany, he obtained a position at the University of Calgary. He has conducted significant studies of mountain sheep that evolved from observation through application of principles. Dr. Geist has authored the keystone books on mountain sheep and numerous scientific articles. He retired in 1995, but remains active in the wildlife arena.

CHARLES P. HIBLER received a B.S. in biology from New Mexico State University in 1956, an M.S. in zoology from Utah State University in 1959, and a Ph.D. from Colorado State University in 1962. Currently he is a professor emeritus in pathology at Colorado State University. He was the director of the Wild Animal Disease Center, College of Veterinary Medicine, Colorado State University and served as the editor of the *Journal of Wildlife Disease* for over seven years.

ABOUT THE CONTRIBUTORS

ROBERT S. HOFFMANN received a B.S. in zoology from Utah State University in 1950, and an M.S. and a Ph.D. in zoology from the University of California, Berkeley, in 1954 and 1955, respectively. He spent 13 years at the University of Montana, Missoula, as an instructor, assistant professor, and associate and full professor in the Department of Zoology. From 1968 to 1986 he served at the University of Kansas in the Department of Zoology and the Department of Systematics and Ecology while acting as curator of mammals at the university's Museum of Natural History. He is currently a senior scientist at the Smithsonian's National Museum of Natural History and served as director of the National Museum of Natural History, assistant secretary for science, and acting provost of the Smithsonian. His research interests have been extensive in the international scientific arena, and he has considerable experience in science policy. His research interests include the evolution of Holarctic mammals, specializing in the Arctic and mountainous regions of the world, and the mammals of USSR, China, and Central Asia.

PAUL R. KRAUSMAN received a B.S. in zoology from the Ohio State University in 1968, an M.S. in wildlife management from New Mexico State University in 1971, and a Ph.D. in wildlife science from the University of Idaho in 1976. He taught at Auburn University before moving to the University of Arizona, where he has concentrated his research on ungulates in arid environments. He has published numerous articles on desert bighorn sheep and other ungulates. Dr. Krausman was editor of *Desert Bighorn Council Transactions* for seven years and also served as editor of the *Journal of Wildlife Management*. His current research interests include habitat relationships between large mammals and their environments in the Southwest, Mexico, and India.

WILLIAM R. LANCE received a B.S. in agriculture from Oklahoma State University in 1962, a D.V.M. from Oklahoma State University in 1965, an M.S. in wildlife biology from Colorado State University in 1971, and a Ph.D. in pathology and wildlife disease from Colorado State University in 1980. From 1977 to 1981 he was a senior research scientist in wildlife diseases at the Wild Animal Disease Center, College of Veterinary Medicine, Colorado State University. He conducted research on diseases of Rocky Mountain bighorn sheep and the role of disease on the decline of mule deer herds in northern New Mexico. In 1981, Dr. Lance founded Wildlife Pharmaceuticals, which is dedicated to research, development, and acquisition of FDA-approved pharmaceuticals for wild and exotic species.

DAVID M. LESLIE JR. is the unit leader of the Oklahoma Cooperative Fish and Wildlife Research Unit and an adjunct professor of zoology at Oklahoma State

University. He received a B.A. in biology from Prescott College in 1974, an M.S. from the University of Nevada in 1977, and a Ph.D. from Oregon State University in 1982. While in Nevada, he and C. L. Douglas studied the desert bighorn sheep in the River Mountains of Lake Mead National Recreation Area, arguably the most managed population of wild sheep in North America. He has served as an associate editor of the *Journal of Wildlife Management* and the *Journal of Mammalogy* and is on the editorial board of the *Wildlife Society Bulletin*. He is currently the editor of the *Journal of Mammalogy*.

CHARLES F. NADLER received an A.B. from Dartmouth College in 1951 and an M.D. from Northwestern University Medical School in 1955. His postgraduate training was done at Washington University, St. Louis, Missouri, and at the National Institutes of Health, University of Colorado Medical Center. He is currently an associate professor of medicine at Northwestern Medical School, Chicago. His research interests include mammalian cytogenetics and biochemical genetics of Asian and North American rodents and ungulates (sheep) and their relationships with the Bering land bridge. Dr. Nadler has conducted fieldwork in Canada, Alaska, Iran, and the USSR.

LYMAN NICHOLS received his B.S. and M.S. in wildlife management from Colorado State University. He is now retired from the Alaska Department of Fish and Game, where he worked as a biologist and pilot and conducted research on Dall's sheep and Rocky Mountain goats. Prior to that, he worked for the Hawaii Division of Fish and Game on axis deer and wild boar research on the islands of Lanai and Hawaii. Following retirement, he owned and operated a small, air-taxi business that specialized in wildlife surveys, but retired again in early 1998. He is happily married and has three grown children and five grandchildren. He and his wife live in the village of Cooper Landing, Alaska, where from their windows they can watch Dall's sheep and mountain goats on the mountains above their home.

ANDREW V. SANDOVAL has been employed by the New Mexico Department of Game and Fish for 23 years. He received a B.S. in wildlife science from New Mexico State University and an M.S. in wildlife science from Colorado State University. For his graduate work, he studied the desert bighorn sheep population in the San Andres Mountains, New Mexico. He then became the bighorn sheep project leader for the New Mexico Department of Game and Fish, gauging the suitability of historic desert bighorn range for the reintroduction of the species. He subsequently studied desert bighorn range for the reintroduction of the species. He has done consulting work on desert bighorn sheep for the Republic of Mexico and the White Mountain Apache Tribe in Arizona and has written over 25 publications on bighorn sheep. Mr. Sandoval is currently the chief of the New Mexico Department of Game and Fish Conservation Services Division, where he continues to oversee desert bighorn restoration efforts. His interests include making rustic furniture and collecting antique firearms.

DAVID M. SHACKLETON earned a B.S. at the University of Leicester, England, before moving to Canada, where he received an M.S. from the University of Western Ontario and a Ph.D. from the University of Calgary. His master's work dealt with differences in social behavior between bison populations and his Ph.D. research investigated differences in behavior and morphology in bighorn sheep populations. His postdoctoral work was carried out in paleobiology at the University of Calgary and at the Paleobiology Division of the National Museums of Canada. Dr. Shackleton is currently an associate professor in the Department of Animal Science, University of British Columbia, and was editor of the IUCN/SSC Caprinae Action Plan dealing with the status and conservation of wild sheep, goats, and their relatives throughout the world.

CHRISTOPHER C. SHANK was educated at Earlham College, Colorado State University, and the University of Calgary. His master's work dealt with social behavior of feral goats and his doctoral research examined the ecological differences between the sexes of bighorn sheep. His postdoctoral work at the University of Munich focused on sex differences in habitat use by chamois. Dr. Shank worked for the United Nations examining the potential for hunting programs, nature reserves, and national parks in Afghanistan and was employed by the Government of the Northwest Territories to work on raptors, protected areas, and endangered species. He is currently working on forest biodiversity concerns for the Alberta Department of Environmental Protection.

TERRY R. SPRAKER received a D.V.M. from Colorado State University in 1972 and a Ph.D. in pathology from Colorado State University in 1978. He was accredited by the American College of Veterinary Pathologist in 1981. Dr. Spraker is currently an associate professor in the Department of Pathology, College of Veterinary Medicine, Colorado State University. His research interests are diseases of free-ranging terrestrial and marine mammals, in particular spongiform encephalopathy in free-ranging ruminants, pneumonia and stress in free-ranging bighorn sheep, and diseases of free-living northern fur seals and Stellar sea lions.

RAUL VALDEZ is a professor in the Department of Fishery and Wildlife Sciences at New Mexico State University. He has field research experience with wild sheep in the United States, Mexico, Iran, Mongolia, Pakistan, Kirghizstan, and Turkmenistan. His interest in wild sheep studies began as a young teenager after he read the Mongolian adventures of Roy Chapman Andrews, the model for Indiana Jones. He graduated with a Ph.D. in wildlife management from Texas A&M University, and his major research interest is ungulate multispecies grazing systems in arid zones.

BRIAN M. WIKEEM received a B.Ed. from the University of Calgary, Alberta, and then taught high school science and math in Australia. He returned to Canada to earn a B.S. in agricultural sciences at the University of British Columbia, where he majored in wildlife. He then earned a Ph.D. in ecology and range management at the University of British Columbia, where he studied the diet of California bighorn sheep in relation to plant community responses. Dr. Wikeem was employed by the British Columbia Ministry of Forests as a range scientist, working on habitat management of bighorn; the biology and ecology of weeds; integrating habitat use between cattle and wildlife; and agroforestry working with cattle, domestic sheep, and forest regeneration before becoming the provincial noxious weed specialist. Dr. Wikeem is now a natural resource management consultant.

ELIZABETH S. WILLIAMS received a B.S. in zoology from the University of Maryland in 1972, a D.V.M. from Purdue University in 1981, and a Ph.D. in pathology from Colorado State University in 1981. She is currently a professor of veterinary science, Department of Veterinary Sciences, University of Wyoming. Her research interests include infectious diseases of wildlife.

INDEX

Numbers for illustrations are in italics; those for tables are in bold.

abortion, 35, 128–129, 258
accidents: and adaptive strategies, 192, 208; desert bighorn, 177–178; Rocky Mountain and California bighorn, 136; thinhorn, 67–68. *See also* injury
activity: budgets, 106–108; patterns of desert bighorn, 161–164; patterns of Rocky Mountain and California bighorn, **45**, 106–109; patterns of thinhorn, 44–46, **45**
adaptation, 11, 18, 241, 258–259, 261; behavioral, 70–71, 187, 192–193, 197–201, 207–208; to forage constraints, 97; morphological, 205; to water constraints, 141, 149, 173
aesthetic value: of burros, 187; of mountain sheep, 3, 180
age estimation: with horns, 7, 40–41, *104*, 104–105, 158–159; with pelage, 161; with teeth, 5, **7**, 40, 105, 159–160
age structure, 128, 169, 172, 175–177, 247, 253. *See also* social organization
aircraft: disturbance caused by, 70–71, 184, 200; use of, for surveying or capture, 58, 171, 176, 236, 251–252
Alamo Hueco Mountains, 141
Alaska Range, 29–30, 33, 44, 57, 61, 65–66, 68, 75–76
albinos, 160
allomothering, 126
altitude. *See* elevation
amensalism, 184, 187
anaplasmosis, 219
anatomy: overview of, 5–9; taxonomic utility of, 11, *12*, 13–15. *See also* growth; horns; morphology; pelage

antelope, pronghorn, 194, 203, 227, 233
antipredator strategies, 110–111, 122, 165, 199–203, 205
aoudad, 11, 22, 186–187, 198, 229
Aravaipa Canyon, 154, 178, 185
argali, 11, 193–194, 196, 199, 201–203, 262
aspect. *See* slope and aspect
ataxic syndrome, 233–234, 236
Atigun Canyon, 70
Audubon's bighorn, 21, 209, 276
automobiles. *See* vehicles
avalanches, 22, 47, 67, 68, 136

Badlands bighorn, 21, 209, 276
Badlands National Park, 170
Banff National Park, 10, 109, 133, 136, 230
Barbary sheep, 11, 22, 186–187, 198, 229
bear: black, 66–67, 194; grizzly, 194, 242; polar, 64; short-faced, 195, 199
bedding: desert bighorn, 143–145, 162–164; and disease, 216–217, 225, 230; Rocky Mountain and California bighorn, **45**, 83, 107–109, 125; thinhorn, **45**, 49, 64
behavior, 190–191; aberrant, 258; adaptive, 192–208; amensal, 184; babysitting, 49; bleating, 124, 181; depressed, 227, 233; distressed, 123; dominance-establishing, 50, 162–163, 181–182; foraging, 28–29, **45**, 97, 145, 190; human influence on, 183–184; as an impediment to interbreeding, 13; lamb, 107, 124–126; lambing, 56, 57, 122–123, 168–169; lamb-rearing, 49, 160; at mineral licks, 35; and population dynamics, 181; reproductive, 17, 51–53, 60, 105,

115–120, 128, 166–168; rigidity of, 183; running, 107, 118, 119, 236; staggering, 227; taxonomic utility of, 276; vigilant, 110, 200; vocal, 200; watering, 141. *See also* activity; bedding; blocking; butting; clashing; coursing; courtship; dominance; fighting; herding; kicking; lip curling; low stretching; male altruism; mounting; playing; rape chasing; tending
Bering land bridge or Beringia, *16,* 16–17, 197
Bering Strait, 266
Big Hatchet Mountains, 141, 148, 150, 154–155, 168, 185, 189
Bill Williams Mountain, 187
bioenergetics, 108–109. *See also* activity
birthing. *See* lambing
birth weight: and adaptive strategies, 202, 206; Rocky Mountain and California bighorn, *99,* 123, 131; thinhorn, 36
bison, 69, 195, 197, 199–200, 203, 205
Black Gap, 188
Black Mountains, 178, 187
black-tailed deer. *See* deer, mule and black-tailed
blocking, 119
blood: and disease, 212, 232, 234, 236; genetic aspects of, 23, 267, **274**, 275; sodium as main constituent of, 155
bluetongue, 227–228, 246
boar, wild, 194
bobcat, 188, 258
body size: affected by glacial activity, 18, **196**, 204–207; of Dall's sheep, 36–39, *37, 38, 39,* **203**; of desert bighorn, *158;* and habitat requirements, 242–243; limited by escape strategy, 202–203; and maturation, 5, 17, 170; of Rocky Mountain and California bighorn, 98–101, *99, 100, 101;* and time spent foraging, 45. *See also* birth weight; growth

bonding (maternal), 122–123, 160
bone marrow, 31, 67, 273
botflies, 126, 219, 221–223
brain abscess, 221
brain size, 193–194
Brooks Range, 23, 27, 29–33, 35, 42–47, 49, 51, 54, 59, 64, 69–70, 204
Buckskin Mountains, 150
burned areas, 28, 32, 71, 81, 257. *See also* fire
burros, 155, 180, 186–188, 253
butting, 5, 49–51, 116–118, *117,* 159, 252

Cabeza Prieta National Wildlife Refuge, 144, 149, 163, 165
calcium, 34–35, 97, 155
canals, 22, 180
Capitol Reef National Park, 142
captivity: desert bighorn in, 151, 155, 164, 166–167, 169, 173, 175, 184; and disease, 210–212, 223–224, 227, 231; interbreeding of thinhorn and bighorn in, 13; red deer in, 126; Rocky Mountain and California bighorn in, 98, 99–100, 106–107, 109, 115–117, 121–123, 132; thinhorn in, 54. *See also* enclosures
capture myopathy, 209, 233, 235–237, 251, 258
carbohydrates: and disease, 211–212; in thinhorn diets, 27, 29
caribou (and reindeer), 64, 68–69, 109, 128, 194, 204
carrying capacity, 29–30, 32, 60, 76–77, 145, 185, 242, 246, 249–250
Cataract Canyon, 184
cattle, 21, 136, 154, 185–186, 256, 271; and disease, 219–220, 227–228, 232. *See also* livestock
censusing, 64, 129, 171, 213, 239, 241, 246–247. *See also* surveys
cestodes, 224–226
Chadwick, L. S., 9

Chemehuevi Mountains, 187
Chilicote Ranch, 188
chloride, 155
chromosomes, 263–266, *264–265*
Chugach Mountains, 35, 43
Chugach National Forest, 32
clashing, horn, 5, 49–51, 67, 116–118, *117*, 159
classification. *See* taxonomy
climate: and adaptive strategies, 192–195, 202–203; in desert bighorn range, 144–145, 148; in Rocky Mountain and California bighorn range, 80–81; in thinhorn range, 23–26; and transplanting considerations, 259
colonization, 18, 187, 189–190, 195–196, 198, 241, 247–249, 253
Colorado River, 178, 213
coloration: anemic, 225; of desert bighorn, 160–161; horn, 159; of sheep with pneumonia, 214; taxonomic utility of, 13; of thinhorn 13, *24, 25*
competition, interspecific, 22; with desert bighorn, 151, 154–155, 184–187, 189, 198; and disease, 209, 214, 216, 224–226; and management considerations, 239, 241, 253; with Rocky Mountain and California bighorn, 136, 138; with thinhorn, 68–69
competition, intraspecific, 73, 110, 111, 163–164, 181
competition, sperm, 118
conception, 52–53, 128, 249, 258. *See also* pregnancy
conservation, future prospects for, 19–22, 69, 139, 183, 212, 239, 260–262
Cooper Landing Closed Area, 59, 70–71
copulation: desert bighorn, 170; Rocky Mountain and California bighorn, 118–120; thinhorn, 51–52. *See also* mounting

corridors, movement, 189, 191, 239, 245–246, 248, 257–258
Cottonwood Mountains, 149
cougar, 136, 177, 178, 188–189, 199, 258, 261
coughing, 210, 214
coursing, 118–119
courtship: 117–119, 128, 162, 182, 205. *See also* behavior, reproductive
cover, security. *See* escape terrain
cover, thermal, 83, 145
Coxcomb Mountains, 149
coyote, 66–67, 131–132, 136, 178, 188, 194, 199, 200, 226
Crescent Mountain, 59
crowding, 207, 211–214, 216, 224, 236

Dall, W. H., 10
Death Valley, 143, 163, 187, 242
deer, 86, 99–100, 102, 109, 185, 219, 221, 241; mule and black-tailed, 22, 69, 86, 136, 149, 155, 194, 219, 227, 230; white-tailed, 69, 125, 194, 223, 227, 233. *See also* caribou; elk; moose
Denali National Park, 34, 59, 65, 70
density dependence, 76, 136–138, 165, 180–182. *See also* population density
dentition. *See* teeth
Desert, 140; Colorado, 148; Great Basin, 140, 146, 170; Mojave, 144, 148, 151, 156; Painted, 146; Sonoran, 146, 148, 156, 168, 170
Desert Bighorn Council, 180
Desert Bighorn Sheep Society, 180
Desert National Wildlife Refuge, 156, 160, 161, 172, 176, 186
development. *See* growth
diarrhea, 225, 227, 228
die-offs: desert bighorn, 180; disease-related, 212–214, 217, 231; historical, 22; and management considerations, 239, 241, 261; Rocky Mountain and California bighorn, 137–138; thin-

horn, 62, 63, 77. *See also* epizootics; famine
diet: and adaptive strategies, 192; of desert bighorn, 148–149, 151–155, 158, 162, 185, 187; and management considerations, 241; overlapping with other species, 136, 151, 155, 185; of Rocky Mountain and California bighorn, 88–97, **90–91, 92, 93**, 114, 136; of thinhorn, 26–31, 34, 68–69, 76. *See also* forage; nutrition
digestibility, 86, 97, 151–152, 200. *See also* forage; nutrition
digestive system, 5, 141, 241
disease, 22, 209–237, **222**, 267, 273; and adaptive strategies, 197–198; in desert bighorn, 161, 169, 173, 178, 182–184, 186, 189–190; and management considerations, 239, 241–242, 246, 251–253, 256, 258, 260–261; in thinhorn, 68, 70, 76, 131, 136–137. *See also* epizootics; parasites
dispersal: and adaptive strategies, 193, 195–196, 208; of desert bighorn, 150, 166, 181, 184, 186–187, 189–191; and management considerations, 241–243, 247; of Rocky Mountain and California bighorn, 111, 132, 138; of thinhorn, 47–48. *See also* colonization; movement
distribution: current, of mountain sheep, *4,* **19**; of desert bighorn, 139–140; historic, of mountain sheep, 3, *4,* 19, 240; of Rocky Mountain and California bighorn, 78–79; of thinhorn, 23–26
disturbance, human, 22, 183–184; and adaptive strategies, 200; of desert bighorn, 171, 183–184, 189; and management considerations, 258; of Rocky Mountain and California bighorn, 131; of thinhorn, 70–71
disturbance, interspecific, 110

disturbance, of vegetation, 26. *See also* fire
dogs, 200, 211–212, 223, 226
Dome Rock Mountains, 153
dominance, 163; among ewes, 49; among rams, 46, 50–51, 53, 59, 116–117. *See also* behavior, dominance-establishing; social status
drought, 146, 161, 182, 231, 248, 250, 255
drowning, 67, 178, 200. *See also* accidents
dust, 69, 211, 252
dystocia, 123, 206

eagles, 65, 66, 122, 188
economic growth, 22, 239. *See also* conservation; habitat, fragmentation; housing developments; roads
ecthyma, 137, 209, 230–232, 241, 253
edema, 211, 227–229, 233, 235
Eldorado Mountains, 160
elevation: and adaptive strategies, 193, 195, 204–205, 208; and disease, 217; and genetics, 267; and management considerations, 242, 244, 245, 255, 257; used by desert bighorn, 142–144, 146, 150, 152, 167; used by Rocky Mountain and California bighorn, 79, 86–87; used by thinhorn, 28, 46–49, 60, 64, 69. *See also* topography
elk, 22, 66, 69, 126, 131, 136, 198, 206, 219, 230, 233
emigration. *See* dispersal; movement
enclosures, 106, 123, 141–142, 153, 188, 210, 212, 231, 236, 252. *See also* captivity
Endangered Species Act, 239
energy: balance, 31, 108; conservation, 50, 163, 181, 193; demands, 44–45, 107, 109; expenditure, 52, 61, 109, 120, 145, 205–206; intake, 100, 109, 151; storage, 60, 120, 193

epizootics, 48, 76, 137, 138, 157, 180, 217, 220, 231, 241. *See also* die-offs
escape terrain: and adaptive strategies, 201, 206–207; desert bighorn, 141–143, 145, 188; and management considerations, 243, 254–256; Rocky Mountain and California bighorn, 80, 122, 130, 136; thinhorn, 23, 25, 32, 46, 60, 65–66, 69
estrus: desert bighorn, 167–169, 171, 181–182, 201; Rocky Mountain and California bighorn, 114–120, 127–128; thinhorn, 50–53, 55
evolution: and adaptations of North American sheep, 15–18, 192–208; molecular, 166
exploitation competition, 184, 187
extinction, 79, 209, 239–242; Pleistocene, 194, 197, 199, 205
extirpation, 13, 21, 138, 180, 184, 186–187, 189–190, 239–242, 247–249, 276

famine, 62, 180, 182, 224, 233
Fannin's sheep, 13
fat: and body condition, 53, 67, 99, 195; as insulation, 161; metabolism of, 31, 60–61, 108; storage, 31, 36, 60–61, 108
fecal analysis, 88, 152, 187
fecundity, 127–129, *127,* 137–138, 169–170, 202, 208, 241
fences, 22, 177–178, 189, 243–244, 246, 257–258
fibrosis, 211, 235
fighting, 43, 116, *117,* 162, 177, 182. *See also* butting; horn clashing; kicking
fire, 26, 32, 42, 71, 82; suppression, 143, 145, 180, 184. *See also* burned areas
fitness, 114, 126, 164, 186, 190, 192–193, 239
forage: of desert bighorn, 151–155; intake and tooth wear, 69; management and monitoring, 31–32, 254–257, 259, 261; and precipitation, 249–250; production and body weight, 100, 206; requirements during gestation, 206, 259; of Rocky Mountain and California bighorn, 81–83, 88–97, **90–91, 92, 93**; selection, 94–97, 255; of thinhorn, 23–28; value, 31; winter, as limiting factor on, 136. *See also* activity; behavior, foraging; diet; forage availability; forage conditions
forage availability, as influenced by: exotic ungulates, 186–187; human disturbance, 183, 200; livestock grazing, 184–186, 246; snow cover, 61–63, *63,* 81, 83; topography, 83–84. *See also* competition, intraspecific
forage conditions: and horn growth, 41–42, 205–206; and lamb survival, 30, 76, 120–121, 131, 259; and population density, 29–30, 60–61, 73–76; and sheep movement, 47–48, 86–88
Fort Peck, 129
Fort Yukon, 10
founder population, 198, 248, 253
fox: arctic, 64; gray, 188; red, 67
Fraternity of the Desert Bighorn, 180

gene pool, 183, 266, 268–269, 271
genetic diversity, 132, 175, 198, 241, 244, 248, 261, 271, 275
genetic drift, 241, 248, 259, 269
genetic isolation, 268, 271
genetics, 263–276; and adaptive strategies, 193, 198; desert bighorn, 160, 170, 175, 183, 187; and horn growth, 102; and management considerations, 240–241, 244, 248, 253, 259–261; molecular, 275; Rocky Mountain and California bighorn, 128–129, 132, 138; thinhorn, 34, 39, 43, 50. *See also* gene pool; genetic diversity; genetic drift
geographic information system, 85, 141, 245
gestation: forage requirements during,

206, 259; metabolic cost of, 202; period, **6**, 53–54, 60, 121, **167**, 169, 203; poisoning during, 35; and timing of weather patterns, 68, 175; ultrasound diagnosis of, 159
glacial periods, 16–18, 193–199
gnats, 227, 256
goat, Persian wild, 186–187
goats, domestic, 186, 219, 223, 275
Grand Canyon, 140, 160, 187
Granite Mountains, 149
Greybrell River, 186
ground cover, 145, 154
ground squirrel, arctic, 68
group integrity, 164–165, 253. *See also* social organization
group size: desert bighorn, 145, 163, 165–166; Rocky Mountain and California bighorn, 110–115, *111, 113, 114;* thinhorn, 49, 55; and transplanting, 253. *See also* social organization
growth: of desert bighorn, 156–160; and morphology, 17; of Rocky Mountain and California bighorn, 97–101, *99, 100, 101;* of thinhorn, 36–39. *See also* body size; horn growth; teeth, formulae

habitat: components, 141–142; of desert bighorn, 141–149, 189; fragmentation, 22, 180, 191, 239–243, 246, 262; and management considerations, 242–246; of Rocky Mountain and California bighorn, 78–88; of thinhorn, 23–26. *See also* cover; elevation; escape terrain; forage; mineral licks; topography; vegetation; water
hair, 64, 108, 160–161, 195, 225
handling (of sheep), 251–252
harems, 51–53, 201, 205
Harquahala (and Little Harquahala) Mountains, 145, 149, 153, 155, 162–163, 165, 173–174, 177
Hart Mountain, 85

heartbeat and heart rate, 35, 108, 184, 200–201, 208, 252
heart disease, 182
helicopters. *See* aircraft
hellebore, false, 35–36
hemoglobins, 267–269
herding (of females by males), 53, 119
herds: condition of, 29, 75, *75;* high-quality, 29–30, 73, 119, 129, 174, 175, 241, 249–250; low-quality, 29, 73, 76, 103–104, 173, 250; selfish, 200
heterozygosity, 132, 198, 241, 271, 275
Hexie Mountains, 149
homeothermy, 100, 109
home range: and adaptive strategies, 202, 206–208; of desert bighorn, 149–151, 164–165, 171, 178, 190; fidelity, 5; and management considerations, 242–244, 256; of Rocky Mountain and California bighorn, 85–86, 110, 113
Hoover Dam, 178
horn clashing, 5, 49–51, 116–118, *117,* 159, 252
horn growth: in desert bighorn, 157–159, *157, 158;* general, 7; and lamb production, 201; in Rocky Mountain and California bighorn, 101–105, *102, 104;* in thinhorn, *24, 25,* 39–44, *41, 42, 43*
horns: and age estimates, 7, 40–41, *104,* 104–105, 158–159; and antipredator strategy, 205–206; brooming of, 5, 43, 103, 105, *157,* 159; entanglement of, 177–178; female compared to male, *106;* and habitat quality, 207; and hunting regulations, 43–44; population effects on, 103–105; record lengths of, **8**, 9, 40; reproductive utility of, 205; and social structure, 44; and survival, 42; taxonomic utility of, 8–9
horses, 69, 136, 253
housing developments, 180, 184, 216, 258

hunting: by aboriginal Americans, 9; and data gathering, 104, 152, 158, 181; and forage production, 30, 76, 133; and full-curl law, 44, 53; illegal, 189; management of, 238, 249–250; unregulated, 22
hunting effects: on behavior, 70–71, 200; on horn size, 60, 76, 106; on lambing, 55, 60; on migration, 48; on sex ratio, 59–60, 120, 128, 172
hybridization, 131–132, 241, 248, 250, 259
hyperplasia, 211
hyperthermia, 233, 236
hypothermia, 61, 122, 131, 173, 203

ibex, 103, 200–201
ice, 67, 85. *See also* snow
Ice Age, 15–18, 193–200, 205–206
immigration, 242, 260. *See also* movement
immobilization, 233, 251–252
inbreeding, 131–132, 241, 248, 250, 259
infection, 137, 182–183, 211–212, 214–216, 218, 220–221, 223–225, 228–232
injury, 50, 177, 190, 193, 232, 251–253. *See also* accidents
interbreeding: by populations of mountain sheep, 247, 257; by species and subspecies of wild sheep, 11–13, 230, 265–266, 276; by wild sheep and domestic sheep, 15–16, 268–269
intrinsic rate of increase, 129, 169, 180
irruptions, population, 73, 76

jaguar, 188
Jasper National Park, 98
Johne's disease, 228–230

Kenai Mountains, 27–28, 31, 34–36, 42, 51–53, 59, 62–64, 66–68
Kenai Peninsula, 29, 32, 34, 47, 51, 54–55, 62, 65, 67, 70, 72–73

keratin, 103, 158
kicking, 117, 119
kidneys, 234, 236
Kluane National Park, 29, 57, 59, 61, 66–67, 69, 71, 73, 76
Kofa Mountains, 153, 165
Kofa National Wildlife Refuge, *140,* 144, 156
Kootenay region, 87, 137

labor. *See* parturition
lactation: and adaptive strategies, 202; desert bighorn, 164, 168, 171; Rocky Mountain and California bighorn, 107, 120–121, 125–126, 129; thinhorn, 42, 44–45, 53, 56. *See also* lambing
Lake Bonneville, 17, 196
lambing: by desert bighorn, *167,* 168–169; and phenology, **54**; by Rocky Mountain and California bighorn, 120–123; seasons, **55**, *167;* by thinhorn, 54–58, **54, 55**; and translocation, 259
lamb survival: and fecundity, *174;* and forage conditions, 30, 76, 120, 259; and lambing rates, 57–58; and maternal bonding, 123–124, 160; and maternal condition, 128–129, 138, 203, 258–259; overview of, 172–175, 178–180; and population dynamics, 58–59, 73–75, *75;* and ram behavior, 207; and suckling, 124–126; and timing of conception, 53, 120–121, 131, 168–169
latitude: and adaptive strategies 195–199; of desert bighorn range, 146, 167, 170; and management considerations, 259; of Rocky Mountain and California bighorn range, 79, 98, 120–121; of thinhorn range, 23, 44, 48, 51, 54
life expectancy, 59–60, 69, 73, 103, 133, 138, 176–177, 233, 250

life history, 6
life tables, 175–177
limiting factor, 69, 136, 144, 180, 188
lip curling, 117–118
liver disorders, 182, 233
livestock, 21–22, 69, 178, 180, 184–185, 189–190, 194, 197; and management considerations, 241, 244, 246, 251, 253, 255–258, 260–261. *See also* cattle; horses
low stretching, 117
lumpy jaw, 221, 232–233
lungworm, 32, 125, 130, 137, 211–219
lying. *See* bedding
lynx, 66–67

Mackenzie Mountains, 29, 33, 40, 44, 48, 57, 62, 70–71
magnesium, 34, 155
male altruism, 207
mark-recapture technique, 246
marmots, 35, 68
marrow, 31, 67, 273
mating. *See* copulation; courtship; reproduction; rut
McGuillivray, D., 10
metabolic rate, 108–109
metabolism: anaerobic, 236; carbohydrate, 212; oxidative, 147; protein, 212
metapopulations, 240, 244, 247, 253, 260, 261
migration: of desert bighorn, 150; inherited, 207–208; of Rocky Mountain and California bighorn, 79, 86–88, 110; of thinhorn, 48. *See also* movement
milk, 36, 61, 103, 107, 124–126, 138, 182, 205, 228, 271. *See also* lactation
mineral licks: false, 35–36; used by desert bighorn, 150, 155; used by Rocky Mountain and California bighorn, 84–85, 87; used by thinhorn, 33–35, 47, 50. *See also* salt

minimum viable population, 138, 240, 247, 259
mites. *See* scabies
molting, 64, 161, 223
moose, 69, 194, 198, 205, 233
morphology, **196**, 197, 201, **203**, 261, 263, 276. *See also* body size; horns; phenotypes
mortality factors: of desert bighorn, 176–180; of Rocky Mountain and California bighorn, 130–137; of thinhorn, 60–71. *See also* abortion; accidents; crowding; disease; disturbance; drowning; famine; mortality rates; overgrazing; parasites; predation
mortality rates: of desert bighorn, 172–177, *174, 179;* and disease, 229; of Rocky Mountain and California bighorn, 132–135, **134**, *135;* of thinhorn, 57–58, 71–72, 72. *See also* die-offs
mouflon, 11, 181, 229, 230, 263, 265, 268, 275
mountain goat, 34, 48, 68, 69, 136, 197
mountain lion, 136, 177, 178, 188–189, 199, 258, 261
Mount Grant, 188
mounting, 51, 183, 205. *See also* copulation
Mount McKinley, 34, 59, 65, 70
movement: and adaptive strategies, 208; corridors, 189, 191, 239, 245–246, 248, 257–258; of desert bighorn, 149–151, 155, 163–164, 183, 189, 190; and disease, 236; and management considerations, 239, 243–244, 248, 253, 260; restricted by snow, 24, 46–47, 62, 81, 109, 144, 207; of Rocky Mountain and California bighorn, 85–88; of thinhorn, 46–49. *See also* migration
mule deer. *See* deer, mule and black-tailed
musk ox, 195, 205

National Bison Range, 129
Native Americans, 9, 198–200
necropsy, 210, 226–227, 233–234
nematodes, 214, 218, 224–226
neoteny, 17
niches, 186, 200, 203
numbers of mountain sheep: historic, 18–22; in 1991, 19–20, **19, 20, 21**
nursing: by desert bighorn, 160, 173; impaired by disease, 231; and management considerations, 259; by thinhorn, 47–49, 55–57, *56*. *See also* behavior, lambing; lactation
nutrition: desert bighorn, 141, 151, 162, 164, 170, 173, 178; and disease, 211, 214, 224–225, 231; and management considerations, 258–259; Rocky Mountain and California bighorn, 84–86, 91, 97–98, 107, 109–110, 115, 124–125, 130–131, 133, 137–138; thinhorn, 29–32, 39, 41–43, 52–53, 57–58, 61–62, 66, 72, 73, 76. *See also* diet; famine; forage

ocelot, 188
osteomyelitis, mandibular, 221, 232–233
overcrowding, 207, 211, 212, 214, 216, 224, 236
overgrazing, 180, 184–185, 246, 257
ovulation, 52, 105, 115
Owyhee River, 186

palatability, 91, 96, 141, 153–154, 160, 217
paleoecology, 193–197
Palliser Range, 133, 136
Panamint Mountains, 187, 242
parasites, 260, 267; and adaptive strategies, 192, 195, 198; of desert bighorn, 182, 186; ecto-, 219–224; gastrointestinal, 224–227; of Rocky Mountain and California bighorn, 98; of thinhorn, 68

paratuberculosis, 228–230
parturition: desert bighorn, 168–170, 173; and management considerations, 241; Rocky Mountain and California bighorn, 87, 120–124; thinhorn, 47, 50, 55–57, 60–61
pasteurellosis, 210–213, 217
pedogenesis, 17, 105
pelage, 13, *24, 25,* 64, 108, 160–161, 195, 214, 225
Peloncillo Mountains, 141, 164, 185
phenology, 86, 153, 161, 168, 170, 173
phenotypes, 138, 195, 196, 208, 268, 269, 271, 273, 275
phosphorous, 97, 155
photoperiod, 51, 53, 115, 258
Pikes Peak, 216–217
Pinto Mountains, 149
Pintwater Range, 160
placenta, ingestion by mother, 123, 160
playing, 49, 56, 107–108, 163
Pleistocene epoch, 15–18, *16,* 193–200, 205–206
pleuritis, 211
Plomosa (and North Plomosa) Mountains, 153, 177
pneumonia, 130–131, 137, 190, 210–219, 227, 231
poaching, 21, 178, 184, 189
polygyny, 201, 205
population density, 48; and forage, 29–30, 60, 73–76; and herd condition, 29, 75; and horn size, 29; and hunting regulations, 249–250; and lamb survival, 131, 175–180; and matriarchal leadership, 165; and movements, 149; and timing of birth, 131; and topography, 142–143. *See also* density dependence; population dynamics
population dynamics: desert bighorn, 169–191; Rocky Mountain and California bighorn, 126–138; thinhorn, 57–60, 72–77, *74, 75. See also* die-

offs; dispersal; fecundity; immigration; mortality rates; population density; survival
population size: effective, 138, 240; of mountain sheep, **20, 21**
postmortem analysis, 210, 226–227, 233–234
potassium, 34, 155, 236
precipitation, 18; in desert bighorn range, 144, 146, 148, 153–154, 161, 168–169, 174–175, 179; and management considerations, 249–250, 259; in Rocky Mountain and California bighorn range, 80, 103; in thinhorn range, 24–25, 41. *See also* ice; rainfall; snow
predation: and adaptive strategies, 199–200, 205–206; and disease, 233; and habitat selection, 83, 121–122; and human disturbance, 183; and migration, 48; and movements, 80; and population dynamics of desert bighorn, 169–170, 177, 181, 182, 183, 188–189; and population dynamics of Rocky Mountain and California bighorn, 130–132, 136, 138; and population dynamics of thinhorn, 57–58, 60, 65–67, 76–77; and social structure, 110–111, 126, 164; and transplant efforts, 247, 252–253
predators. *See* bear; bobcat; coyote; dogs; eagles; lynx; mountain lion; predation; wolverine; wolves
pregnancy: and adaptive strategies, 206; desert bighorn, 159, 170, 182; and disease, 215, 236; and management considerations, 236, 259; Rocky Mountain and California bighorn, 80, 87, 113, 128; thinhorn, 36, 42, 44, 52–53, 57, 61. *See also* conception
productivity, 58, 76, 87, 104, 120, 173, 229, 247, 259. *See also* fecundity; population dynamics; reproduction; survivorship

progesterone, 159
pronghorn antelope, 194, 203, 227, 233
protein: in desert bighorn diets, 151; and disease, 211–212, 234; and genetic analysis, 267–269, 273, 275; in Rocky Mountain and California bighorn diets, 97; in thinhorn diets, 27–29, 31
protozoans, 5, 219, 224–226
puberty, 53, 105
Pusch Ridge, 143, 153, 184
pyothorax, 211

radio telemetry, 58, 86, 135, 149–150, 164, 189–190, 229, 244, 253, 261
rainfall: in desert bighorn range, 144, 150, 153, 171; and management considerations, 257; in thinhorn range, 61
range management, 31, 185–186
rape chasing, 118–119
rate of increase, intrinsic, 129, 169, 180
ratios: birth/death, 249; calcium/phosphorus, 97; lamb/ewe, of desert bighorn, 171, 176, 178, 181; lamb/ewe, and management considerations, 250, 254; lamb/ewe, of Rocky Mountain and California bighorn, 130, 136; lamb/ewe, of thinhorn, 32, 57–59, 71, 75, 75; power/mass, 203; sex, of desert bighorn, 170–172, 177, 182; sex, and management considerations, 253; sex, of Rocky Mountain and California bighorn, 118–120, 127–130, 132; sex, of thinhorn, 59; weight/height, 53; yearling/ewe, 58, 176
recognition, by ewes and lambs (of each other), 123–124
recreation, 22, 180, 184, 216
recruitment. *See* lamb survival; survivorship
reindeer. *See* caribou
reintroduction, 79, 175, 180, 185, 188–

189, 198, 208, 252. *See also* translocation
reproduction: age of first, 57, 127, 137, 169–170, 249; and body size, 103; and climate, 258–259; and competition, 184; costs of, 36, 100, 129, 135, 193; of desert bighorn, 167–168, *167;* and escape terrain, 243; and forage quality, 30, 193; and horn size, 205; and lambing rate, 57; and lamb survival, 169; and photoperiod, 51, 53, 115, 258; of Rocky Mountain and California bighorn, 115–120; of thinhorn, 51–53. *See also* courtship; lambing; population dynamics; twinning
resting. *See* activity
Rhode, C., 62
River Mountains, 147, 165, 172, 174, 177, 250
roads, 22, 35, 70, 178, 180, 189, 211, 241, 245, 258
Rocky Mountain National Park, 155, 186
rodents, 68, 243
rumen analysis, 26, 88, 152, 154
running, 107, 118, 119, 236
ruptured muscle syndrome, 233, 235
rut: and adaptive strategies, 201, 206–207; desert bighorn, 162, 164, 167–168, 169, 177, 181–182; and management considerations, 243–244, 257; Rocky Mountain and California bighorn, 85, 100, 105, 109, 113–120, *114, 117;* thinhorn, 40, 46, 47, 51, 52, 59–60
rutting. *See* behavior, dominance-establishing; behavior, reproductive; rut

saber-toothed cat, 199
salt, 22, 34, 84–85, 155–156, 230. *See also* mineral licks
San Andres Mountains, 141, 143–145, 147–148, 154–156, 162, 166, 168–169, 171, 181, 186, 188–189, 220, 241
San Gabriel Mountains, 141, 144–146, 155, 168, 173, 183, 243, 245, 254
Sangamon interglacial period, 17. *See also* glacial periods
Santa Rosa Mountains, 143, 146–147, 173, 175
scabby mouth, 137, 209, 230–232, 241, 253
scabies, 157, 186, 190, 209, 219–220, 241
Seward Highway, 35
sexual dimorphism, 207; of horns, 7
sexual maturity, 42, 57, 73, 105–106, 166–167, 215
sexual segregation, 48, 85, 110–112, 116, 150, 154, 164–165, 206–208. *See also* competition, intraspecific; social organization
sheep, domestic, 15–16, 21–22; and desert bighorn, 151, 156, 182, 186, 189–190, 194; diseases of, 212–213, 216, 219–221, 223–231; genetics of, 266–269, 271, 273, 275; management of, 256; and Rocky Mountain and California bighorn, 108, 115–117, 120–125, 127, 131, 137; and thinhorn, 31, 35, 52–53
Sheep Mountain, 27, 36–37, 44, 52, 55, 69, 71
shipping fever, 182
shock, 182–183, 233, 236
Siberian snow sheep, 11, 15, 17, 23, 27, 205, 263, 266
sinusitis, chronic, 190, 209, 221–224, **222**
size. *See* birth weight; body size; brain size; group size; horns; population size
slope and aspect: and diseases, 217; and management considerations, 242, 243, 255; preferred by desert bighorn, 142, 143, 145, 185; preferred by Rocky Mountain and California

bighorn, 79, 83–84, 96; preferred by thinhorn, 28, 61. *See also* topography

snails, 214–217

snow: depth and hardness, *63;* and desert bighorn, 144; and disease, 211; and forage availability, 61–63, *63,* 81, 83; and predation, 66, 201–202; and restriction of movements, 24, 46–47, 62, 81, 109, 144, 207; and Rocky Mountain and California bighorn, 81, 83, 108, 109; and thinhorn, 60–64, *63;* as tracking aid, 88; as water source, 32–33, 85. *See also* ice

social organization: of desert bighorn, 164–166; of Rocky Mountain and California bighorn, 109–115, *111, 113, 114;* of thinhorn, 49–50

social pressure, 181

social status, 114–115; of ewes, 108; of rams, 50, 116–118, 127–128. *See also* behavior, dominance-establishing; social organization

Society for the Conservation of Bighorn Sheep, 180

sodium, 34–35, 155–156, 236

sore-mouth, 137, 209, 230–232, 241, 253

spermatogenesis, 105, 116, 166–167

Spring Mountains, 160

standing, **45**. *See also* activity; behavior

starvation, 62, 180, 182, 224, 233

sterility, 181

Stone, A. J., 11

stress: and adaptive strategies, 193; in desert bighorn, 156, 164, 178, 181–183; and disease, 210–212, 214, 217, 226, 228, 230–231, 233–234, 236–237; and genetics, 273; and management considerations, 251–252, 255, 258–260; in Rocky Mountain and California bighorn, 135, 137–138; thermal, 83, 107, 120, 145, 148–149, 170, 252; in thinhorn, 53, 59–60; water, 148, 243, 258

suckling, 124–126, 128, 173

sudden death syndrome, 233–234

summer range: of desert bighorn, 146, 149; and disease, 217; of Rocky Mountain and California bighorn, 81, 85, 87–88, 116; of thinhorn, 46–49, 68

Superstition Mountains, 143, 153

Surprise Mountain, 59, 71, 73, 75

surveys: aerial, 58, 171, 176; foot, 171

survival: density dependent, 137–138; differential, of sexes, 59–60, 129, 132, 137; and horn growth, 42; population, 240–242, 246–251; prehistoric sheep, 9; ram, *179;* species, 22, 186, 262; and weather, 60–62, 80, 145, 259. *See also* lamb survival; life tables; population dynamics; predation

survivorship, 72, 129–130, 133, 176–177, *179*

Tanana Hills, 10

tapeworm, 226

taxonomy, 5, 11–15, *12,* **14**, **15**, 197, 208; genetic, 263–276; and management considerations, 261; of Rocky Mountain and California bighorn, 78; of thinhorn, 23

teeth: and age estimates, 5, **7**, 40, 105, 159–160; formulae, **6, 7**; infections of, 221, 232; wear and longevity, 69–70

temperature, ambient: and adaptive strategies, 201, 204–205; in desert bighorn range, 18, 144–145, 147–148, 166; and management considerations, 251; in Rocky Mountain and California bighorn range, 80, 83, 107–109, 115; in thinhorn range, 23–24, 33, 63–64. *See also* thermal stress

temperature, body, 260

tending, 118–119

testosterone, 109, 116

Texas bighorn, 15, 209

thermal stress, 107, 120, 145, 170, 233, 236

ticks, 219

topography, 200–201, 245; and desert bighorn, *140,* 141–144, 149, 151, 186;

and Rocky Mountain and California bighorn, 79–80, 85, 96, *111;* and thinhorn, 28, 46, 60. *See also* escape terrain; slope and aspect
trailers, transport, 252
Trans-Alaska Pipeline, 32, 35, 70
transferrins, 269–275, **270**, **272**
translocation, 21, 188–189; and disease, 218, 231, 237–238; and management considerations, 239, 242, 247–251, 254–261; techniques, 251–253. *See also* handling; reintroduction; transportation; trapping
transplantation. *See* reintroduction; translocation
transportation (of sheep), 231, 234–237, 252–253, 258
Trap Cave, 195–196
trapping (of sheep), 251
trophy rams, 8, 60, 106, 156
twinning, 5, 54, 127–128, 169, 203

ultrasound, 128, 159
urial, 11, 15, 194, 263

vegetation: of desert bighorn range, 140–146, 151, 155, 166; and digestion, 5; and disease, 214, 216–217; and lambing period, 121, 170; and management considerations, 242–246, 255–257, 261; and movements, 47; of Rocky Mountain and California bighorn range, 81–83, 85–97; of thinhorn range, 26–32, 60; as a water source, 5. *See also* escape terrain; forage; habitat
vehicles, 22, 35, 70, 177–178, 200, 258; and transport of sheep, 236, 252
Venegas, M., 10
vigilance, 110, 200
Virgin Mountains, 153
visibility, 83, 140–141, 145–146, 159, 165, 200, 256
vocalization, 124, 181, 200

water, 5, 212, 216, 223, 236; availability and predation, 9; needs of desert bighorn, 139, 141, 143–144, 147–151, 153; needs of Rocky Mountain and California bighorn, 84–85, 126; needs of thinhorn, 32–33; and sheep management, 238, 243, 253–258
waterhole counts, 161, 171–172
weaning, 57, 120, 126, 173, 259
weather: and disease, 210, 214; and forage selection, 96; and horn growth, 41; and lambing sites, 122; and lamb survival, 57, 131, 173; and movements, 47, 258; and population dynamics, 72, 76–77, 138, 169, 250, 254; and predation, 65; and transplant success, 260; variables, as limiting factors for thinhorn, 60–64
weight. *See* birth weight; body size; growth
weight loss, 156; in Rocky Mountain and California bighorn, 100–101, 109, 128; in thinhorn, 36–37, 52
white-tailed deer. *See* deer, white-tailed
wilderness value of mountain sheep, 3, 22, 141, 241
Wildhorse Island, 129
winter range: of desert bighorn, 149, 166; and disease, 217; and management considerations, 243; of Rocky Mountain and California bighorn, 81, 83, 85–88, 110, 114–116, 122, 131, 136; of thinhorn, 28–30, 32, 46–47, 49, 60–63, 68, 73, 75, 76
wolverine, 66–67, 194
wolves, 65–67, 77, 194, 198–199, 200, 202, 226
Wrangell Mountains, 43, 64

Yosemite National Park, 242

Zion National Park, 164, 173, 188